T0281426

Computer Methods in
Chemical Engineering

Computer Methods in Chemical Engineering

Second Edition

Nayef Ghasem

CRC Press
Taylor & Francis Group
Boca Raton London New York

CRC Press is an imprint of the
Taylor & Francis Group, an **informa** business

Second edition published 2022
by CRC Press
6000 Broken Sound Parkway NW, Suite 300, Boca Raton, FL 33487-2742

and by CRC Press
2 Park Square, Milton Park, Abingdon, Oxon, OX14 4RN

© 2022 Taylor & Francis Group, LLC
First edition published by CRC Press 2015

CRC Press is an imprint of Taylor & Francis Group, LLC

Library of Congress Cataloging-in-Publication Data

Names: Ghasem, Nayef, author.
Title: Computer methods in chemical engineering / Nayef Ghasem.
Description: Second edition. | Boca Raton: CRC Press, [2022] | Includes
    bibliographical references and index. | Summary: "This textbook presents
    the most commonly used simulation software, along with the theory
    involved. It covers chemical engineering thermodynamics, fluid
    mechanics, material and energy balances, mass transfer operations,
    reactor design, and computer applications in chemical engineering. The
    Second Edition is thoroughly updated to reflect the latest updates in
    the featured software and has added a focus on real reactors, introduces
    AVEVA Process Simulation software, and includes new and updated
    appendixes. It gives chemical engineering students and professionals the
    tools needed to solve real-world problems"—Provided by publisher.
Identifiers: LCCN 2021031719 (print) | LCCN 2021031720 (ebook) | ISBN
    9780367765255 (hbk) | ISBN 9780367765248 (pbk) | ISBN 9781003167365 (ebk)
Subjects: LCSH: Chemical engineering—Data processing.
Classification: LCC TP184 .G48 2022 (print) | LCC TP184 (ebook) | DDC 660—dc23
LC record available at https://lccn.loc.gov/2021031719
LC ebook record available at https://lccn.loc.gov/2021031720

ISBN: 978-0-367-76525-5 (hbk)
ISBN: 978-0-367-76524-8 (pbk)
ISBN: 978-1-003-16736-5 (ebk)

DOI: 10.1201/9781003167365

Typeset in Times
by KnowledgeWorks Global Ltd.

Access the Support Materials: https://www.routledge.com/9780367765255.

Contents

Preface

PURPOSE OF THE BOOK

In industry, complicated problems are often not solved by hand for two reasons: human error and time constraints. Many different simulation programs are used in industry, depending on field, application, and desired simulation products. When software is used to its full capabilities, it is a potent tool for a chemical engineer in various fields, including oil and gas production, refining, chemical and petrochemical processes, environmental studies, and power generation. Although most software packages are user-friendly, considerable effort is required to master these software packages.

Software packages, such as UniSim/Hysys, PRO/II, Aspen Plus, SuperPro, and recently Aveva Process Simulation, have been developed to perform rigorous solutions of most unit operations in chemical engineering. However, as a design engineer, one always needs to know the fundamental theory and calculation methods to enable one to make decisions about the validity of these black box packages to verify the results. Most software packages are interactive process simulation programs. They are user-friendly and powerful programs that can solve various kinds of chemical engineering processes. However, solving an issue with using software packages requires multiple conditions and options, which requires good knowledge to use the program efficiently. This book aims to introduce chemical engineering students to the most commonly used simulation software packages and the theory to cover core chemical engineering courses. The book is helpful in understanding parts applied in various basic chemical engineering subjects such as chemical engineering thermodynamics, fluid mechanics, material and energy scales, mass transfer processes, reactor design, computer applications in chemical engineering, and industrial applications for graduate projects.

The second edition of *Computer Methods in Chemical Engineering* contains a theoretical description of process units followed by numerous examples solved by manual calculations. The entire book is in SI units. The book includes the recent versions of five software packages, UniSim/Hysys, PRO/II, Aspen Plus, SuperPro Designer, and Aveva Process Simulatin, through step-by-step instructions. The book is perfect for students and professionals and gives them the tools to solve real problems involving mainly thermodynamics and fluid phase equilibria, fluid flow, material and energy balances, heat exchangers, reactor design, distillation, absorption, and liquid-liquid extraction.

The most commonly used simulation software packages listed as follows and a brief introduction to each software is available in the appendices.

1. **UniSim/Hysys (Appendix A)**

    UniSim is software package for process design and simulation. It is user-friendly, excellent for petroleum refining and petrochemicals; it includes excellent Hysys to UniSim converter. A friendly graphical user interface, pretty similar to Hysys. Hysys and UniSim have almost identical capabilities. Hysys is older, whereas UniSim is comparatively new.

http://hwll.co/UniSim

2. **PRO/II simulation (Appendix B)**

Aveva PRO/II is a steady-state process simulator for process design and operational analysis for process engineers in the chemical processing and polymer industries. It includes a chemical component library, thermodynamic property prediction methods, and unit operations such as heat exchangers, compressors, distillation columns, and reactors as found in the chemical processing industries. It can perform steady-state mass and energy balance calculations for modeling continuous processes.

https://www.aveva.com/en/products/pro-ii-simulation/

3. **Aspen Plus (Appendix C)**

Aspen Plus is the leading chemical process simulator in the market. Software that will allow the user to build a process model and then simulate it using complex calculations (models, equations, math calculations, regressions, etc.).

http://www.Aspentech.com/

4. **SuperPro Designer (Appendix D)**

SuperPro Designer is a process simulator dealing with environmental issues such as wastewater treatment, air pollution control, waste minimization, pollution prevention. SuperPro provides a single umbrella modeling of manufacturing and end-of-pipe treatment processes, project economic evaluation, and environmental impact assessment.

https://www.intelligen.com/

5. **Aveva Process Simulation (Appendix E)**

The software was formally known as SimCentral, a new platform to manage how to engineer processes across their entire lifecycle. Using the Aveva process Simulation software package, users can simplify the design of the operation tool work together for process development, simplify modeling complexity, reduce time and cost, an appealing user experience for the next generation of engineers, and accelerated process simulation and design.

https://www.aveva.com/en/products/process-simulation/

COURSE OBJECTIVES

Students should be able to:

1. Recognize what chemical engineers do.
2. Explain out the functions of the basic chemical engineering unit operations.
3. Solve unit process system that faces chemical engineers.
4. Simulate the fundamental chemical processes using software packages such as UniSim/Hysys, PRO/II, Aspen Plus, SuperPro Designer, and Aveva Process Simulation
5. Design chemical process units manually and with software packages.

COURSE-INTENDED OUTCOMES

The relationship of the covered subjects to the program outcomes (based on ABET criteria). By completing the topics covered in this book, students should be able to:

1. Know the features of the best five software packages used in process simulation and flow sheeting; UniSim/Hysys, PRO/II, Aspen Plus, SuperPro Designer, and Aveva Process Simulation (1,2).
2. Verify manual calculations with available software packages to simulate pipes, pumps, compressors, heaters, air coolers, and shell and tube heat exchangers (1,2).
3. Design and simulate unit process systems such as chemical reactors, distillation columns, absorption, and extraction (1,2).
4. Work in teams, including a beginning ability to work in multi-disciplinary teams (5).
5. Communicate effectively through presentations and class participation (3).

Acknowledgments

The author would like to thank Allison Shatkin, acquiring editor for this book, and Gabrielle Vernachio, the editorial assistant, for their help and cooperation. The author would also like to express thanks to Mihaela Hahne and Julien de Beer (Aveva global academic program manager) and the Aveva support engineers for their kind cooperation and for providing the educational license of the Aveva Process Simulation software, and Honeywell for giving the academic permission of UniSim. The author would like to thank engineer Abdul Raouf from the United Arab Emirates University for sharing his experience and comments while teaching the subject for a few years. The author appreciated the comments and suggestions of the reviewers.

Author

Nayef Ghasem is a professor of chemical engineering at the United Arab Emirates University, where he teaches undergraduate courses in process modeling and simulation, natural gas processing, reactor design in chemical engineering, and graduate and undergraduate courses in chemical engineering. He has published primarily in modeling and simulation, bifurcation theory, polymer reaction engineering, advanced control, and CO_2 absorption in gas-liquid membrane contactors. He is the author of *Principles of Chemical Engineering Processes: Material and Energy Balances* (CRC Press, 2015), *Computer Methods in Chemical Engineering* (CRC Press, 2009), *Modeling and Simulation of Chemical Process Systems* (CRC Press, 2018). He is a senior member of the American Institute of Chemical Engineers (AIChE).

1 Thermodynamics and Fluid-Phase Equilibria

At the end of this chapter, students should be able to:

1. Estimate the vapor pressure of pure components.
2. Determine the boiling point and dew point of a mixture.
3. Estimate the molar volume using the equation of state (EOS).
4. Plot the effect of temperature versus density.
5. Use UniSim/Hysys, Aspen Plus, PRO/II, SuperPro, and Aveva Process Simulation software packages to estimate physical properties.

1.1 INTRODUCTION

Phase-equilibrium thermodynamics deals with the relationships that govern the distribution of a substance between gas and liquid phases. When a species is transferred from one phase to another, the transfer rate decreases with time until the second phase is saturated with the species, holding as much as it can hold at the prevailing process conditions. When concentrations of all species in each phase cease to change, the phases are at phase equilibrium. When two phases are in contact, a redistribution of the components of each phase occurs through evaporation, condensation, dissolution, and precipitation until a state of equilibrium is reached in which the temperatures and pressures of both phases are the same, and the compositions of each phase no longer change with time. A species' volatility is the degree to which the species tends to be transferred from the liquid phase to the vapor phase. The vapor pressure of a species is a measure of its volatility. Estimation of vapor pressure can be carried out by empirical correlation.

When a liquid is heated slowly at constant pressure, the temperature at which the first vapor bubble forms is called bubble point temperature. When the vapor is cooled slowly at constant pressure, the temperature at which the first liquid droplet forms is known as dew point temperature.

1.2 BOILING POINT CALCULATIONS

When heating a liquid consisting of two or more components, the bubble point is where the first formed bubble of vapor. Given that vapor will probably have a different composition in the liquid, the bubble point and the dew point at different compositions provide valuable data required to design of the distillation column. For single-component mixtures, the bubble point and the dew point are the same

DOI: 10.1201/9781003167365-1

1

and the same as the boiling point. At the bubble point, the following relationship holds:

$$\sum_{i=1}^{n} y_i = \sum_{i=1}^{n} K_i x_i = 1.0 \tag{1.1}$$

where

$$K_i = \frac{y_i}{x_i} \tag{1.2}$$

K_i is the distribution coefficient or K factor, defined as the ratio of mole fraction in the vapor phase y_i to the mole fraction in the liquid phase x_i at equilibrium. When Raoult's law and Dalton's law hold for the mixture, the K factor defined as the ratio of the vapor pressure to the total pressure of the system [1]:

$$K_i = \frac{P_{v,i}}{P} \tag{1.3}$$

1.3   DEW POINT CALCULATION

The dew point is the temperature at which a given parcel of air must be cooled, at constant barometric pressure, for water vapor to condense into water. The condensed water is called dew. Dew point is a saturation point. The basic equation for the dew point is as follows:

$$\sum_{i=1}^{n} x_i = \sum_{i=1}^{n} \frac{y_i}{K_i} = 1.0 \tag{1.4}$$

1.4   VAPOR PRESSURE CORRELATIONS

One of the most successful correlations is the Antoine equation, which uses three coefficients, A, B, and C, depending on the analyzed substance. Antoine equation is as follows:

$$\log(P_v) = A - \frac{B}{T+C} \tag{1.5}$$

If Raoult's law and Dalton's law hold, values of K_i are calculated from the vapor pressure (P_v) and the total pressure (P) of the system.

$$K_i = \frac{P_v}{P} \tag{1.6}$$

1.5 RELATIVE VOLATILITY

The K factors are strongly temperature dependent because of the change in vapor pressure, but the relative volatility of K for two components changes only moderately with temperature. The ratio of K factors is the same as the relative volatility (α_{ij}) of the components

$$\alpha_{ij} = \frac{y_i/x_i}{y_j/x_j} = \frac{K_i}{K_j} \tag{1.7}$$

when Raoult's law applies,

$$\alpha_{ij} = \frac{P_{v,i}}{P_{v,j}} \tag{1.8}$$

Example 1.1: Bubble Point

Find the bubble point temperature for a mixture of 35 mol% n-hexane, 30% n-heptane, 25% n-octane, and 10% n-nonane at 1.5 total atm pressure. Compare the manual calculation with the predicted results obtained from the following software packages: UniSim, PRO/II, and Aspen Plus, SuperPro, and Aveva process simulation.

SOLUTION

HAND CALCULATION

Assume a temperature (e.g., T = 110°C), calculate the vapor pressure using the Antoine equation, and then calculate the summation of y_i; if 1, then the temperature is the boiling point temperature, and if not, consider different temperatures (Tables 1.1 and 1.2).

At 110°C, the summation of $\Sigma K_i x_i = 1.127$, and at T = 100°C, $\Sigma K_i x_i = 0.862$; by interpolation at $\Sigma K_i x_i = 1.0$, the bubble point is 105.2°C.

UNISIM CALCULATIONS

In case of UniSim, add all components involved in the mixture, i.e., hexane, heptane, octane, and nonane, and their compositions, i.e., 0.35, 0.3, 0.25, and 0.1, respectively. Select Antoine as the fluid package, and then enter the simulation

TABLE 1.1
Bubble Point Calculation at Assumed T = 110°C

Component	x_i	P_v (110°C), atm	$K_i = P_v/1.5$	$y_i = K_i x_i$
n-hexane	0.35	3.11	2.074	0.726
n-heptane	0.30	1.385	0.923	0.277
n-octane	0.25	0.623	0.417	0.104
n-nonane	0.10	0.292	0.1945	0.020
			$\Sigma K_i x_i = 1.127$	

TABLE 1.2

Bubble Point Calculation at Assumed T = 100°C

Component	x_i	P_v (100°C), atm	$K_i = P_v/1.5$	$y_i = K_i x_i$
n-hexane	0.35	2.42	1.61	0.565
n-heptane	0.30	1.036	0.69	0.207
n-octane	0.25	0.454	0.303	0.076
n-nonane	0.10	0.205	0.137	0.0137
			$\Sigma K_i x_i = 0.862$	

environment. Click on stream in the object pallet, click on any place in the simulation area, double click on stream 1, and enter each component's molar compositions. While on the conditions page, set the vapor/phase fraction = 0. The calculated temperature (which is the boiling point temperature at the given pressure of 1.5 atm) is 106.2°C, as shown in Figure 1.1.

PRO/II CALCULATION

Start Aveva PRO/II and create a new file, File>New, which leads to the simulation environment. It is called process flow diagram (PFD) screen. Enter the components involved in the system and choose the correct equation of state. First, click on the Component Selection button on the toolbar (Benzene ring icon). In the popup menu, type in the desired components' names or select them from a list already inside the PRO/II. After choosing all species, click OK, then OK to return to the PFD. Next, click on the Thermodynamics Data button, select Peng–Robinson for the estimation of physical properties. Add all components involved in the mixture. Click on Streams in the object pallet, then click anywhere in the simulation area. Double click the stream S1 and specify the pressure as 1.5 atm. For the second specification, select Bubble Point. The calculated bubble point is 106.762°C (Figure 1.2).

ASPEN PLUS CALCULATIONS

The easiest way to estimate the bubble point temperature with Aspen Plus is to build a simple mixing process with feed stream S1 and exit stream S2. The property

Worksheet	Stream Name	1
Conditions	Vapour / Phase Fraction	0.0000
Properties	Temperature [C]	106.2
Composition	Pressure [kPa]	152.0
K Value	Molar Flow [kgmole/h]	100.0
User Variables	Mass Flow [kg/h]	1.016e+004
Notes	Std Ideal Liq Vol Flow [m3/h]	14.76
Cost Parameters	Molar Enthalpy [kJ/kgmole]	-2.071e+005
	Molar Entropy [kJ/kgmole-C]	150.8
	Heat Flow [kJ/h]	-2.071e+007
	Liq Vol Flow @Std Cond [m3/h]	14.69

FIGURE 1.1   UniSim calculates the bubble point temperature (T_{bp} = 106.2°C) for the case described in Example 1.1.

Stream Name Stream Description		S1
Phase		Liquid
Temperature Pressure	C ATM	106.762 1.500
Flowrate	KG-MOL/HR	0.454
Composition HEXANE HEPTANE NONANE OCTANE		0.350 0.300 0.100 0.250

FIGURE 1.2 PRO/II calculates the bubble point temperature (T_{bp} = 106.76°C) of the liquid mixture described in Example 1.1.

estimation method is Peng–Robinson. Double click on S1 and fill in pressure and composition. Since the bubble, point temperature is to be determined, set the vapor-to-phase ratio to 0.0. The system is ready to run. The bubble point temperature is 379.9 K (106.7°C) (Figure 1.3).

AVEVA PROCESS SIMULATION

Create a new simulation (e.g., Example 1.1) in Aveva Process Simulation.

1. Copy DefFluid to the Example 1.1 model library, right-click on the DefFluid icon, and select Edit.
2. Click on Methods, and for system select Peng–Robinson.

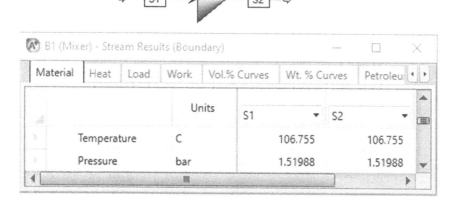

FIGURE 1.3 Aspen Plus calculates the bubble point temperature (T_{bp} = 106.75°C) of the liquid mixture described in Example 1.1.

FIGURE 1.4 Aveva Process Simulation calculates the bubble point temperature (T_{bp} = 106.76°C) for the case described in Example 1.1.

3. Drag the Source icon to the canvas and set P = 1.5 atm, W = 1 kg/s, and VF = 0.
4. For the Fluid Type, from the pull-down menu, select Example 1.1/DefFluid.
5. Once completed, Aveva Process Simulation is squared and solved. The calculated bubble point temperature is 106.76°C (Figure 1.4).

Example 1.2: Dew Calculation

Find the dew point temperature for a mixture of 35 mol% n-hexane, 30% n-heptane, 25% n-octane, and 10% n-nonane at 1.5 total atm pressure.

SOLUTION

MANUAL CALCULATIONS

Assume the temperature, calculate the vapor pressure using the Antoine equation, and then calculate the summation of x_i; if 1, then the temperature is the dew point temperature, and if not, consider other temperatures. To make use of the previously assumed temperature, assume T = 110°C (Table 1.3). Increase the assumed temperature to T = 130°C (Table 1.4). By interpolation, the dew point is 127.27°C.

TABLE 1.3

Dew Point Calculation at Assumed T = 110°C for the Case in Example 1.2

Component	y_i	P_v (110°C), atm	$K_i = P_v/1.5$	$x_i = y_i/K_i$
n-hexane	0.35	3.11	2.074	0.169
n-heptane	0.30	1.385	0.923	0.325
n-octane	0.25	0.623	0.417	0.600
n-nonane	0.10	0.292	0.1945	0.514
				$\Sigma y_i/K_i = 1.608$

TABLE 1.4
Dew Point Calculation at Assumed T = 130°C for the Case in Example 1.2

Component	y_i	P_v (130°C), atm	$K_i = P_v/1.5$	$x_i = y_i/K_i$
n-hexane	0.35	4.94	3.29	0.106
n-heptane	0.30	2.329	1.553	0.193
n-octane	0.25	1.12	0.747	0.335
n-nonane	0.10	0.556	0.371	0.27
				$\Sigma y_i/K_i = 0.94$

UNISIM CALCULATION

UniSim determines the dew point temperature by setting the vapor/phase fraction to 1.0 (Figure 1.5). The calculated temperature is the dew point temperature (T_{dp} = 125.6°C). Peng–Robinson is the suitable fluid package for the hydrocarbon gases listed in Table 1.4.

PRO/II CALCULATION

In a recent case in PRO/II, add all components involved (hexane, heptane, octane, and nonane); for the fluid package, select the Peng–Robinson EOS, click on stream in the object pallet, and then click anywhere in the simulation area. Double click on the stream S1 and specify pressure as 1.5 atm; as a second specification, select Dew Point from the pull-down menu. Double click on flow rate and identify the molar composition of all streams. Enter any value for total flow rate, for example, 1.0 kgmol/h. The PRO/II predicted dew point temperature is 125.811°C (Figure 1.6).

ASPEN CALCULATION

Start Aspen Plus and create New case. Add all the components involved, i.e., hexane, heptane, octane, and nonane, and their molar fractions, i.e., 0.35, 0.3, 0.25, and 0.1, respectively. The property estimation method is Peng–Robinson. The dew point temperature is to be determined; set the vapor-to-phase ratio to 1, and the pressure to 1.5 atm. The system is ready to run. The calculated dew point temperature is 399 K (125.85°C) (Figure 1.7).

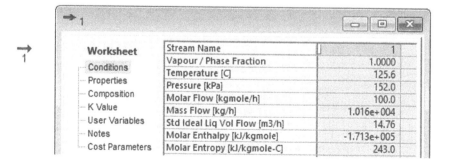

FIGURE 1.5   UniSim calculates the dew point temperature (T_{dp} = 125.6°C) for the hydrocarbon gas mixture in Example 1.2.

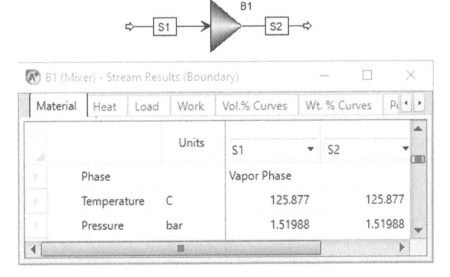

Stream Name Stream Description		S1
Phase		Vapor
Temperature Pressure	C ATM	125.811 1.500
Flowrate	KG-MOL/HR	0.454
Composition HEXANE HEPTANE NONANE OCTANE		0.350 0.300 0.100 0.250

FIGURE 1.6 PRO/II calculates the dew point temperature ($T_{dp} = 125.811°C$) for the hydrocarbon vapor mixture presented in Example 1.2.

AVEVA PROCESS SIMULATION

Create a new simulation in the Aveva Process Simulation (e.g., Example 1-2).

1. Copy DefFluid to the Example 1.2 model library, right-click on the DefFluid icon, and select Edit.
2. Click on Methods, and for system select Peng–Robinson.
3. Drag the Source icon to the canvas and set P = 1.5 atm, W = 1 kg/s, and VF = 1.

B1

⇨— S1 —▶ ◀— S2 —⇨

(A) B1 (Mixer) - Stream Results (Boundary)   — ☐   ✕

| Material | Heat | Load | Work | Vol.% Curves | Wt. % Curves | P ◀ ▶ |

	Units	S1 ▼	S2 ▼
Phase		Vapor Phase	
Temperature	C	125.877	125.877
Pressure	bar	1.51988	1.51988

FIGURE 1.7 Aspen Plus calculates the dew point temperature 399 K ($T_{dp} = 125.85°C$) for the vapor hydrocarbon mixture presented in Example 1.2.

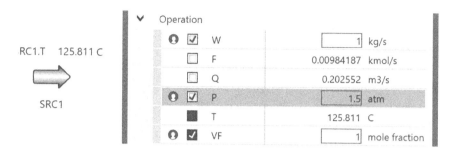

FIGURE 1.8 Aveva Process Simulation calculates the dew point temperature (T_{dp} = 125.811°C) for the vapor mixture described in Example 1.2.

4. For the Fluid Type, from the pull-down menu, select Example 1.2/DefFluid.
5. Once completed, Aveva Process Simulation is squared and solved. The calculated dew point temperature is 125.81°C (Figure 1.8).

Example 1.3: Vapor Pressure of Gas Mixture

Find the mixture's vapor pressure composed of 35 mol% n-hexane, 30% n-heptane, 25% n-octane, and 10% n-nonane at 130°C. Compare manual results with predicted results obtained from the available commercial software packages.

SOLUTION

HAND CALCULATION

Use the Antoine equation for n-hexane and the other components listed in Table 1.5.

$$\log(P_v, mmHg) = A - \frac{B}{(T+C)} = 6.84 - \frac{1168.72}{(130+224.21)} = 3722 \text{ mmHg } (4.94 \text{ atm})$$

UNISIM CALCULATIONS

Create the stream with specified conditions and compositions; select the Antoine equation for the fluid package.

TABLE 1.5

Vapor Pressure of the Gas Mixture at T = 130°C for the Case in Example 1.3

Component	y_i	P_v (130°C), atm
n-hexane	0.35	4.94
n-heptane	0.30	2.33
n-octane	0.25	1.12
n-nonane	0.10	0.556

Mixture vapor pressure = $(0.35 \times 4.94) + (0.3 \times 2.329) + (0.25 \times 1.12) + (0.1 \times 0.556) = 2.76$

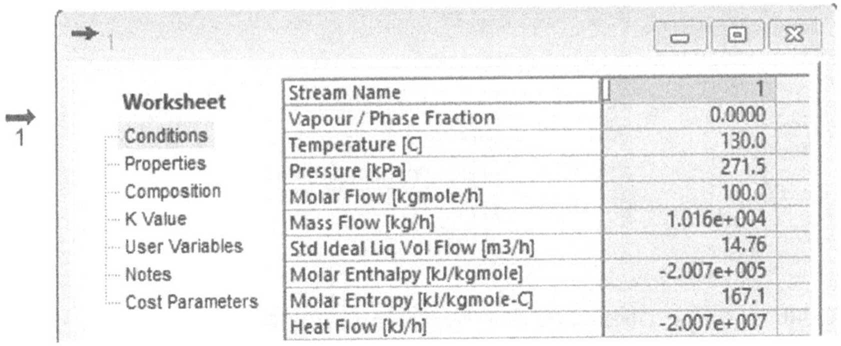

Stream Name	1
Vapour / Phase Fraction	0.0000
Temperature [C]	130.0
Pressure [kPa]	271.5
Molar Flow [kgmole/h]	100.0
Mass Flow [kg/h]	1.016e+004
Std Ideal Liq Vol Flow [m3/h]	14.76
Molar Enthalpy [kJ/kgmole]	-2.007e+005
Molar Entropy [kJ/kgmole-C]	167.1
Heat Flow [kJ/h]	-2.007e+007

Worksheet
- Conditions
- Properties
- Composition
- K Value
- User Variables
- Notes
- Cost Parameters

FIGURE 1.9 UniSim calculates the liquid mixture's vapor pressure at 130°C (271.5 kPa) for the case described in Example 1.3.

Set the vapor/phase fraction to 0.0 and the temperature to the desired temperature to find the vapor pressure. Then UniSim will calculate the pressure; the calculated pressure is the vapor pressure (or more precisely, bubble point pressure = vapor pressure) at the specified temperature. To calculate the dew point, set the vapor fraction to 1. The result depends on many parameters, for example, the selection of fluid package and components in a mixture. Using the Antoine equation, the vapor pressure of pure n-hexane at 130°C is 500.2 kPa, and the vapor pressure of pure n-heptane, n-octane, and n-nonane is 236 kPa, 113.4 kPa, and 56.33 kPa, respectively. The vapor pressure using Peng–Robinson EOS at 130 is 496 kPa, the same as obtained by the Antoine equation. For the gas mixture at 130°C, the vapor pressure is 271.5 kPa (Figure 1.9).

PRO/II CALCULATION

PRO/II estimates the gas mixture's vapor pressure specified in Example 1.3 by setting the temperature at which vapor pressure is to be calculated (in this case, 130°C). Peng–Robinson is the suitable fluid package. The second specification is the bubble point. The gas mixture has calculated 2.644 atm (267.9 kPa) vapor pressure (Figures 1.10).

ASPEN PLUS CALCULATIONS

To calculate the gas mixture's vapor pressure at 130°C with Aspen Plus, set the vapor fraction to zero. The vapor pressure is 2.644 atm (267.498 kPa) (Figure 1.11).

AVEVA PROCESS SIMULATION

Create a new simulation (e.g., Example 1.3) in Aveva Process Simulation.

1. Copy DefFluid to the Example 1.3 model library, right-click on the DefFluid icon, and select Edit.
2. Click on Methods, and for system select Peng-Robinson.
3. Drag the Source icon to the canvas and set P = 1.5 atm, W = 1 kg/s, and VF = 1.
4. For the Fluid Type, from the pull-down menu, select Example 1.3/DefFluid.
5. Once completed, Aveva Process Simulation is squared and solved. The calculated dew vapor pressure at 130°C is 267.593 kPa (Figure 1.12).

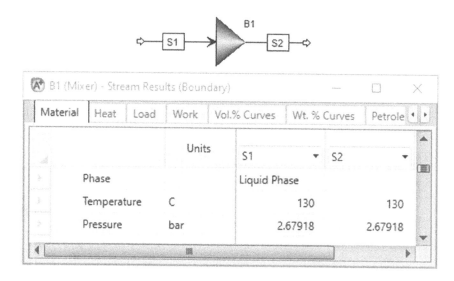

Stream Name Stream Description		S1
Phase		Liquid
Temperature Pressure	C ATM	130.000 2.644
Flowrate	KG-MOL/HR	0.454
Composition HEXANE HEPTANE NONANE OCTANE		0.350 0.300 0.100 0.250

FIGURE 1.10 PRO/II calculates the vapor pressure (267.9 kPa) of the liquid mixture at 130°C for the case described in Example 1.3.

1.6 EQUATIONS OF STATE

The equation of state (EOS) describes the required expression to relate gases' specific volume to temperature and pressure. EOS relates the molar quantity and gas volume to temperature and pressure. EOS is used to predict p, V, n, and T for real gases, pure components, or mixtures. The simplest example of an EOS is the ideal gas law [2]. EOS is formulated by collecting experimental data and calculating the coefficients in a proposed equation using statistical fitting. The literature presented numerous EOSs, the equations involving two or more coefficients. Cubic EOSs such as Redlich–Kwong

B1

S1 → ▷ → S2

A B1 (Mixer) - Stream Results (Boundary)		— □ ✕

Material	Heat	Load	Work	Vol.% Curves	Wt. % Curves	Petrole ◄ ►

	Units	S1 ▼	S2 ▼
Phase		Liquid Phase	
Temperature	C	130	130
Pressure	bar	2.67918	2.67918

FIGURE 1.11 Aspen Plus calculates the vapor pressure (267.43 kPa) of the liquid mixture at 130°C for the case described in Example 1.3.

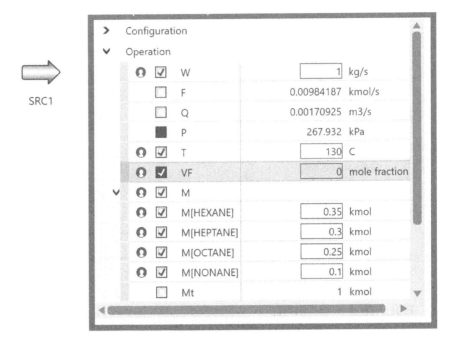

FIGURE 1.12 Aveva Process Simulation calculates the vapor pressure (267.593 kPa) of the liquid mixture at 130°C for the case described in Example 1.3.

equation of state, Soave–Redlich–Kwong, and Peng–Robinson can have an accuracy of 1–2% over an extensive range of conditions of many compounds. For solving n or V, one must solve a cubic equation that might have more than one real root. For example, Peng–Robinson EOS can easily be solved for p if V and T are given [3].

$$p = \frac{RT}{V-b} - \frac{a}{V(V+b)+b(V-b)} \tag{1.9}$$

The constants a and b are determined as follows:

$$a = 0.45724\left(\frac{R^2 T_c^2}{p_c}\right)\left[1 + m\left(1 - T_r^{1/2}\right)\right] \tag{1.10}$$

$$b = 0.07780\left(\frac{RT_c}{p_c}\right) \tag{1.11}$$

$$m = 0.37464 + 1.54226\omega - 0.26992\omega^2 \tag{1.12}$$

$$T_r = \frac{T}{T_c} \tag{1.13}$$

where ω is acentric factor, T_c and p_c are critical temperature and critical pressure, respectively, and V is specific volume.

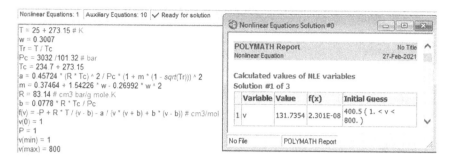

FIGURE 1.13 Polymath programs determined the specific volume (0.131 m³/kg mol) calculated for the case in Example 1.4.

Example 1.4: Specific Molar Volume of N-Hexane

Estimate the specific molar volume of n-hexane at 1 atm and 25°C. Compare the manual calculation with those predicted from the available software packages.

SOLUTION

POLYMATH CALCULATION

Equations 1.9–1.13 can be easily solved using the polymath program (Figure 1.13). The calculated molar volume is 131 cm³/mol (0.131 m³/kg mol).

UNISIM SIMULATION

UniSim calculates the molar volume of pure n-hexane at 1 atm and 25°C. Peng–Robinson is a suitable fluid package for estimating the fluid properties of the given components. Select a material stream, specify the temperature, pressure, flow rate (e.g., 100 kgmol/h), and composition. From Worksheet/Properties menu (Figure 1.14), the UniSim predicted molar volume is 0.131 m³/kgmol.

PRO/II SIMULATION

Set the stream conditions with pure n-hexane. Peng–Robinson is the suitable fluid package to estimate the physical properties of hexane. Specific molar volume is the inverse of the molar density (mass density divided by the molecular weight

Worksheet		
Conditions	Kinematic Viscosity [cSt]	0.4540
Properties	Liq. Mass Density (Std. Cond) [kg/m3]	666.2
Composition	Liq. Vol. Flow (Std. Cond) [m3/h]	0.1293
K Value	Liquid Fraction	1.000
User Variables	Molar Volume [m3/kgmole]	0.1312
Notes	Mass Heat of Vap. [kJ/kg]	336.5
Cost Parameters	Phase Fraction [Molar Basis]	0.0000
	Surface Tension [dyne/cm]	17.86
	Thermal Conductivity [W/m-K]	0.1142

FIGURE 1.14 Molar volume determined by UniSim for the case described in Example 1.4.

Stream Name		S1
Stream Description		
Phase		Vapor
Temperature	C	130.0000
Pressure	ATM	1.0000
Total Molar Comp. Concentrations	KG-MOL/M3	
HEXANE		7.7103

FIGURE 1.15   PRO/II determines the molar volume (1/7.7103 = 0.13 m³/kgmol) at 130°C for the case described in Example 1.4.

of n-hexane). Generate from the molar concentration report by clicking on the Stream Property Table in the toolbar, click anywhere in the simulation area, then double click on the generated table, and from the "Property List to be used," select the "Molar Conc. Report;" the result should look like Figure 1.15.

ASPEN CALCULATIONS

Aspen Plus determines the molar volume of pure n-hexane (7.63 kmol/m³), and converts to molar volume (1/7.63 = 0.131 m³/kmol) at 25°C and 1 atm for the case in Example 1.4. The Aspen Plus calculates molar volume from Results Summary/ Streams, and molar density is 0.131 (Figure 1.16).

SUPERPRO CALCULATION

Using the SuperPro designer, select pure component and then select n-hexane (molecular weight = 86.18). At a temperature of 25°C and pressure = 1 atm, the density of pure n-hexane can be calculated using the following equation obtained from the pure component property windows:

$$\text{Density (g/L)} = 924.33 - 0.8999T \text{ (K)}$$

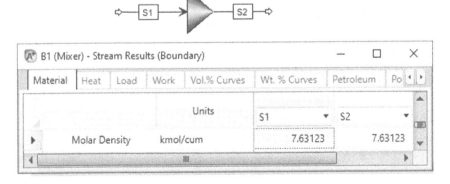

FIGURE 1.16   Aspen Plus determines the molar volume of pure n-hexane at 25°C and 1 atm for the case in Example 1.4.

SRC1.State.VI 0.1313 m3/kmol

SRC1

		Operation		
❶ ☑	W		1	kg/s
☐	F		0.260092	Nm3/s
☐	Q		0.00152361	m3/s
❶ ☑	P		1	atm
❶ ☑	T		25	C
☐	VF		-0.789061	mole fraction
❶ ☑	M			

FIGURE 1.17 Aveva Process Simulation determines the molar volume (0.131 m³/kg-mol) of pure n-hexane at 25°C and 1 atm for the case in Example 1.4.

$$\text{Density (g/L)} = 924.33 - 0.8999(298) = 656 \text{ g/L or } 656 \text{ kg/m}^3$$

Accordingly, the molar volume is 0.131 m³/kgmol.

AVEVA PROCESS SIMULATION

Create a new simulation (e.g., Example 1.4) in Aveva Process Simulation.

1. Copy DefFluid to the Example 1.4 model library, right-click on the DefFluid icon, and select Edit.
2. Click on Methods, and for system select Peng-Robinson.
3. Drag the Source icon to the canvas and set P = 1.0 atm, T = 25°C, W = 1 kg/s.
4. For the Fluid Type, from the pull-down menu, select Example 1.4/DefFluid.
5. Once completed, Aveva Process Simulation is squared and solved.
6. Right-click on the source and select Full Properties, click on SRC1.State.
7. Drag the molar volume (V1) to the canvas.
8. The calculated molar volume is 0.1313 m³/kgmol (Figure 1.17).

1.7 PHYSICAL PROPERTIES

Many correlations are available in the literature to measure physical properties such as density, viscosity, and specific heat as a function of temperature.

1.7.1 LIQUID DENSITY

For saturated-liquid molar volume, the Gunn and Yamada method is used [4].

$$\frac{V}{V_{sc}} = V_r^0 \left(1 - \omega\Gamma\right) \tag{1.14}$$

where V is the liquid-specific volume, and V_{sc} is the scaling parameter at $T_r = 0.6$.

$$V_{sc} = \frac{V_{0.6}}{0.3862 - 0.0866} \tag{1.15}$$

where $V_{0.6}$ is the saturated-liquid molar volume at a reduced temperature of 0.6. If $V_{0.6}$ is not available, then approximately V_{sc} can be estimated by

$$V_{sc} = \frac{RT_c}{P_c}(0.2920 - 0.0967\omega) \tag{1.16}$$

In most cases, V_{sc} is close to V_c. However, if the saturated-liquid molar volume is available at any temperature, V_{sc} can be eliminated, as shown later in Equation 1.17. In Equation 1.17, $V_r^{(0)}$ and Γ are functions of reduced temperature, and ω is the acentric factor.

For $0.2 \le T_r \le 0.8$

$$V_r^{(0)} = 0.33593 - 0.33953T_r + 1.5194T_r^2 - 2.02512T_r^3 + 1.11422T_r^4 \tag{1.17}$$

For $0.8 < T_r < 1.0$

$$V_r^{(0)} = 1 + 1.3(1 - T_r)^{1/2}\log(1 - T_r) - 0.50879(1 - T_r) - 0.91534(1 - T_r)^2 \tag{1.18}$$

For $0.2 \le T_r < 1.0$

$$\Gamma = 0.29607 - 0.09045T_r - 0.04842T_r^2 \tag{1.19}$$

where, $T_r = T/T_c$.

In the absence of experimental data, one may assume volume or mass additivity to calculate mixture densities from pure components.

$$\bar{\rho} = \sum_{i=1}^{n} x_i\rho_i \tag{1.20}$$

$$\frac{1}{\rho} = \sum_{i=1}^{n} \frac{x_i}{\rho_i} \tag{1.21}$$

Equation 1.21 is more accurate than Equation 1.20.

Example 1.5: Estimation of the Benzene Density

Estimate benzene's density as a function of temperature at 1 atm pressure and 0–70°C temperature range. Compare the manual results with the available software packages' predicted values.

SOLUTION

MANUAL CALCULATIONS, POLYMATH

The set of equations in Section 1.5 is solved using a polymath nonlinear equations solver, as shown in Table 1.5. Figure 1.18 shows the density of liquid benzene at 70°C calculated by polymath. Densities at different temperatures can be found by changing the value of T.

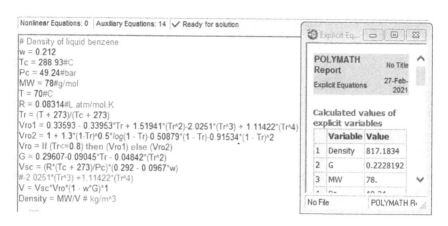

# Density of liquid benzene
w = 0.212
Tc = 288.93#C
Pc = 49.24#bar
MW = 78#g/mol
T = 70#C
R = 0.08314#L.atm/mol.K
Tr = (T + 273)/(Tc + 273)
Vro1 = 0.33593 - 0.33953*Tr + 1.51941*(Tr^2)-2.0251*(Tr^3) + 1.11422*(Tr^4)
Vro2 = 1 + 1.3*(1-Tr)^0.5*log(1 - Tr)-0.50879*(1 - Tr)-0.91534*(1 - Tr)^2
Vro = If (Tr<=0.8) then (Vro1) else (Vro2)
G = 0.29607-0.09045*Tr - 0.04842*(Tr^2)
Vsc = (R*(Tc + 273)/Pc)*(0.292 - 0.0967*w)
#-2.0251*(Tr^3) +1.11422*(Tr^4)
V = Vsc*Vro*(1 - w*G)*1
Density = MW/V # kg/m^3

POLYMATH Report — No Title

Explicit Equations — 27-Feb-2021

Calculated values of explicit variables

	Variable	Value
1	Density	817.1834
2	G	0.2228192
3	MW	78.

No File | POLYMATH R

FIGURE 1.18 Polymath calculates the liquid density using Equations 1.14 to 1.19 at 70°C, and the calculated value is 817.1834 kg/m³.

UNISIM SIMULATION

Generate a material stream, specify temperature, pressure composition, then use: Tools ≫ Utilities ≫ Property Table. Then click on Add Utility and then click on View Utility. Click on Select Stream to select stream 1, and fill in the popup menu as shown in Figure 1.19. Click on Calculate. From the performance page, select press Plot. Figure 1.20 shows the change in mass density as a function of temperature.

PRO/II SIMULATION

With PRO/II, it is easy to find the physical properties of components such as density, viscosity, and surface tension as a function of temperature. After opening PRO/II, click on the Input button in the toolbar and while in the input menu, click on Launch TDB (Thermo Data Manager). Under Data Bank Type, select SIMSCI and select benzene from the components (Figure 1.21). Click on TempDep, and then choose the density, followed by liquid (Figure 1.21). Choose the variables (Temperature for this example) with which the physical properties will vary. For example, to find density as a temperature function at 1 atm pressure, the variable

FIGURE 1.19 Property table is used in UniSim to measure the density's property as a function of temperature for the case present in Example 1.5.

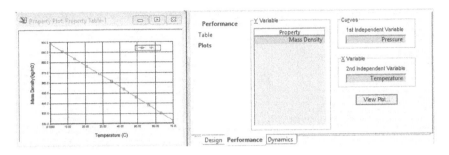

FIGURE 1.20 UniSim generated the plot of density versus temperature for the case described in Example 1.5.

would be temperature. Before we can continue to calculate the physical properties, we need to vary our temperature over a range. Highlight the temperature by clicking on the box in front of the chosen temperature. Once done, an arrow appears next to the selected temperature, and the box is highlighted. Then click on the Range/List at the bottom of the screen (Figure 1.22).

Aspen Plus Simulation

1. Start Aspen Plus, click New, then create a blank case; benzene is the only targeted component for this example.
2. Click on Method, then Specifications on the arrow pointing downward in the Base method box, scroll down, and from the list of options, choose NRTL.
3. Click on the Next button in the toolbar to continue. A popup screen appears; click on OK to continue.
4. Click on the Pure button in the toolbar. In the new menu, for the property, select RHO, and for component, select benzene and click ">" to move benzene from available to selected components (Figure 1.23).

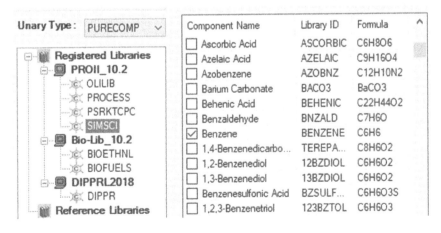

FIGURE 1.21 PRO/II selection of benzene component using the Input/Launch TDM for the case described in Example 1.5.

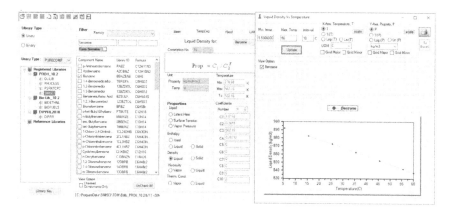

FIGURE 1.22   PRO/II calculates benzene's density as a function of temperature for the case described in Example 1.5.

5. Select kg/cum as units to RHO and C as units of temperature.
6. Click on Run Analysis (Figure 1.24).

SuperPro Designer

With SuperPro, it is much easier to plot physical properties (density in this case) versus adjustable variables (temperature). Select the components involved (benzene), double click on the components name, and click on the Physical (T-dependent) button. Click on the plot of any T Dependent Property. Then click on Show Graph. The graph should appear like that in Figure 1.25.

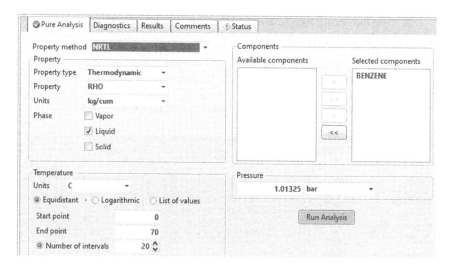

FIGURE 1.23   Aspen Plus pure analysis menu is used to measure pure component physical properties (density of benzene versus temperature) for the case described in Example 1.5.

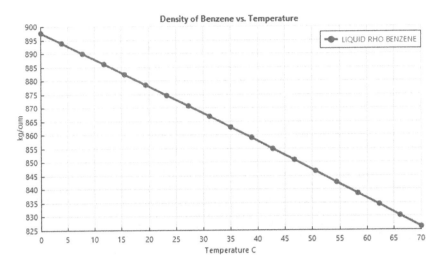

FIGURE 1.24 Aspen Plus generates benzene density versus temperature for the case described in Example 1.5.

AVEVA PROCESS SIMULATION

To estimate benzene's density as a function of temperature at 1 atm pressure and 0–70°C temperature range, create a new simulation (e.g., Example 1.5) in Aveva Process Simulation.

1. Copy DefFluid to the Example 1.5 submodel library, right-click on the DefFluid icon, and select Edit.
2. Click on Methods, and for system select Peng-Robinson.
3. Drag the Source icon to the canvas and set P = 1.0 atm, W = 1 kg/s.
4. For the Fluid Type, from the pull-down menu, select Example 1.5/DefFluid.
5. To create a new curve, go to Model Library; go to Simulation tab, i.e., Example 1.5, right-click and select Create New Curve.
6. Then follow the instructions manual "Aveva Process Simulation Building guide.pdf" under chapter 12 "Curve Editor".
7. Figure 1.26 shows the effect of temperature on benzene density.

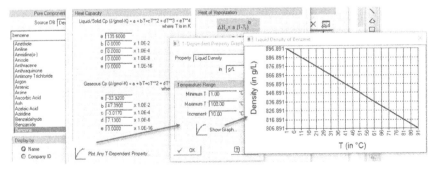

FIGURE 1.25 SuperPro generated the plot of density versus temperature for the case described in Example 1.5.

FIGURE 1.26 Aveva Process Simulation generated the plot of density versus temperature (0–70°C) for the case described in Example 1.5.

Example 1.6: Estimate Density of Liquid Mixture

The density of 50 wt% H_2SO_4 in water at 25°C and 1 atm is 1.39 kg/m³. Estimate the liquid mixture's density using the following densities of pure H_2SO_4 and water and then compare it with the experimentally obtained value. Density of H_2SO_4 at 25°C = 1.834 g/cm³, and the density of H_2O at 25°C = 0.998 g/cm³. Compare the manual calculation with UniSim, PRO/II, Aspen Plus, SuperPro, and Aveva Process Simulation.

SOLUTION

HAND CALCULATIONS

The density of liquid mixture (method 1)

$$\bar{\rho} = \sum_{i=1}^{n} x_i \rho_i = 0.5 \times 0.998 + 0.5 \times 1.834 = 1.42 \text{ g/cm}^3 \left(1420 \frac{\text{kg}}{\text{m}^3} \right)$$

The density of liquid mixture (method 2)

$$\frac{1}{\rho} = \sum_{i=1}^{n} \frac{x_i}{\rho_i} = \frac{0.5}{0.998} + \frac{0.5}{1.834} \rightarrow \bar{\rho} = 1.29 \text{ g/cm}^3 \left(1290 \frac{\text{kg}}{\text{m}^3} \right)$$

The percent error using the first and second equation is 7.3% and 1.5%, respectively.

UNISIM SIMULATION

In a new case in UniSim, select the two components (H_2SO_4 and water) and Peng–Robinson–Stryjek–Vera (PRSV) equation of state for the property estimation. Select the material stream, specify the temperature as 25°C, and set the pressure to 1 atm. The basis of the assumption is 100.0 kmol/h of the mixture. Figure 1.27 shows the predicted result from the stream properties (1,391 kg/m³).

PRO/II SIMULATION

Start a new case in PRO/II and add the two components involved in the mixture (H_2SO_4 and water). Select PRSV for the property estimation method. Select a material stream, and specify its temperature as 25°C and pressure as 1 atm. The basis of the assumption is 1.0 kmol/h of the mixture. Figure 1.28 shows the simulated result.

Worksheet	Stream Name	1	A ^
Conditions	Molecular Weight	30.44	
Properties	Molar Density [kgmole/m3]	45.71	
Composition	Mass Density [kg/m3]	1391	
K Value	Act. Volume Flow [m3/h]	2.188	
User Variables	Mass Enthalpy [kJ/kg]	-1.274e+004	
Notes	Mass Entropy [kJ/kg-C]	0.8211	
Cost Parameters	Heat Capacity [kJ/kgmole-C]	111.6	
	Mass Heat Capacity [kJ/kg-C]	3.666	

FIGURE 1.27 UniSim calculates 50 wt% H_2SO_4 in water liquid mixture density (1,391 kg/m³) for the case described in Example 1.6.

ASPEN SIMULATION

Create a new case in Aspen Plus, add the two components (H_2SO_4 and water), and select the Peng–Robinson for property estimation. Select a material stream, specify the temperature as 25°C and pressure as 1 atm. The basis of the assumption is 1.0 kmol/h of the mixture. Figure 1.29 shows the Aspen Plus predicted density of the liquid mix (1,408.45 kg/m³).

AVEVA PROCESS SIMULATION

Start Aveva Process Simulation, create a new simulation, and rename it (e.g., Example 1.6).

1. Copy DefFluid to the Example 1.6 model library, right-click on the DefFluid icon, and select Edit.
2. Click on Methods, and for the system, select NRTL.
3. Click on Component List and add H_2SO_4, delete N_2 and O_2.
4. Drag the Source icon to the canvas, specify P = 1.5 atm, T = 25°C, and W = 1 kg/s.
5. For the Fluid Type, from the pull-down menu, select Example 1.6/DefFluid.
6. Once completed, Aveva Process Simulation is squared and solved. The calculated density is 1,288.77 kg/m³ (Figure 1.30).

Example 1.7: Use of Henry's Law

A gas containing 1.0 mol% of ethane and the remaining nitrogen comes into direct contact with water at 20.0°C and 20.0 atm. Estimate the mole fraction of dissolved ethane in the water stream.

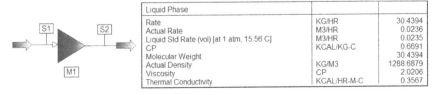

Liquid Phase		
Rate	KG/HR	30.4394
Actual Rate	M3/HR	0.0236
Liquid Std Rate (vol) [at 1 atm, 15.56 C]	M3/HR	0.0235
CP	KCAL/KG-C	0.6691
Molecular Weight		30.4394
Actual Density	KG/M3	1288.6879
Viscosity	CP	2.0206
Thermal Conductivity	KCAL/HR-M-C	0.3567

FIGURE 1.28 PRO/II calculates the liquid mixture density composed of 50 wt% H_2SO_4 in water (1,289 kg/m³) for the case described in Example 1.6.

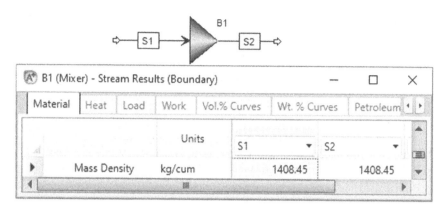

FIGURE 1.29 Aspen Plus calculates the mixture density composed of 50 wt% H_2SO_4 in water (1,408.45 kg/m³) using the case described in Example 1.6.

SOLUTION

MANUAL CALCULATIONS

Hydrocarbons are relatively insoluble in water, so ethane's solution is likely to be very dilute. We should therefore assume that Henry's law applies [5] and look up Henry's constant for ethane in water:

$$y_A P = x_A H_A(T) \rightarrow x_A = \frac{y_A P}{H_A(T)} = \frac{(0.100)(20.0 \text{ atm})}{2.63 \times 10^4 \text{ atm}} = 7.60 \times 10^{-6} \frac{\text{mol } C_2H_6}{\text{mol}}$$

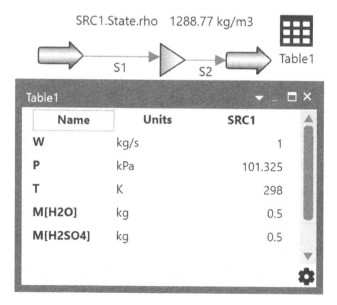

FIGURE 1.30 Aveva Process Simulation calculates the mixture density of 50% H_2SO_4 in water using the case described in Example 1.6.

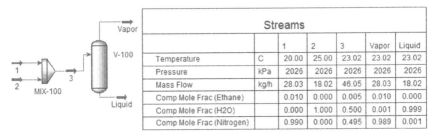

Streams			1	2	3	Vapor	Liquid
Temperature	C		20.00	25.00	23.02	23.02	23.02
Pressure	kPa		2026	2026	2026	2026	2026
Mass Flow	kg/h		28.03	18.02	46.05	28.03	18.02
Comp Mole Frac (Ethane)			0.010	0.000	0.005	0.010	0.000
Comp Mole Frac (H2O)			0.000	1.000	0.500	0.001	0.999
Comp Mole Frac (Nitrogen)			0.990	0.000	0.495	0.989	0.001

FIGURE 1.31   UniSim generates ethane mole fractions saturated with water vapor at 20°C and 20 atm for the case described in Example 1.7.

UniSim Simulation

In a recent case in UniSim, add the components ethane and water, and select the appropriate fluid package (NRTL). Enter the simulation environment and mix the two streams. The Workbook is used to display the stream summary table below the process flowsheet. Click on Workbook in the toolbar; once the Workbook appears, click on Setup in the Workbook menu, and then click on Add to add the required variables from the list of variables. Once all necessary information is added to the Workbook, right-click anywhere in the PFD area and select Add Workbook Table (Figure 1.31).

PRO/II Simulation

Start PRO/II and create a new case, add Ethane, Nitrogen, and Water, and select NRTL as the appropriate fluid package. Connect two feed streams to a mixer and the outlet stream to a flash unit. Figure 1.32 shows the PFD and stream summary generated by PRO/II for the case described in Example 1.7.

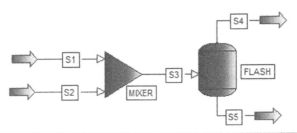

Stream Name		S1	S2	S4	S5
Phase		Vapor	Liquid	Vapor	Liquid
Temperature	C	20.00	20.00	19.63	19.63
Pressure	KG/CM2	20.66	20.00	20.00	20.00
Flowrate	KG-MOL/HR	1.00	1.00	0.98	1.02
Total Molar Comp. Rates	KG-MOL/HR				
ETHANE		0.010	0.000	0.010	0.000
NITROGEN		0.990	0.000	0.969	0.021
WATER		0.000	1.000	0.001	0.999

FIGURE 1.32   PRO/II generates ethane mole fractions saturated with water vapor at 20°C and 20 atm for the case described in Example 1.7.

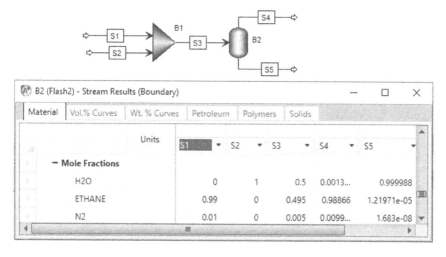

FIGURE 1.33 Aspen Plus generates ethane mole fractions saturated with water vapor at 20°C and 20 atm for the case described in Example 1.7.

ASPEN PLUS SIMULATION

Start Aspen Plus and create a new case, add the components ethane, nitrogen, and water, and select Peng–Robinson as the appropriate fluid package. Connect the two feed streams to a mixer and then the mixer's outlet stream to a flash unit. Figure 1.33 shows the Aspen Plus generated PFD and stream summary.

AVEVA PROCESS SIMULATION

Start Aveva Process Simulation, create a new simulation, and rename it (e.g., Example 1-7). Copy DefFluid to the Example 1-7 model library, right-click on the DefFluid icon and select Edit, and then follow the steps:

1. Click on Methods, and for the system, select NRTL.
2. Click on Component List and add ethane, delete O_2 from the DefFluid default components.
3. Drag to the canvas and connect the two Source icons, the mixer, and the flash Drum.
4. Set P = 20 atm, T = 20°C, W = 1 kg/s for the feed streams.
5. For the Fluid Type, from the pull-down menu, select Example 1-7/ DefFluid.
6. Once completed, Aveva Process Simulation is squared and solved.
7. The calculated mole fraction of the vapor phase is shown in Figure 1.34.

Example 1.8: Raoult's Law for Hydrocarbon Mixtures

An equal molar mixture of benzene and toluene is in equilibrium with its vapor at 30.0°C. What is the mole fraction of benzene and toluene in the vapor phase?

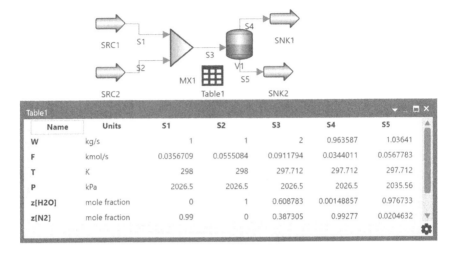

FIGURE 1.34 Aveva Process Simulation generates ethane mole fractions saturated with water vapor at 20°C and 20 atm for the case described in Example 1.7.

SOLUTION

HAND CALCULATIONS

Assuming Raoult's law applies [5],
The vapor pressure of benzene, P_B^*

$$\log\left(p_B^*\right) = 6.906 - \frac{1211}{T + 2208} \xrightarrow{T=30°C} p_B^* = 119 \text{ mmHg } \left(15.86 \text{ kPa}\right)$$

Vapor pressure of toluene, P_T^*

$$\log\left(p_T^*\right) = 6.9533 - \frac{1343.9}{T + 219.38} \xrightarrow{T=30°C} p_T^* = 36.7 \text{ mmHg } \left(4.89 \text{ kPa}\right)$$

The partial pressure of benzene, P_B

$$p_B = y_B P = x_B P_A^*\left(T\right) = 0.5\left(119\right) = 59.5 \text{ mmHg } \left(7.93 \text{ kPa}\right)$$

The partial pressure of toluene, P_T

$$p_T = y_T P = x_T p_A^*\left(T\right) = 0.5\left(36.7\right) = 18.4 \text{ mmHg } \left(2.45 \text{ kPa}\right)$$

The total pressure, P

$$P = 59.5 + 18.4 = 77.9 \text{ mm Hg } (10.4 \text{ kPa})$$

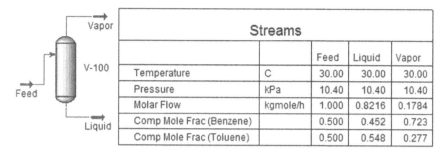

Streams			Feed	Liquid	Vapor
Temperature	C		30.00	30.00	30.00
Pressure	kPa		10.40	10.40	10.40
Molar Flow	kgmole/h		1.000	0.8216	0.1784
Comp Mole Frac (Benzene)			0.500	0.452	0.723
Comp Mole Frac (Toluene)			0.500	0.548	0.277

FIGURE 1.35 UniSim calculates benzene and toluene's mole fractions for the case described in Example 1.8.

Mole fraction of benzene in the vapor phase, y_B

$$y_B = \frac{p_B}{P} = 0.764$$

Mole fraction of toluene in the vapor phase, y_T

$$y_T = \frac{p_T}{P} = 0.236$$

UniSim Simulation

Start UniSim and in a new case, add benzene and toluene components and select Peng–Robinson equation of state is a proper fluid package for hydrocarbons. Select the Separator from the object palate, then connect the feed to two exit streams. The feed stream is at 30°C, and the vapor fraction is zero. Figure 1.35 shows the PFD and stream summary generated by UniSim for the case described in Example 1.8.

PRO/II Simulation

In a recent case in PRO/II, add benzene and toluene components, Peng–Robinson, as a proper fluid package for hydrocarbons. Select a flash unit and connect two feed streams and two product streams; the feed stream is 30°C and the calculated manual pressure is 10.4 kPa. Figure 1.36 shows the process flowsheet and stream summary generated by PRO/II for the case described in Example 1.8.

Aspen Plus Simulation

In a recent case in Aspen Plus, add benzene and toluene components and select the appropriate fluid package (Peng–Robinson) for hydrocarbons. The feed stream is at 30°C and the vapor fraction is zero. The mole fraction of benzene and toluene in the vapor phase is 0.70 and 0.30, respectively. Figure 1.37 shows the process flowsheet and stream summary generated with Aspen Plus.

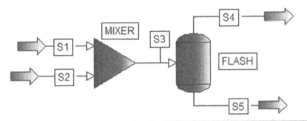

Stream Name		S1	S2	S3	S4	S5
Phase		Liquid	Liquid	Liquid	Vapor	Liquid
Temperature	C	30.0	30.0	15.0	30.0	30.0
Pressure	atm	0.2	0.0	0.0	0.1	0.1
Flowrate	KG-MOL/HR	1.0	1.0	2.0	0.0	2.0
Composition						
BENZENE		1.00	0.00	0.50	0.76	0.50
TOLUENE		0.00	1.00	0.50	0.24	0.50

FIGURE 1.36 PRO/II calculates benzene and toluene's mole fractions at equilibrium for the case described in Example 1.8.

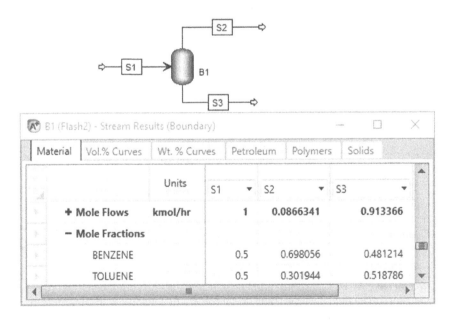

FIGURE 1.37 Aspen Plus calculates benzene and toluene's mole fractions at equilibrium for the case described in Example 1.8.

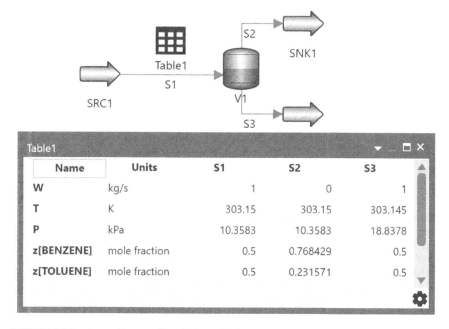

Table1					▼ ☐ ✕
Name	**Units**	**S1**	**S2**	**S3**	
W	kg/s	1	0	1	
T	K	303.15	303.15	303.145	
P	kPa	10.3583	10.3583	18.8378	
z[BENZENE]	mole fraction	0.5	0.768429	0.5	
z[TOLUENE]	mole fraction	0.5	0.231571	0.5	

FIGURE 1.38 Aveva Process Simulation calculates benzene and toluene's mole fractions at equilibrium for the case described in Example 1.8.

Aveva Process Simulation

Start Aveva Process Simulation and create a new simulation (e.g., Example 1-8)

1. Copy DefFluid to the Example 1-8 model library, right-click on the DefFluid icon, and select Edit.
2. Click on Methods, and for the system, leave the default selected system SRK.
3. Drag to the canvas and then connect Source icon, Drum, and two Sinks.
4. Click on the source stream and set VF = 0, T = 30°C, and W = 1 kg/s.
5. Compositions of benzene and toluene (equal molar) M: 0.5 kmol each.
6. For the Fluid Type, from the pull-down menu, select Example 1.8/DefFluid.
7. Once completed, Aveva Process Simulation is squared and solved. The calculated stream compositions are shown in Figure 1.38.

PROBLEMS

1.1 DEW POINT CALCULATION

Calculate the temperature and composition in equilibrium with a gas mixture containing 10.0 mol% benzene, 10.0 mol% toluene, and balance nitrogen (considered none condensable) at 1 atm. The recommend fluid package is Peng–Robinson.

1.2 COMPRESSIBILITY FACTORS

Fifty cubic meters per hour of methane flow through a pipeline at 40.0 bar absolute and 300.0 K. Estimate the mass flow rate. The appropriate fluid package for hydrocarbon compounds is Peng–Robinson.

1.3 USE OF RAOULT'S LAW

A liquid mixture contains 40% (mole percent) benzene, and the remaining is toluene in equilibrium with its vapor at 30.0°C. Use UniSim, PRO/II, and Aspen Plus. The appropriate fluid package for liquid compounds is NRTL. What are the system pressure and the composition of the vapor?

1.4 USE OF RAOULT'S LAW

A liquid mixture contains an equal molar of benzene and toluene equilibrium with its vapor at 0.12 atm. What are the system temperature and the composition of the vapor phase? Use UniSim, PRO/II, and Aspen Plus. The applicable fluid package for liquid compounds is NRTL-ideal.

1.5 USE OF HENRY'S LAW

A gas containing 1.00 mol% of ethane and the remaining methane contacted water at 20.0°C and 20.0 atm. Estimate the mole fraction of dissolved ethane and methane using UniSim, PRO/II, and Aspen Plus. The recommend fluid package is Peng–Robinson.

1.6 USE OF HENRY'S LAW

A gas containing 15.00 mol% of CO_2 and the balance is methane comes into direct contact with water at 20.0°C and 20.0 atm. Estimate the mole fraction of dissolved CO_2 in water using UniSim, PRO/II, and Aspen Plus. The recommend fluid package is Peng–Robinson.

1.7 DEW POINT CALCULATION

Find the dew point temperature for a mixture of 45 mol% n-hexane, 30% n-heptane, 15% n-octane, and 10% n-nonane at 2.0 total atm pressure. The recommend fluid package is Peng–Robinson.

1.8 BUBBLE-POINT CALCULATION

Find the bubble point temperature for a mixture of 45 mol% n-hexane, 30% n-heptane, 15% n-octane, and 10% n-nonane at 5.0 total atm pressure. The recommend fluid package is Peng–Robinson.

1.9 VAPOR PRESSURE OF GAS MIXTURE

Find the vapor pressure for the binary mixture of 50 mol% n-hexane and 50% n-heptane at 120°C. The recommend fluid package is Peng–Robinson.

1.10 VAPOR PRESSURE OF GAS MIXTURE

Find the vapor pressure for the pure components and the mixture of 35 mol% n-hexane, 30% n-heptane, and 35% n-octane at 150°C. The recommend fluid package is Peng–Robinson.

REFERENCES

1. McCabe, W. L., J. C. Smith and P. Harriot, 2005. Unit Operations of Chemical Engineering, 7th edn, McGraw-Hill, New York, NY.
2. Riggs, J. B. and D. M. Himmelblau, 2012. Basic Principles and Calculations in Chemical Engineering, 8th edn, Prentice-Hall, Englewood Cliffs, NJ.
3. Peng, D. Y. and D. B. Robinson, 1976. A new two-constant equation of state, Industrial and Engineering Chemistry: Funding, 15, 59.
4. Reid, R. C., J. M. Prausnitz and T. K. Sherwood, 2001. The Properties of Gases and Liquids, 5th edn, McGraw-Hill, New York, NY.
5. Smith, J. M., H. C. Vaness, M. M. Abbott and M. Swihart, 2018. Introduction to Chemical Engineering Thermodynamics, 7th edn, McGraw-Hill, New York, NY.

2 Fluid Flow in Pipes, Pumps, and Compressors

At the end of this chapter, students should be able to:

1. Define the type of flow regime in pipes, pumps, and compressors.
2. Determine pressure drop in the pipeline, the pipe inlet rate, and pipe length.
3. Define the valuable power input needed to overcome the friction losses in a pipeline.
4. Calculate Brake Power for pumps and compressors.
5. Verify the manual calculations with UniSim/Hysys, PRO/II, Aspen Plus, and Aveva Process Simulation software.

2.1 FLOW IN PIPES

This section introduces the most general form of Bernoulli's equation for steady incompressible flows. Bernoulli's equation is composed of kinetic energy, potential energy, and internal energy. Equation 2.1 represents the energy balance equation between the inlet at point 1 and the pipeline's exit at point 2 (Figure 2.1). The process flow diagram (PFD) represents sudden contraction (i.e., the connection of the exit of the first tank to the inlet of pipe 1), two 90° elbows, and sudden expansion (i.e., the connection of the exit of pipe four and the inlet of the second tank). The energy equation for incompressible fluids [1, 2]:

$$\frac{P_1}{\rho} + gz_1 + \frac{V_1^2}{2} = \frac{P_2}{\rho} + gz_2 + \frac{V_2^2}{2} + W_s + \sum F \qquad (2.1)$$

where P_1 is the pressure at point 1 and P_2 is the pressure at point 2, ρ is the average fluid density, z_1 is the height at point 1, z_2 is the height at point 2, V_1 is the inlet velocity, V_2 is the exit velocity, W_s is the shaft work, and ΣF is the summation of friction losses. The friction loss is due to pipe skin friction, expansion losses, contraction losses, and fitting losses [3].

The summation of friction loss:

$$\sum F = f \frac{L}{D} \frac{V_2^2}{2} + \left(K_{exp} + K_c + K_f \right) \frac{V_2^2}{2} \qquad (2.2)$$

where K_{exp} is the expansion loss, K_c is the contraction loss, and K_f is the fitting loss. Fitting losses include losses due to elbows (K_e), tees (K_T), and globe valves (K_G). Expansion loss (K_{exp}) is determined using

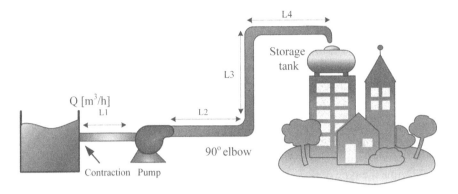

FIGURE 2.1   Water transport from the first tank to the elevated storage tank.

$$K_{exp} = \left(1 - \frac{A_1}{A_2}\right)^2 \tag{2.3}$$

where A_1 and A_2 are the cross-sectional areas at the inlet and the exit, respectively. The contraction loss (K_c):

$$K_c = 0.55\left(1 - \frac{A_2}{A_1}\right) \tag{2.4}$$

For turbulent flow, $K_e = 0.75$ (for 90° elbow), $K_T = 1.0$ (tee), $K_G = 6.0$ (globe valve), $K_C = 2.0$ (check valve). For a horizontal pipe with the same inlet and exit diameter and incompressible fluid, $V_1 = V_2$. To calculate the pressure drop between the inlet and the exit of a horizontal pipe, calculate the average velocity and then use the Reynolds number (Re) to determine the flow regime (i.e., laminar, transient, or turbulent). The average velocity expressed in terms of the flow rate:

$$V = \frac{Q}{A_c} = \frac{Q}{\pi D^2 / 4} \tag{2.5}$$

where V is the average velocity, A_c is the pipe's inner cross-sectional area, Q is the inlet fluid volumetric flow rate, and D is the pipe's inner diameter. Reynolds number is expressed as follows:

$$Re = \frac{\rho\, VD}{\mu} \tag{2.6}$$

2.1.1 LAMINAR FLOW

In fully developed laminar flow (Re < 4,000) in a circular horizontal pipe, the pressure loss and the head loss are given by

$$\frac{\Delta P}{\rho} = \frac{P_1 - P_2}{\rho} = \Delta P_L = f \frac{L}{D} \frac{V^2}{2} \qquad (2.7)$$

The friction factor

$$f = \frac{64}{Re} \qquad (2.8)$$

Under laminar flow conditions, the friction factor, f, is directly proportional to viscosity and inversely proportional to the velocity, pipe diameter, and fluid density. The friction factor is independent of pipe roughness in the laminar flow because the disturbances caused by surface roughness are damped by viscosity [4]. The pressure drop in laminar flow for a circular horizontal pipe is

$$\Delta P_L = \frac{32\mu L V}{D^2} \qquad (2.9)$$

When the flow rate and the average velocity are held constant, the head loss becomes proportional to viscosity. The head loss, h_L, is related to the pressure loss by

$$h_L = \frac{\Delta P_L}{\rho g} = \frac{31\mu L V}{\rho g D^2} \qquad (2.10)$$

2.1.2 TURBULENT FLOW

When the flow is turbulent, the relationship becomes more complex and is best shown by the graph because the friction factor is a function of both Reynolds number and roughness [5, 6]. The degree of roughness is designated as the sand grain diameter ratio to the pipe diameter (ε/D). The pipe roughness determines the relationship between the friction factor and Reynolds number. From these relationships, it is apparent that roughness is essential in determining the friction factor's magnitude for rough pipes. At a high Reynolds number, the friction factor depends entirely on roughness and the friction factor obtained from the rough pipe laws. In fully developed turbulent flow (Re > 4,000) in a circular pipe, the pressure drop for turbulent flow is

$$\Delta P = \Delta P_L = f \frac{L}{D} \frac{\rho V^2}{2} \qquad (2.11)$$

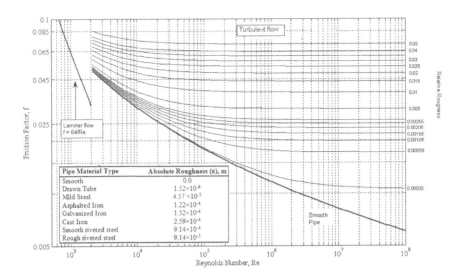

FIGURE 2.2 Moody diagram measures the pipe friction factor at various Reynolds number and the pipe material.

The friction factor, f, can be found from the Moody diagram (Figure 2.2) based on the Colebrook equation in the turbulent regime [7].

$$\frac{1}{\sqrt{f}} = -2\log\left(\frac{\varepsilon/D}{3.7} + \frac{2.51}{Re\sqrt{f}}\right) \tag{2.12}$$

Alternatively, the explicit equation for the friction factor derived by Swamee and Jain estimates the absolute roughness (Equation 2.13).

$$f = \frac{0.25}{\left[\log\left((\varepsilon/3.7\ D) + \left(5.74/Re^{0.9}\right)\right)\right]^2} \tag{2.13}$$

Head loss for turbulent flow

$$h_L = \frac{\Delta P_L}{\rho g} \tag{2.14}$$

The pipe's relative roughness is ε/D, where ε is pipe roughness and D is the pipe's inner diameter. Moody diagram (Figure 2.2) determines the friction factor using Equation 2.13. The valuable power input is the amount needed to overcome the frictional losses in the pipe.

$$\dot{W}_{pump} = Q\Delta P \tag{2.15}$$

Example 2.1: Pressure Drop in a Horizontal Pipe

Water is flowing in a 10-m horizontal smooth pipe at 4.0 m/s and 25°C. The water density is 1,000 kg/m³, and the viscosity of water is 0.001 kg/m/s. The tube is Schedule 40, 1-inch nominal diameter (2.66 cm insider diameter [ID]). Water inlet pressure is 2 atm. Calculate pressure drop in the pipe using manual calculations, compare the results with those obtained using UniSim, PRO/II, Aspen Plus, and Aveva Process Simulation.

SOLUTION

HAND CALCULATIONS

Reynolds number determines the flow regime:

$$Re = \frac{\rho VD}{\mu} = \frac{\left(1000\frac{kg}{m^3}\right)\left(4\frac{m}{s}\right)(0.0266 \text{ mm})}{\left(0.001\frac{kg}{m.s}\right)} = 1.064 \times 10^5$$

Since Reynolds number is more significant than 4,000, the flow is turbulent. The relative roughness of the smooth pipe is

$$\varepsilon/D = \frac{0}{0.04 \text{ m}} = 0$$

The friction factor, f, can be determined from the Moody chart (Figure 2.2) or Swamee and Jain alternative equation,

$$f = \frac{0.25}{\left[\log\left((0/3.7\ D)+\left(5.74/\left(1.064\times10^5\right)^{0.9}\right)\right)\right]^{0.9}} = 0.0176$$

The calculated friction factor, f = 0.0176. Then the pressure drop

$$\Delta P = P_1 - P_2 = \Delta P_L = f\frac{L}{D}\rho\frac{V^2}{2}$$

$$\Delta P = (0.0176)\left(\frac{10m}{0.0266m}\right)\frac{\left(1000\frac{km}{m^2}\right)\left(4\frac{m}{s}\right)^2}{2}\left(\frac{1\ kN}{1000\ kg\cdot m/s}\right)\left(\frac{1\ kPa}{1\ kN/m^2}\right)$$

$$= 52.93 \text{ kPa}$$

The head loss

$$h_L = \frac{\Delta P_L}{\rho g} = f\frac{L}{D}\frac{V^2}{2g} = 0.0176\frac{10 \text{ m}}{0.0266 \text{ m}}\frac{(4m/s)^2}{2(9.81 \text{ m/s}^2)} = 5.13 \text{ m}$$

The volumetric flow rate

$$Q = VA = (4\text{m/s})\left(\frac{\pi(0.0266 \text{ m})^2}{4}\right) = 2.22 \times 10^{-3} \text{ m}^3/\text{s}$$

The power input is needed to overcome the frictional losses in the pipe.

$$\dot{W}_{pump} = Q\Delta P = \left(2.22 \times 10^{-3} \text{m}^3/\text{s}\right)(50.4 \text{ kPa})\left(\frac{1 \text{ kW}}{1 \text{ kPa} \cdot \text{m}^3/\text{s}}\right) = 0.12 \text{ kW}$$

Therefore, a power input of 0.12 kW is needed to overcome the pipe's frictional losses.

UniSim Simulation

In UniSim, the object palette's pipe segment offers three calculation modes: pressure drop, flow rate, and pipe length. The information provided helps in selecting the appropriate model. The UniSim/Hysys simulation of fluid flow in a pipe:

1. Start a new case in UniSim and use the SI units from the Tools menu, Preferences, and Variables. Choose water as the component flowing in the pipe, and the American Society of Mechanical Engineers (ASME) Steam as Property packages and click Enter the simulation Environment.
2. Select a material stream by double-clicking on the blue arrow from the top of the object palette and fill in the Stream Name: Inlet.
3. Specify the volumetric feed rate, Q, based on the velocity of 4 m/s and the inner pipe diameter of 0.0266 m,

$$Q = A_c \times V = \frac{\pi(0.0266 \text{ m})^2}{4} \times \frac{4 \text{ m}}{\text{s}}\left(\frac{3600 \text{ s}}{1 \text{ h}}\right) = 8.03\frac{\text{m}^3}{\text{h}}$$

4. Enter values for feed pressure, temperature, and volumetric flow rate (Figure 2.3). In the composition menu, enter the mole fraction as 1 for water.
5. Add the pipe segment by double-clicking on the pipe segment in the object palette. Click on the Rating tab and then on Add Segment. The pipe length is 10 m; specify the Pipe Material as "smooth" by choosing this value from the drop-down list (see Figure 2.4).
6. Click on View Segment and select Schedule 40. Then click on the nominal diameter entry and select 1 inch diameter. To choose one of the options, click on 25.4 mm (1 inch) and select Specify (Figure 2.5).
7. Double click on the product stream and enter 25°C for its temperature (isothermal operation). Display the stream summary table below the PFD in the simulation area by clicking on the toolbar's Workbook icon. Once the Workbook appears from the workbook menu, click on the Setup command, click on Add in the workbook tabs group, and select the variable that needs to appear in the table. Right click on the PFD area below

Worksheet	Stream Name	Inlet
·· Conditions	Vapour / Phase Fraction	0.0000
·· Properties	Temperature [C]	25.00
·· Composition	Pressure [kPa]	202.6
·· K Value	Molar Flow [kgmole/h]	443.3
·· User Variables	Mass Flow [kg/h]	7987
·· Notes	Std Ideal Liq Vol Flow [m3/h]	8.003
···· Cost Parameters	Molar Enthalpy [kJ/kgmole]	-2.850e+005
	Molar Entropy [kJ/kgmole-C]	6.613
	Heat Flow [kJ/h]	-1.264e+008
	Liq Vol Flow @Std Cond [m3/h]	7.995
	Fluid Package	Basis-1
	Phase Option	Multiphase

FIGURE 2.3 Feed stream conditions required by UniSim for the case described in Example 2.1.

the PFD and click on Add Workbook Table; the result seems as given in Figure 2.5. While on the Design page, click on the parameters; the UniSim calculated pressure drop is 51.38 kPa (see Figure 2.6).

PRO/II SIMULATION

Start PRO/II and click on New to create a new simulation. Click on the component selection icon (the benzene ring in the toolbar), select water. Click on the Thermo button and select the stream fluid package (BWRS01). Click on the pipe segment in the object palette, and then click anywhere in the PFD area to place the pipe. Click on Streams in the object palette, then generate the inlet stream (S1) and the exit stream (S2). Double click on stream S1, specify inlet temperature, and pressure feed stream. Click on Flow rate and composition, determine inlet flow rate, and stream composition (Figure 2.7).

Double click on the Pipe icon in the PFD and specify nominal pipe diameter, pipe length, elevation change, and K factor when available. For a smooth pipe, the

Rating	Length - Elevation Profile		Pipe Info: PIPE-100	
Sizing	Segment	1	Pipe Parameters	
Pipeline Profile	Fitting/Pipe	Pipe	Pipe Schedule	Schedule 40
Heat Transfer	Length [m]	10.00	Nominal Diameter [mm]	25.4000
	Elevation Change [m]	0.0000	Inner Diameter [mm]	26.6446
	Outer Diameter [mm]	33.40	Pipe Material	Smooth
	Inner Diameter [mm]	26.64	Roughness [mm]	0.000e-01
	Wall Thickness [mm]	3.378	Pipe Wall Conductivity [W/m-K]	45.000
	Material	Smooth		
	Increments	5	Available Nominal Diameters	

[mm]	[mm]	[mm]
25.40	152.4	406.4
38.10	203.2	457.2
50.80	254.0	508.0
76.20	304.8	609.6
101.6	355.6	

Append Insert Multiple Add 1 Delete...   Specify

☐ Transpose Table

Design **Rating** Worksheet Performance Dynamics Deposition

FIGURE 2.4 Pipe specifications (sizing/rating) and pipe parameters inside diameter (ID)/outside diameter (OD) required by UniSim for the piping system presented in Example 2.1.

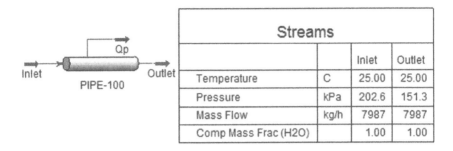

FIGURE 2.5 Pipe process flowsheet and stream summary generated by UniSim for the case described in Example 2.1.

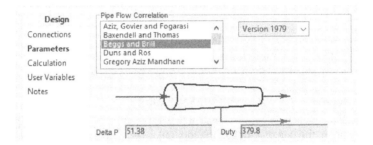

FIGURE 2.6 Pipe Delta P calculated by UniSim (Design/Parameters) for the case described in Example 2.1.

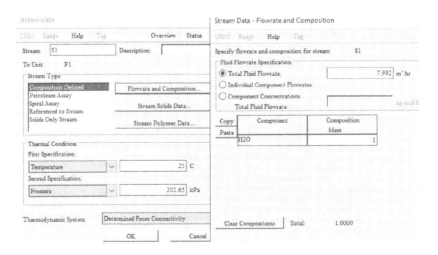

FIGURE 2.7 PRO/II requires inlet stream conditions (temperature, pressure, and flow rate) to specify the feed stream for the case described in Example 2.1.

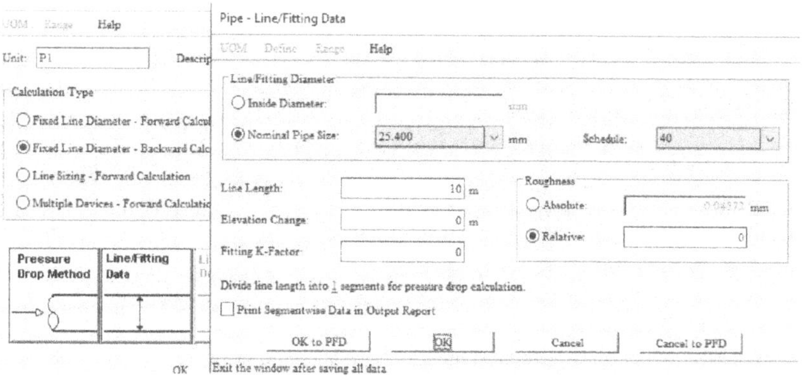

FIGURE 2.8 Pipe segment input menu and nominal pipe size required by PRO/II for the case described in Example 2.1.

relative roughness is zero (Figure 2.8). Click on the run or the small arrow in the toolbar. After the run is successfully converged, generate the results report. Figure 2.9 shows the converged process flowsheet and stream summary. The calculated total pressure drop is 52.1 kPa.

ASPEN SIMULATION

Start the Aspen Plus and create a new simulation case. Add the component (water) and select the fluid package (Steam-TA). Select pipe under the Pressure Changers tab from the Equipment Model Pallete, and then click on the flowsheet window where the piece of equipment appears. To add material streams to the simulation, select the Material stream from the Model Pallete. When the material stream option is selected, several arrows will appear on each of the unit operations. Red arrows indicate a required stream, and blue arrows indicate an optional stream. Click on the process flowsheet where one would like the Stream to begin and click again to end the Stream. In a fashion similar to that of the equipment, each click will add a new stream to the process flowsheet until clicking on Select Mode (the arrow at the left button corner). For this example, add one Stream into the pipe, and one product stream leaves the line. The process flowsheet should be complete at this point, and it should somewhat resemble the one shown in Figure 2.10.

Stream Name Stream Description		S1	S2
Phase		Water	Water
Temperature Pressure	C KPA	25.000 202.650	25.000 254.387
Flowrate	KG-MOL/HR	443.186	443.186
Composition H2O		1.000	1.000

FIGURE 2.9 PRO/II generated process flowsheet and pressure drop for water flow in the pipe case described in Example 2.1.

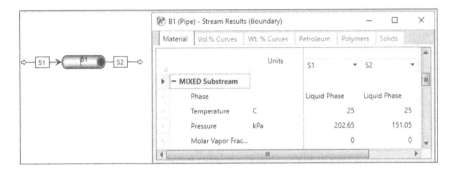

FIGURE 2.10 Aspen Plus generated process flowsheet and pressure drop for water flow in the pipe case described in Example 2.1.

Aspen Plus has a toolbar tool that will automatically take the user through the required data input stepwise. The button that does this is the blue N with the arrow next to it in the toolbar. An alternative method is to double click on the material stream, specify the feed stream conditions (25°C, 2 atm), and then double click on the pipe segment and specify pipe conditions (10 m length, 1-inch Schedule 40, zero elevation, and zero roughness). After the feed stream is established and the pipe segment is defined, the simulation status changes to Required Input is Complete. There are few ways to run the simulation; the user could select either Next in the toolbar to say whether the required inputs are completed and ask whether to run the simulation.

AVEVA PROCESS SIMULATION

The pipe model is a flow-based equipment model. It uses a constant-density Darcy equation designed to model pipes with moderate pressure drop (Figure 2.11).

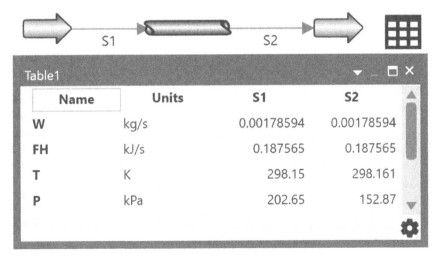

FIGURE 2.11 Aveva Process Simulation generated a process flowsheet and pressure drop for water flow in the pipe case described in Example 2.1. The pressure difference is 49.78 kPa.

The pipe handles up to 10% pressure drop. If the pressure drop is more significant than 10%, multiple pipes in series should be used. Essential variables are nominal pipe size (NPS) or diameter nominal. Default pipe roughness is 0.00015 ft for commercial steel pipe.

1. Copy the DefFluid from the Process library to the created simulation named: Example 2.1, library, right click on DefFluid and select Edit, click on Methods and select SRK from the pull-down menu. Click on Component List and in the Enter the component name, type water.
2. Drag the pipe to the simulation area (Canvas), connect feed (Source) and exit streams (Si)
3. Fully specify the feed stream (S1): F = 0.0022 m³/s, P = 202.6 kPa, T = 25°C.
4. Double click on the pipe and enter the following variables: the pipe size is 1.0 inch, the pipe is a Schedule (Sch) 40, length (L) is 10 m, and the pipe diameter margin is 0.0247 m.
5. To generate the table below the pipe system, drag the Table from Tools to the simulation area. Right click on the stream S1 and click on table > Add all process. Streams > Table1.
6. Double click on Table1, click on Expand the configuration section (the star at the bottom right corner of the table), select and delete (-) unwanted rows.

Conclusions

The comparison of pressure drop values calculated by manual calculation (52.93 kPa) and various software packages, UniSim (51.38 kPa), PRO/II (51.737 kPa), Aspen Plus (51.6 kPa), and Aveva Process Simulation (49.78 kPa), reveals that there is a slight deviation between manual calculations and simulation predictions. The discrepancy in the manual-calculated value is due to the assumption made by taking the inlet conditions in calculating the Reynolds number, while the average of inlet and exit streams should be considered to have better results.

Example 2.2: Pressure Drop of Natural Gas in Horizontal Pipe

Natural gas contains 85 mol% methane and 15 mol% carbon dioxide (density, $\rho = 2.879$ kg/m³ and the viscosity, $\mu = 1.2 \times 10^{-5}$ kg/m/s) is pumped through a horizontal pipe Schedule 40, 6-inch-diameter cast iron pipe at a mass flow rate of 363 kg/h. If the pipe inlet's pressure is 3.45 bars and 25°C, the pipe length is 20 km downstream, assuming incompressible flow. Calculate the pressure drop across the pipe. Is the assumption of incompressible flow reasonable?

Compare the manual calculation with the predictions from the accessible software package (e.g., UniSim/Hysys, PRO/II, Aspen Plus, and Aveva Process Simulation).

SOLUTION

Hand Calculations

The energy equation for the pipe flow is

$$gz_1 + \frac{p_1}{\rho} + \frac{V_1^2}{2} = gz_2 + \frac{p_2}{\rho} + \frac{V_2^2}{2} + f\frac{L}{D}\frac{V_2^2}{2} + \sum F + W_s$$

Since the pipe is horizontal $z_1 = z_2$, and the flow is assumed to be incompressible (the assumption is not suitable for gases) with a constant diameter $V_1 = V_2$, the pressure drop, in this case, can be calculated by using the following equation:

$$\frac{p_1 - p_2}{\rho} = f \frac{L}{D} \frac{V_2}{2}$$

The nominal pipe diameter is 6 inches. Schedule 40; consequently, the inner pipe diameter is 0.154 m, and the velocity is

$$V = \frac{\dot{m}}{\rho A} = \frac{\dot{m}}{\rho \left(\pi D_i^2 / 4 \right)} = \frac{\left(363 \frac{kg}{h} \right) \left(\frac{h}{3600s} \right)}{\left(2.879 \frac{kg}{m^3} \right) \pi \left(0.15m \right)^2 / 4} = \frac{1.98\ m}{s}$$

Reynolds number,

$$Re = \frac{\rho VD}{\mu} = \frac{\left(2.849\ kg/m^3 \right) \left(1.98\ m/s \right) \left(0.154\ m \right)}{1.2 \times 10^{-5}\ kg/ms} = 7.14 \times 10^4$$

Since Reynolds number is greater than 4,000, the flow is turbulent and the roughness factor for the Cast iron pipe is $\varepsilon = 0.00026$ m (Table 2.1). The relative roughness of a cast iron pipe is $(\varepsilon/D) = (0.00026\ \text{m})/(0.15\ \text{m}) = 0.0017$ m. From the relative roughness and $Re = 7.14 \times 10^4$, we can find the Friction factor, $f = 0.024$ (Figure 2.2). Using the calculated friction factor, the pressure drop is calculated.

$$P_1 - P_2 = \rho \left(f \frac{L}{D} \frac{V_2}{2} \right) = \left(2.879\ kg/m^3 \right) \left((0.024) \frac{20,000\ m}{0.154\ m} \frac{\left(1.98\ m/s \right)^2}{2} \right)$$
$$= 17,590\ kg/m/s^2$$

TABLE 2.1

Roughness Factors Used by UniSim

Pipe Material Type	Absolute Roughness (ε), m
Smooth	0.0
Drawn tube	1.52×10^{-6}
Mild steel	4.57×10^{-5}
Asphalted iron	1.22×10^{-4}
Galvanized iron	1.52×10^{-4}
Cast iron	2.59×10^{-4}
Smooth riveted steel	9.14×10^{-4}
Rough riveted steel	9.14×10^{-3}

Converting the pressure drop to the units of kPa,

$$P_1 - P_2 = \left(17{,}590\,\frac{kg}{m/s^2}\right)\left(\frac{N}{1\,kg/m/s^2}\right)\left(\frac{1\,kPa}{1{,}000\,N/m^2}\right) = 17.59\;kPa$$

If the initial pressure is 3.45 bars, the downstream pressure (P_2) is:

$$P1 - P2 = 17.59\;kPa \Rightarrow P2 = 345 - 17.59\;kPa = 327\;kPa$$

UNISIM SIMULATION

Open a blank case in UniSim; select methane and carbon dioxide as components and Peng–Robinson as the fluid package, respectively. Specify the feed stream conditions and compositions. Specify the temperature of the exit stream as that for the inlet stream. Click on the pipe segment and then on Append Segment in the rating page; then specify pipe length as 20 km. The inlet and the exit of the pipe are at the same level; the elevation is zero. Click on View Segment, and specify the pipe schedule and pipe material.

Figure 2.12 displays the result by clicking on the Workbook icon in the toolbar. The Workbook appears. From the Workbook, click on the Setup button. Once the setup view appears, click on Add in the workbook tabs group and select the variable desired in the table. Right click on the PFD area below the PFD and click on Add Workbook Table.

SIMULATION WITH PRO/II

Repeating the same procedure used to construct the process flowsheet of Example 2.1 with PRO/II, the pressure drop is 17.224 kPa, as shown in Figure 2.13 (red: results with warnings).

ASPEN SIMULATIONS

Following the same procedure used in constructing the process flowsheet of Example 2.1 with Aspen, remember that the pipe has no fittings. Figure 2.14 depicts the simulation results.

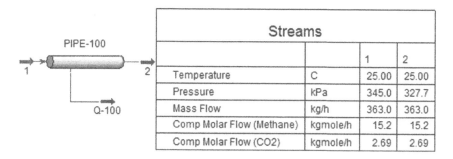

Streams			1	2
Temperature	C		25.00	25.00
Pressure	kPa		345.0	327.7
Mass Flow	kg/h		363.0	363.0
Comp Molar Flow (Methane)	kgmole/h		15.2	15.2
Comp Molar Flow (CO2)	kgmole/h		2.69	2.69

FIGURE 2.12 UniSim generates the pressure drop through the pipeline (17.3 kPa); the case described in Example 2.2.

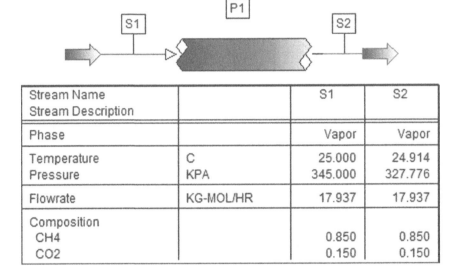

Stream Name Stream Description		S1	S2
Phase		Vapor	Vapor
Temperature Pressure	C KPA	25.000 345.000	24.914 327.776
Flowrate	KG-MOL/HR	17.937	17.937
Composition CH4 CO2		0.850 0.150	0.850 0.150

FIGURE 2.13 PRO/II calculated the pressure drop across the pipe for the case described in Example 2.2.

AVEVA PROCESS SIMULATION

The pipe model is a flow-based equipment model. It uses a constant-density Darcy equation designed to model pipes with moderate pressure drop.

The single pipe is suitable for up to 10% pressure drop; otherwise, multiple pipe segments in series are preferable. Essential variables are NPS or diameter nominal. Default pipe roughness is 0.00015 ft for commercial steel pipe.

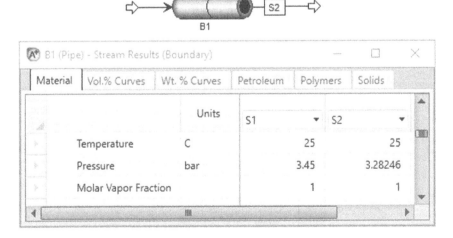

FIGURE 2.14 Aspen Plus generated the pipe process flow diagram and stream summary for the case existing in Example 2.2, where the pressure difference is 16.754 kPa.

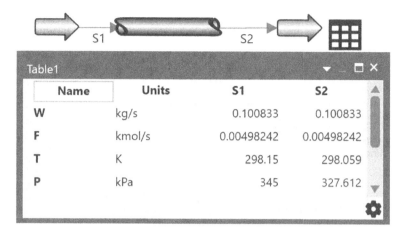

Name	Units	S1	S2
W	kg/s	0.100833	0.100833
F	kmol/s	0.00498242	0.00498242
T	K	298.15	298.059
P	kPa	345	327.612

FIGURE 2.15 Aveva Process Simulation generated pipe process flow diagram and stream summary for the fluid flow in pipes (Example 2.2) where the pressure difference is 17.388 kPa.

1. Copy the DefFluid from the Process library to the created simulation named: Example 2.1, library, right click on DefFluid and select Edit, click on Methods and select SRK from the pull-down menu. Click on Component List and in the Enter the component name, type water.
2. Drag the pipe to the simulation area (Canvas), connect feed (Source) and exit streams (Si)
3. Fully specify the feed stream (S1): F = 0.1 kg/s, P = 345 kPa, T = 25°C.
4. Double click on the pipe and enter the following variables: the pipe size is 1.0 inch, the pipe is a schedule (Sch) 40, length (L) is 20,000 m, and the pipe diameter margin is 0.0247 m.
5. To generate the table below the pipe system, drag the Table from Tools to the simulation area. Right click on the stream S1 and click on table > Add all process. Streams > Table1.
6. Double click on Table1, click on Expand the configuration section (the star at the bottom right corner of the table), select and delete (-) unwanted rows (Figure 2.15).

CONCLUSION

The manually calculated pressure drop is 17.59 kPa, and the pressure drops predicted by UniSim, PRO/II, Aspen, and Aveva are 17.3 kPa, 17.224 kPa, 17.754 kPa, and 17.388 KPa, respectively. Simulation software results were very close to each other; however, there is a discrepancy between simulation results and manual calculation due to physical properties. It is clear from the solution of this example that the density of gases is a function of both temperature and pressure.

Example 2.3: Calculate Pipe Inlet Flow (Given D and ΔP)

Water at 2 atm and 25°C is flowing in a horizontal 10 m in a mild steel pipe; the pipe's pressure drop is equal to 118 kPa. The tube is Schedule 40 and has a 1-inch nominal diameter smooth pipe (ID: 1.049 inches or 0.0266 m). Calculate inlet water velocity and liquid volumetric flow rate. Compare the manual calculation with the predictions from the accessible software package (e.g., UniSim/Hysys, PRO/II, Aspen Plus, and Aveva Process Simulation).

SOLUTION

HAND CALCULATIONS

Fluid velocity and fluid flow rate are calculated from the known values of pipe diameter and pressure drop. Since there is no fitting in the pipe, the pressure drop is a function of pipe skin friction only as of the following:

$$\Delta P = f \frac{L}{D} \rho \frac{V^2}{2}$$

Substituting values into the above equation,

$$118 \times 10^3 \ Pa = f \frac{10 \ m}{0.0266 \ m} \left(1,000 \ kg/m^3\right) \frac{V^2}{2}$$

The friction factor can be found from the Moody diagram or calculated:

$$f = \frac{0.25}{\left[\log\left((\varepsilon/3.7 \ D) + (5.74/Re^{0.9})\right)\right]^2}$$

The friction factor is a function of Reynolds number,

$$f = \frac{0.25}{\left[\log\left(\frac{4.57 \times 10^{-5} \ m}{3.7 \times 0.0266 \ m} + \frac{5.74}{Re^{0.9}}\right)\right]^2}$$

The Reynolds number

$$Re = \frac{\rho V D}{\mu} = \frac{1,000 \ kg/m^3 \ (V) 0.0266 \ m}{0.001 \ kg/m/s}$$

Since Reynolds number is a function of velocity, manually a trial-and-error solution is needed to calculate the inlet velocity. Assume velocity V, and calculate Reynolds number. The Reynolds number calculates the friction factor using the Moody diagram shown in Figure 2.2; otherwise, the equations mentioned above. Calculate the pressure drop and then compare the calculated result with the given value of pressure drop, which is 118 kPa in the example. Repeat until the desired pressure drop is close to 118 kPa. Quickly Polymath calculated the pipe inlet velocity which is 5.29 m/s (Figure 2.16), and volumetric flow rate is $2.94 \times 10^{-3} \ m^3/s$.

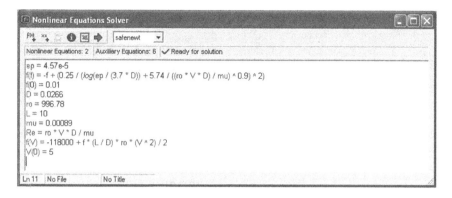

FIGURE 2.16 The polymath program and the generated results solved the set of equations described in Example 2.3.

UniSim Simulation

Select a recent case in UniSim. From the components menu, select the water component, and for the fluid package, select ASME steam, then click on Enter the simulation environment button. Select the pipe segment icon from the object palette and specify the feed stream conditions, stream composition, and product stream temperature. While on the rating page, click on Append Segment and enter the pipe specification shown in Figure 2.17. While on the pipe Design page, click on Parameters, and enter the pressure drop Delta P as 118 kPa. "The pressure drop is greater than 10%" is only a warning and can be ignored.

To view the velocity, click on the Performance tab, then click on View Profile. The UniSim calculated velocity is 5.13 m/s (Figure 2.18). The volumetric flow rate = velocity × pipe cross-section

$$Q = VA = \left(5.13 \text{ m/s}\right)\left(\frac{\pi\left(0.0266 \text{ m}\right)^2}{4}\right) = 2.85 \times 10^{-3} \text{ m}^3/\text{s}$$

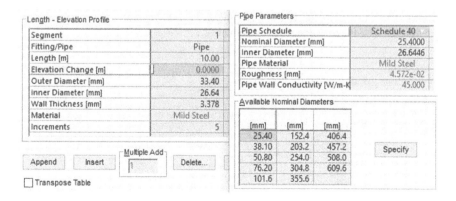

FIGURE 2.17 Pipe length–elevation profile is required by UniSim for the case described in Example 2.3.

Friction Gradient [kPa/m]	Static Gradient [kPa/m]	Accel Gradient [kPa/m]	Liquid Re	Vapour Re	Liquid Velocity [m/s]
11.80	0.0000	0.0000	1.531e+005		5.133
11.80	0.0000	0.0000	1.531e+005		5.133
11.80	0.0000	0.0000	1.531e+005		5.133
11.80	0.0000	0.0000	1.531e+005		5.133
11.80	0.0000	0.0000	1.531e+005		5.133
11.80	0.0000	0.0000	1.531e+005		5.133

FIGURE 2.18   UniSim generated the velocity profile in a 10-m horizontal pipe for the case described in Example 2.3.

SIMULATION WITH PRO/II

The option of having the pressure drop and calculating velocity in UniSim is not available in PRO/II; instead, a case study or assuming flow rate and calculating pressure drop. Figure 2.19 demonstrates the process flowsheet of the pipe using PRO/II. Click the pipe icon in the PFD and specify the pipe length as 10 m, zero elevation, and zero fittings. Mild steel absolute roughness is 4.57×10^{-5}m entered in the PRO/II pipeline/fittings menu, as shown in Figure 2.20.

Stream Name Stream Description		S1	S2
Phase		Water	Water
Temperature Pressure	C KPA	25.00 202.650	25.00 84.569
Total Mass Rate	KG/HR	10318.000	10318.000
Total Weight Comp. Fractions H2O		1.000	1.000

FIGURE 2.19   PRO/II produces the pipe process flow diagram and stream summary for fluid flow in pipes, the case described in Example 2.3.

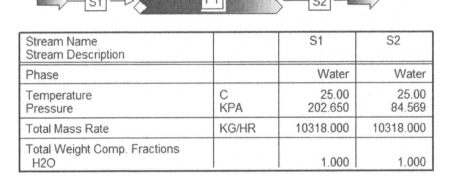

FIGURE 2.20   PRO/II pipeline/fitting data for the case described in Example 2.3.

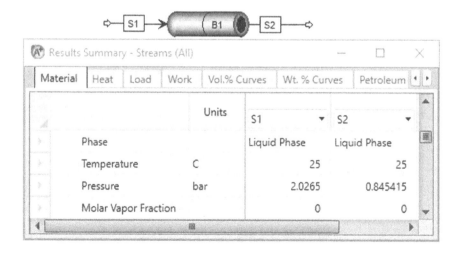

CALC TOTAL PRESSURE DROP, KPA		118.01754	
CALC MAX LINE FLUID VELOCITY, M/SEC		5.15671	
MIXTURE FLOWING FLUID PROPERTIES		INLET	OUTLET
TEMPERATURE, C		25.00000	25.00000
PRESSURE, KPA		202.64999	84.63245
MOLE FRACTION LIQUID		1.00000	1.00000
VELOCITY, M/SEC		5.15671	5.15671
SLIP DENSITY, KG/M3		996.81276	996.81276
SLIP LIQUID HOLDUP FRACTION, (VOL/VOL)		1.00000	1.00000
TAITEL-DUKLER-BARNEA FLOW REGIME		SINGLE PHASE	SINGLE PHASE

FIGURE 2.21 PRO/II calculates inlet fluid velocities (velocity 5.156 m/s), extracted from the generated report leads to a pressure drop of 118 kPa for the case described in Example 2.3.

The trial-and-error procedure estimates the required pressure drop of 118 kPa. Assume the inlet liquid volumetric flow rate, run the simulator, and then check the pressure drop. Keep repeating the process until reaching a pressure drop of 118 kPa. The calculated velocity extracted from the generated report is 5.156 m/s (Figure 2.21).

ASPEN SIMULATION

Aspen Plus required a trial-and-error procedure in the same way as that used with PRO/II. Following the same procedure in Example 2.1. Figure 2.22 depicts the process flowsheet and stream table simulation results. Volumetric flow rate is 0.00287 m³/s (velocity = 5.17 m/s).

	Units	S1	S2
Phase		Liquid Phase	Liquid Phase
Temperature	C	25	25
Pressure	bar	2.0265	0.845415
Molar Vapor Fraction		0	0

FIGURE 2.22 Aspen Plus generated the pipe process flow diagram and stream summary for the case presented in Example 2.3.

Aveva Process Simulation

Start Aveva Process Simulation and generate a new simulation, rename it (e.g., Example 2.3).

1. Copy the DefFluid from the Process library to the created simulation named: Example 2.1, library, right click on DefFluid and select Edit, click on Methods and select SRK from the pull-down menu. Click on Component List and in the Enter the component name, type water.
2. Drag the pipe to the simulation area (Canvas), connect feed (Source), and exit streams.
3. Specify the T and P of feed stream (S1): P = 345 kPa, T = 25°C, M[H₂O] = 1.
4. Double click on the pipe and enter the following variables: the pipe size is 1.0 inch, Schedule (Sch) 40, length (L) is 10 m, D_margin = 0.0001, Dep = 118 kPa, P1 = 202.65 kPa.
5. To generate the table below the pipe system, drag the Table from Tools to the simulation area. Right click on the stream S1 and click on table > Add all process. Streams > Table1.
6. Double click on Table1, click on Expand the configuration section (the star at the bottom right corner of the table), select and delete (-) unwanted rows.
7. The Aveva Process Simulation predicts the PFD and stream summary (Figure 2.23).

Conclusions

The velocity obtained by manual calculations (5.29 m/s) is slightly higher than that obtained by UniSim (5.13 m/s) and closer to those obtained by PRO/II (5.156 m/s), Aspen Plus (5.17 m/s), and Aveva Process Simulation (5.15 m/s).

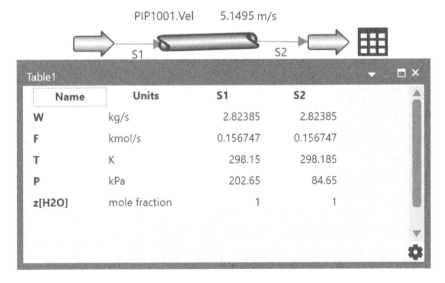

FIGURE 2.23 Aveva Process Simulation generated pipe process flow diagram and stream summary for the fluid flow in pipes, the case described in Example 2.3.

Example 2.4: Effect of Liquid Flow Rate on Pressure Drop

Water is flowing in a pipeline at 25°C. The pipe is 6 inches nominal diameter, Schedule 40 commercial carbon steel pipe (length, L = 1,500 m), pipe inlet pressure, P_1 = 20 atm, exit pressure, P_2 = 2 atm, the change in pipe elevation from pipe entrance to exit is 100 m (height). Plotting the inlet volumetric flow rate vs. pressure drop through the pipe. Compare the manual calculation with the predictions from the accessible software package (e.g., UniSim/Hysys, PRO/II, Aspen Plus, and Aveva Process Simulation).

SOLUTION

HAND CALCULATIONS

Assume an inlet liquid volumetric flow rate of 0.01 m³/s, find the fluid velocity, and then Reynolds number using calculated speed. Use the Moody diagram to calculate the friction factor and then calculate the pressure drop. Repeat for flow rates 0.02, 0.04, and 0.06 m³/s, and plot the calculated pressure drop versus the inlet liquid flow rate using excel.

UNISIM SIMULATION

Start a recent case in UniSim, select water as the pure component, ASME steam for the fluid package, and then enter the simulation environment. Select the pipe segment from the object palette, double click on the pipe, and fill in the connection page. Click on the Worksheet tab; set the feed and product stream temperatures to 20°C (isothermal condition) and the feed pressure to 20 atm. Click on the Rating tab; then click on Append segment and specify the pipe's parameters.

Specify the pipe's outer and inner diameters by double-clicking on the View Segment, select pipe Schedule 40, 6-inch nominal diameter (152.4 mm), and the pipe material as mild steel (Figure 2.24). The example requests the flow rate plot versus the pressure drop with 18 atm as the maximum bound.

A valuable tool in UniSim is the DataBook. From the Tools menu, select DataBook. Click on Insert and add the following variables: feed and actual liquid flow. After that, click on OK. Under the Object column, select PIPE 100, and for the variable, set the pressure drop and then click OK to close. Click on the Case Studies tab at the bottom; choose the pressure drop as the independent variable and actual liquid flow as the dependent variable (Figure 2.25). Rename the current case study to DP versus Q. Click on View and specify the low bound as 10 atm

Length - Elevation Profile			Pipe Info: PIPE-100		
Segment		1			✕
Fitting/Pipe		Pipe	Pipe Parameters		
Length [m]		1500	Pipe Schedule		Schedule 40
Elevation Change [m]		100.0	Nominal Diameter [mm]		152.4000
Outer Diameter [mm]		168.3	Inner Diameter [mm]		154.0510
Inner Diameter [mm]		154.1	Pipe Material		Mild Steel
Wall Thickness [mm]		7.112	Roughness [mm]		4.572e-02
Material		Mild Steel	Pipe Wall Conductivity [W/m-K]		45.000

FIGURE 2.24 UniSim requires pipe segment specifications data for the case described in Example 2.4.

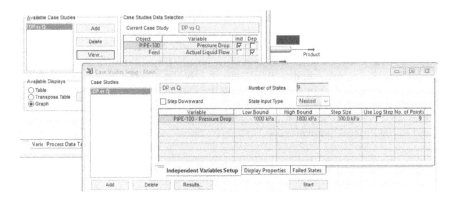

FIGURE 2.25 UniSim-Data Book requires the dependent and independent variables for studying the effect of liquid feed rate on pressure drop for the case described in Example 2.4.

and the high bound as 18 atm, and the Step Size as 0.5. Click on Start. When the calculations are complete, click on Results. Then click on View to see the generated graph (Figure 2.26).

PRO/II SIMULATION

To plot pressure drop versus liquid flow rate using the PRO/II software package, carry out the following steps:

1. Click on the Case Study icon in the toolbar (Input/case study) and rename PAR1 or leave as default; in this case, PAR1 renamed to FLOW.
2. Click on the red Parameter, a menu will pop up; since the inlet stream flow rate is the manipulated variable, select Stream from the pull-down

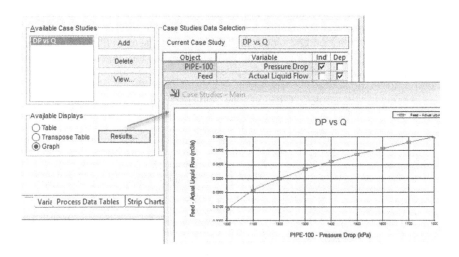

FIGURE 2.26 UniSim generates pressure drop versus actual liquid flow rate for the case described in Example 2.4.

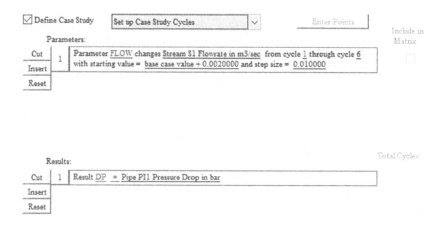

FIGURE 2.27 PRO/II case study parameters selection menu for the case described in Example 2.4.

menu under Stream/Unit. Under Stream Name, select S1 and then click on the Parameter below the Stream/Unit cell and select Flow rate.

3. Specify cycles from 1 through 6. Set the offset of the base case value to 0.002 and the step size to 0.001. After adding the Parameter, the result in the second row is to be specified.

4. Click on a result and rename it as pressure drop DP. Under Stream/Unit, select Pipe, and for the name, select the name of the pipe (PI1).

5. Click on the red word Parameter just below the Stream/Unit cell and select the Parameter as the Pressure drop and the bar units. The final case study and the Parameter menu should appear like that in Figure 2.27. Click on OK to close the case study window.

6. Click on run, select case study under Output in the toolbar, and then select plots. Enter plot name and title and then click on Data. Select the DP as the x-axis and Flow as the y-axis; the final windows should appear like that in Figure 2.28.

7. Click on Preview Plot, or from the Options menu, select Plot Setup, then select "Excel 97 and above." When generating a plot, PRO/II will open Excel and plot one worksheet and the source data on another. The case study tabular output can also be exported to Excel. In the case study setup window, click on View Table and view the table; click on Copy to Clipboard, paste into Excel and use any Excel plot option as per choice.

8. Figure 2.29 reveals that the liquid volumetric flow rate increases as the pressure drop across the pipe increases.

ASPEN SIMULATION

1. Start Aspen Plus and create a new case. Select water component by typing water and then enter. For the fluid package, select STEAM-TA. If one is not sure, click on Method Assistant to find the correct fluid package.

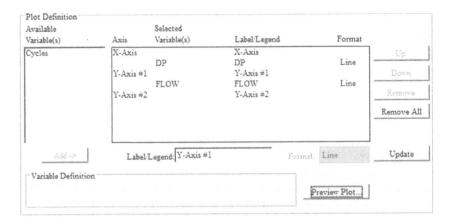

FIGURE 2.28   PRO/II case study – plot definition menu as described in Example 2.4.

2. Next, go to the simulation part. Click Pressure Changers, and select pipe and click somewhere in the simulation area. Click on Material in Model Palette and connect feed stream (S1) and product stream (S2).
3. Click on the feed stream and enter Temperature, Pressure, and flow rate, 25°C, 20 atm, and 10.0 kg/s, respectively. For composition, changer to mass fraction and enter 1.0 under the value.
4. Click on the Pipe object, type the values of pipe length (1,500 m). Use pipe Schedule 40, carbon steel, nominal diameter 6 inches. The elevation is 100 m (flow in uphill if positive, downhill is negative). The pipe has no fittings.
5. Figure 2.30 shows the simulation result generated by Aspen Plus. The total pressure drop is 1,006 kPa.

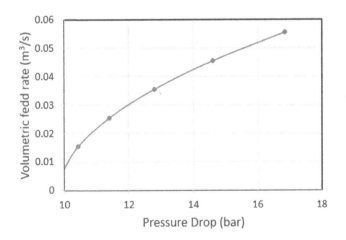

FIGURE 2.29   Liquid volumetric flow rates versus pressure drop generated by PRO/II for the case described in Example 2.4.

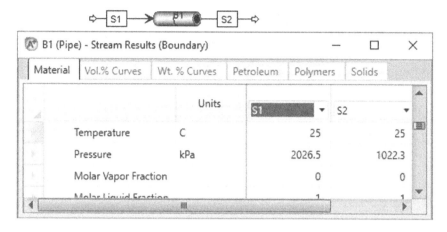

FIGURE 2.30   Aspen Plus generated a process flowsheet and stream summary for the case described in Example 2.4.

6. To plot the pressure drop versus the flow rate, click on the Model Analysis Tool and then Sensitivity. Click on the Define tab and then on the New button, the default name is S-1, click New again, name the first variable for pressure drop as PDROP (Figure 2.31).

7. Click on Vary and New and repeat the same for the liquid volume flow rate (manipulated variable), as shown in Figure 2.32.

8. Click the Tabulate tab, and for Column No, enter 1.0 and PDROP for the tabulated variable.

9. Click on RUN to run the Sensitivity analysis. To plot the figure, click on the result and then click on Custom above Plot. In the popup menu for the X-Axis, select PDROP and for the Select curve to plot, check the box near Vary volumetric flow rate. The results should look like that shown in Figure 2.33.

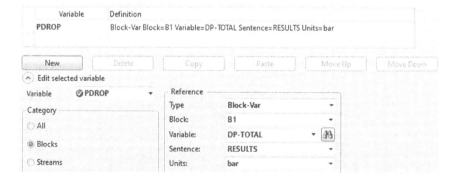

FIGURE 2.31   Specify pressure drop as a variable and blocks as the category for the case described in Example 2.4.

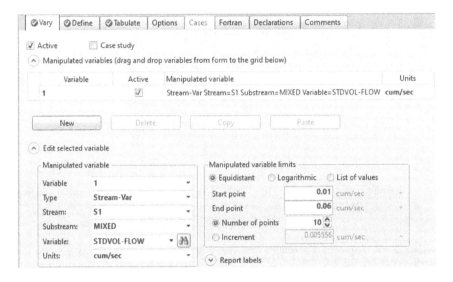

FIGURE 2.32   The setting of the liquid flow rate as a stream manipulated variable for the case study required in Example 2.4.

AVEVA PROCESS SIMULATION

Start Aveva Process Simulation and generate a new simulation, rename it (e.g., Example 2.4).

1. Copy the DefFluid from the Process library to the created simulation named: Example 2.4, library, right click on DefFluid and select Edit, click on Methods and select SRK from the pull-down menu. Click on Component List and in the Enter the component name, type water.
2. Drag the pipe to the simulation area (Canvas), connect the feed stream (Source), and the Sink exit streams.

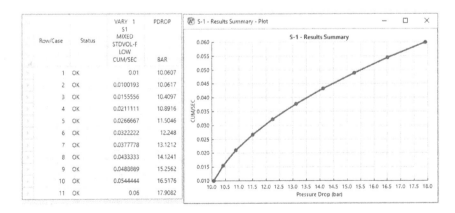

FIGURE 2.33   Selection of pressure drop as the x-axis for the sensitivity analysis for the case described in Example 2.4.

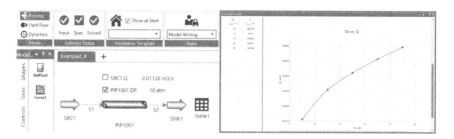

FIGURE 2.34 Aveva Process Simulation generates the process flowsheet and stream summary for the case described in Example 2.4.

3. Specify the T and P of feed stream (S1): P = 345 kPa, T = 25°C, M[H₂O] = 1.
4. Double click on the pipe and enter the following variables: the pipe size is 6.0 inch, Schedule (Sch) 40, length (L) is 10 m, D_margin = 0.0001, Dep = 118 kPa, P1 = 2026.5 kPa.
5. To generate the table below the pipe system, drag the Table from Tools to the simulation area. Right click on the stream S1 and click on table > Add all process. Streams>Table1.
6. Double click on Table1, click on Expand the configuration section (the star at the bottom right corner of the table), select and delete (-) unwanted rows.
7. To create a new curve, go to Model Library, go to Simulation tab, i.e., Example 2.5, right click and select Create New Curve.
8. Then follow the AVEVA Process Simulation Building guide.pdf under chapter 12 "Curve Editor" (Figure 2.34).

Example 2.5: Pipeline with Fitting and Pump

Water at 20°C pumped from the feed tank at 5 atm pressure to an elevated storage tank 15 m high at the rate of 18 m³/h. All pipes are 4-inch, Schedule 40 commercial carbon steel pipes (Figure 2.35). The pump has an efficiency of 65%.

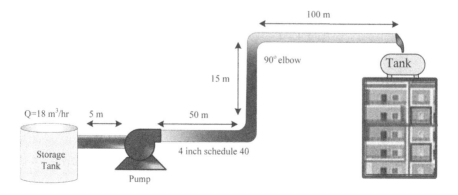

FIGURE 2.35 Process flow diagram for the fluid flowing in a pipe with the pump of the case described in Example 2.5.

Calculate the power needed for the pump to overcome the pressure loss in the pipeline. Compare the manual calculation with the predictions from the accessible software package (e.g., UniSim/Hysys, PRO/II, Aspen Plus, and Aveva Process Simulation).

Useful Data: Density of water ρ = 998 kg/m³, viscosity μ = 1.0 × 10⁻³ kg/m/s, pipe diameter for 4 inches. Schedule 40 is ID = 0.1023 m, and pipe cross-sectional area, A = 8.2 × 10⁻³ m².

SOLUTION

MANUAL CALCULATIONS

In the previous example, there were no pumps installed in the pipelines; however, the driving force provided by a pump in many practical situations. The energy equation,

$$\frac{P_1}{\rho} + gz_1 + \frac{V_1^2}{\rho} = \frac{P_2}{\rho} + gz_2 + \frac{V_2^2}{2} + W_s + \sum F$$

Reynolds number,

$$Re = \frac{\rho VD}{\mu} = \frac{\left(998 \text{ kg/m}^2\right)\left(\dfrac{5 \times 10^{-5} \text{ m}^3/\text{s}}{8.22 \times 10^{-3} \text{m}^2}\right)}{1 \times 10^{-3} \dfrac{\text{kg}}{\text{m/s}}} = 6.18 \times 10^4$$

Since Reynolds number is more significant than 4,000 and the flow is turbulent, from the Moody chart

$$\frac{\varepsilon}{D} = \frac{4.6 \times 10^{-5}}{0.1023} = 0.00045; \text{ hence, } f = 0.02$$

The friction losses

$$\sum F = f \frac{L}{D} \frac{V^2}{2} + \frac{V^2}{2}\left(K_c + 2K_e + K_{ex}\right)$$

Contraction loss at the exit feed tank, $A_2/A_1 = 0$, since $A_1 \gg A_2 = 0$

$$K_c = 0.55\left(1 - \frac{A_2}{A_1}\right) = 0.55(1-0) = 0.55$$

Expansion loss at the inlet of the storage tank, $A_1/A_2 = 0$, since $A_2 \gg A_1 = 0$

$$K_{ex} = \left(1 - \frac{A_1}{A_2}\right)^2 = (1-0) = 1.0$$

Summation of friction losses

$$\sum F = 0.02 \left(\frac{170 \text{ m}}{0.1023 \text{ m}} \right) \frac{\left(0.608 \frac{\text{m}}{\text{s}} \right)^2}{2} + \frac{\left(0.608 \frac{\text{m}}{\text{s}} \right)^2}{2} \left(0.55 + 2(0.75) + 1 \right) = 6.84 \frac{\text{m}^2}{\text{s}^2}$$

The energy equation:

$$0 = \frac{P_2 - P_1}{\rho} + g(z_2 - z_1) + \frac{V_2^2 - V_1^2}{2} + W_s + \sum F$$

Substitute

$$0 = 0 + 9.806 \text{ m/s}^2 (15 - 0) + 0 + W_s + 6.84 \text{ m}^2/\text{s}^2$$

The shaft work is $W_s = -154 \text{ m}^2/\text{s}^2$
  The Pump power (kW) = mW_p
  $W_s = -\eta W_p$, substitution: $-154 = -0.65 W_p$, hence $W_p = 234 \text{ m}^2/\text{s}^2$

$$\text{Brake kW} = mW_p = \left(5 \times 10^{-3} \frac{\text{m}^3}{\text{s}} \right) \left(998 \frac{\text{kg}}{\text{m}^3} \right) \left(234 \frac{\text{m}^2}{\text{m}^2} \right) \left(\frac{1 \text{ kJ}}{1{,}000 \text{ kg/m}^2/\text{s}^2} \right) = 1.17 \text{ kW}$$

$$\text{Pump hp} = 1.17 \text{ kW} = \frac{1 \text{ hp}}{0.7457 \text{ kW}} = 1.57 \text{ hp}$$

UNISIM SIMULATION

Open a blank case in UniSim, select water for component, ASME Steam as a fluid package, and then enter the simulation environment. Build pipe flowsheet. Double click on feed stream and specify feed stream conditions. Double click on the pipe segment on the object palette, switch to the Rating page, click on the Append Segment, and then add pipes and fitting as shown in Figures 2.36.

Click on the Design tab and then Parameters to find the pressure drop across the pipe. The pump horsepower (hp) required is shown in Figure 2.37. The case used the SET logical operator built-in UniSim logical operators.

Length - Elevation Profile

Segment	1	2	3	4	5	6	7	8
Fitting/Pipe	Coupling/Unic	Pipe	Pipe	Elbow 90 Std	Pipe	Elbow 90 Std	Pipe	Coupling/Unic
Number of Fittings/Length [m]	1	5.000	50.00	1	15.00	1	100.0	1
Elevation Change [m]	0.0000	0.0000	0.0000	0.0000	15.00	0.0000	0.0000	0.0000
Outer Diameter [mm]	<empty>	114.3	114.3	<empty>	114.3	<empty>	114.3	<empty>
Inner Diameter [mm]	102.3	102.3	102.3	102.3	102.3	102.3	102.3	102.3
Wall Thickness [mm]	<empty>	6.020	6.020	<empty>	6.020	<empty>	6.020	<empty>
Material	Mild Steel	Mild Steel	Mild Steel	Mild Steel	Mild Steel	Mild Steel	Mild Steel	Mild Steel
Increments	1	5	5	1	5	1	5	1

FIGURE 2.36 Pipe segment specifications for the section from the pipe inlet to the elbow.

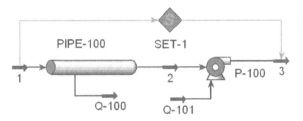

Streams				
		1	2	3
Temperature	C	20.00	20.00	20.02
Pressure	kPa	506.6	352.9	506.6
Mass Flow	kg/h	1.796e+004	1.796e+004	1.796e+004
Comp Mass Frac (H2O)		1.0000	1.0000	1.0000

FIGURE 2.37   The pressure drops across the pipe and pumps duty generated by UniSim for the case described in Example 2.5. Power is 1.585 hp.

SIMULATION WITH PRO/II

Build the process flowsheet shown in Figure 2.38 using PRO/II. Double click on stream S1 to enter the inlet stream conditions (temperature, pressure, total flow rate, and molar composition). Double click on the pipe segment and specify the NPS, length, roughness, elevation, and fitting K factor (Figure 2.39). Double click on the pump and enter the outlet pressure equal to the feed pressure (stream S1) to maintain zero pressure drops in the system, as shown in Figure 2.40. Run the system and generate the output report of 1.59 hp (Shaft power, 1.186 kW).

ASPEN SIMULATION

Follow the same procedure shown in the previous example for constructing the process flowsheet using Aspen Plus. Specify feed stream conditions. Double click

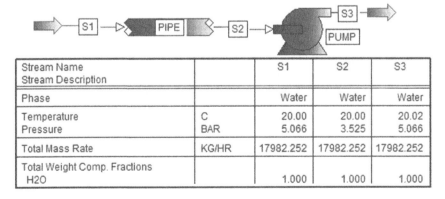

Stream Name Stream Description		S1	S2	S3
Phase		Water	Water	Water
Temperature Pressure	C BAR	20.00 5.066	20.00 3.525	20.02 5.066
Total Mass Rate	KG/HR	17982.252	17982.252	17982.252
Total Weight Comp. Fractions H2O		1.000	1.000	1.000

FIGURE 2.38   PRO/II generated the pressure drops across the pipe and pumps for the case described in Example 2.5.

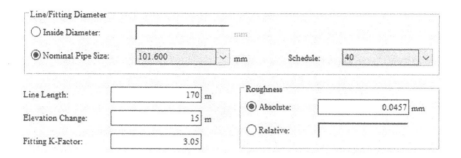

FIGURE 2.39 PRO/II pipeline/fitting data required for the case presented in Example 2.5.

on the pipe segment and enter the pipe rise 15 m. Specify that the pump exit pressure is defined as the pipe inlet pressure (506.3 kPa) to overcome the pipeline's pressure drop. Figure 2.41 shows the Aspen process flowsheet and stream table. Figure 2.42 shows the pump brake horsepower.

AVEVA PROCESS SIMULATION

1. Start Aveva Process Simulation and create a new simulation and rename it to Example 2.5.
2. Copy the Cooling water icon as the suitable method from the Fluids model library to the Example 2.5 model library.
3. Drag the pipe and pump icons from the Process model library to the canvas.
4. Connect inlet and exit stream the pump.
5. Fully specifies the feed stream (T, P, Q).
6. Double click on the pipe and specify the pipe length and diameter.
7. Double click on the pump icon and specify $P_2 = 506$ kPa, eta = 0.65.
8. Figure 2.43 presents the PFD and stream summary.

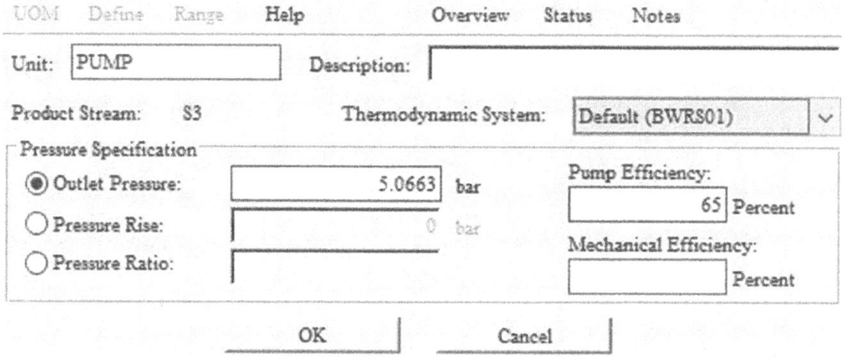

FIGURE 2.40 PRO/II specified the pump exit pressure and percent efficiency for the case described in Example 2.5.

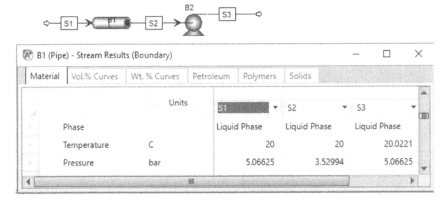

CONCLUSIONS

Pump brake power is calculated manually and simulated using four commercial software packages: UniSim, PRO/II, Aspen Plus, and Aveva Process Simulation, 1.57 hp, 1.585 hp, and 1.186 kW (1.59 hp), 1.585 hp, 1.59 hp, respectively. All results were almost the same.

Example 2.6: Pressure Drop through Pipeline and Fitting

Water at 25°C (density 1,000 kg/m³) and 2.5 atm pressure transferred with a 2.0 hp pump that is 75% efficient at a rate of 15 m³/h to storage tank 20 m high. All the

Summary	Balance	Performance Curve	Utility Usage	Status
Fluid power	0.768207639	kW		
Brake power	1.18186	kW		
Electricity	1.18186	kW		
Volumetric flow rate	300.021	l/min		
Pressure change	1.53631	bar		
NPSH available	35.8186	m-kgf/kg		
NPSH required				
Head developed	15.6904	m-kgf/kg		
Pump efficiency used	0.65			

FIGURE 2.42   Pump Brake power generated by Aspen Plus for the case described in Example 2.5. Power is 1.585 hp.

Name	Units	S1	S2	S3
W	kg/s	4.99553	4.99553	4.99553
Q	m3/s	0.005	0.005	0.005
T	K	293.15	293.187	293.207
P	kPa	506.625	352.742	506.625

FIGURE 2.43 Aveva Process Simulation generates a process flowsheet and stream summary for the piping system for the case presented in Example 2.5.

pipes are 4-inch Schedule 40 mild steel pipes except for the last section, a 2 inch. Schedule 40 steel pipe. There are three 4-inch nominal diameter standard 90° elbows and one reducer to connect the 2 inches pipe (115 m long), as shown in Figure 2.44. Calculate the pressure drop across the system Compare the manual calculation with the predictions from the accessible software package (e.g., UniSim/Hysys, PRO/II, Aspen Plus, and Aveva Process Simulation).

SOLUTION

HAND CALCULATION

The average water density is 1,000 kg/m³, and the moderate viscosity of water is 0.001 kg/m/s. The pipeline consists of two piping diameters:

- ID = 102.3 mm, OD = 114.3 mm (4-inch Schedule 40).
- ID = 52.50 mm, OD = 60.33 mm (2-inch Schedule 40).

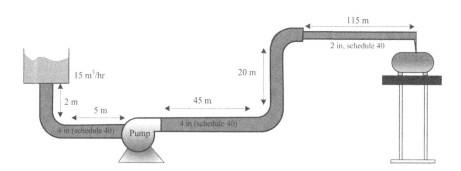

FIGURE 2.44 Process flow diagram of the pumping system described in Example 2.6.

For the mild steel pipe, the roughness, ε, is 4.57×10^{-5} m

$$V_1 = \frac{Q}{A_1} = \frac{Q}{\pi D_1^2/4} = \frac{\left(15 \text{ m}^3/\text{h}\right)\left(\text{h}/3,600 \text{ s}\right)}{\pi \left(0.1023 \text{ m}\right)^2/4} = 0.507 \text{ m/s}$$

Reynolds number for the 4 inches Schedule 40 pipe,

$$Re_1 = \frac{\rho V_1 D_1}{\mu} = \frac{\left(1,000 \dfrac{\text{kg}}{\text{m}^3}\right)\left(0.507 \dfrac{\text{m}}{\text{s}}\right)\left(0.1023\text{m}\right)}{\left(0.001 \dfrac{\text{kg}}{\text{m.s}}\right)} = 5.2 \times 10^4$$

The relative roughness for the 4 inches Schedule 40 pipe,

$$\frac{\varepsilon}{D_1} = \frac{4.57 \times 10^{-5} \text{ m}}{0.1023 \text{ m}} = 4.47 \times 10^{-4}$$

Using the Moody chart (Figure 2.2),

$$f_1 = 0.022$$

The average velocity for the 2 inches Schedule 40 pipe is

$$V_2 = \frac{Q}{A_2} = \frac{Q}{\pi D_2^2/4} = \frac{\left(15 \text{ m}^3/\text{h}\right)\left(\text{h}/3600 \text{ s}\right)}{\pi \left(0.0.0525 \text{ m}\right)^2/4} = 1.93 \text{ m/s}$$

Reynolds number for the 2 inches Schedule 40 pipe is

$$Re_2 = \frac{\rho V_2 D_2}{\mu} = \frac{\left(1000 \text{ kg/m}^3\right)\left(1.93 \text{ m/s}\right)\left(0.0525 \text{ m}\right)}{0.001 \text{ kg/m} \cdot \text{s}} = 1.0 \times 10^4$$

Relative roughness of the pipe (inch), the pipe is schedule 40, is

$$\frac{\varepsilon}{D_2} = \frac{4.57 \times 10^{-5} \text{ m}}{0.0525 \text{ m}} = 8.7 \times 10^{-4}$$

Using the Moody chart (Figure 2.2), $f_2 = 0.028$

Friction loss in the line before the pump, ΣF_1: there is one elbow 90° in this segment

$$\sum F_1 = f_1 \frac{L_1}{D_1} \frac{V_1^2}{2} + K_e \frac{V_1^2}{2}$$

Substituting values,

$$\sum F_1 = (0.022)\frac{(7\,m)}{0.102\,m}\frac{\left(0.507\frac{m}{s}\right)^2}{2} + (0.75)\frac{\left(0.507\frac{m}{s}\right)^2}{2} = 0.29\frac{m^2}{s^2}$$

Pressure drop for the line before the pump is

$$\Delta P_1 = \rho\left[g(z_2 - z_1) + \frac{V_2^2 - V_1^2}{2} + \sum F_1\right]$$

Substitution of known values:

$$\Delta P_1 = \left(1,000\frac{kg}{m^3}\right)\left[\left(9.81\frac{m}{s^2}\right)(0 - 2\,m) + \frac{\left(0.507\frac{m}{s}\right)^2 - \left(0.507\frac{m}{s}\right)^2}{2} + 0.29\frac{m}{s^2}\right]$$

Simplify further,

$$\Delta P_1 = 1,000\ kg/m^3 \,(-19.61 + 0 + 0.29)m^2/s^2 = -19,320\ kg/m/s^2$$

Convert pressure units to kPa

$$\Delta P_1 = -19,320\frac{kg}{m/s^2}\left(\frac{N}{kg/m/s^2}\right)\left(\frac{1\,Pa/m^2}{1\,N}\right)\left(\frac{1\,kPa}{1,000\,Pa}\right) = -19.30\ kPa$$

Friction loss in the line after the pump, ΣF_2: in this line segment, there are two various pipes (4 inches and 2 inches diameter); consequently, two velocities. There are also two 90° elbows and one sudden contraction joining the two pipes.

$$\sum F_2 = f_1 \frac{L_3}{D_2} \frac{V_2^2}{2} + f_2 \frac{L_4}{D_2} \frac{V_2^2}{2} + 2K_e \frac{V_1^2}{2} + K_c \frac{V_2^2}{2}$$

Contraction loss

$$K_c = 0.55\left(1 - \frac{A_2}{A_1}\right) = 0.55\left(1 - \frac{\pi D_2^2/4}{\pi D_1^2/4}\right) = 0.55\left(1 - \frac{D_2^2}{D_1^2}\right)$$

Simplify

$$K_c = 0.55\left(1 - \frac{(0.0525\ \text{m})^2}{(0.1023\ \text{m})^2}\right) = 0.405$$

Friction loss in the line after the pump,

$$\sum F_2 = f_1 \frac{L_3}{D_2} \frac{V_2^2}{2} + f_2 \frac{L_4}{D_2} \frac{V_2^2}{2} + 2K_e \frac{V_1^2}{2} + K_c \frac{V_2^2}{2}$$

Substitute known values,

$$\sum F_2 = (0.022)\frac{(65\ \text{m})}{0.102\ \text{m}} \frac{\left(0.51\frac{\text{m}}{\text{s}}\right)^2}{2} + (0.028)\frac{(115\ \text{m})}{0.0525} \frac{\left(1.93\frac{\text{m}}{\text{s}}\right)^2}{2}$$
$$+ 2(0.75)\frac{\left(0.51\frac{\text{m}}{\text{s}}\right)^2}{2} + (0.405)\frac{\left(1.93\frac{\text{m}}{\text{s}}\right)^2}{2}$$

Solving for $\sum F_2$:

$$\sum F_2 = (1.6 + 114.2 + 0.19 + 0.754)\ \text{m}^2/\text{s}^2 = 116\ \text{m}^2/\text{s}^2$$

Pressure drop in the line after the pump

$$\Delta P_2 = \rho\left[g(z_2 - z_1) + \frac{V_2^2 - V_1^2}{2} \sum F\right]$$

Substituting required values

$$\Delta P_2 = \left(1,000\frac{\text{kg}}{\text{m}^3}\right)\left[\left(9.81\frac{\text{m}}{\text{s}^2}\right)(20\ \text{m} - 2\ \text{m}) + \frac{\left(1.93\frac{\text{m}}{\text{s}}\right)^2 - \left(0.507\frac{\text{m}}{\text{s}}\right)^2}{2} + 116\frac{\text{m}^2}{\text{s}^2}\right]$$

Simplifying further,

$$\Delta P_2 = 1,000\frac{kg}{m^3}(176.51+1.73+116)\frac{m^2}{s^2} = 294,000\frac{kg}{m/s^2}$$

The pressure drop in the second pipe segment in kPa is

$$\Delta P_2 = \left(2.94\times10^5\frac{kg}{m/s^2}\right)\left(\frac{N}{kg/m/s}\right)\left(\frac{1\,Pa/m^2}{1\,N}\right)\left(\frac{1\,kPa}{1,000\,Pa}\right) = 294\ kPa$$

The mass flow rate in kg/s

$$m = \rho Q = \left(1,000\ kg/m^3\right)\left(15\ m/h\right)\left(\frac{1\,h}{3,600\,s}\right) = 4.17\ kg/s$$

The shaft work

$$W_s = \frac{-\eta W_p}{\dot{m}} = -0.75(2\ hp)\left(\frac{0.7457\ kW}{1\ hp}\right)\left(\frac{1\ kJ/s}{1\ kW}\right)\left(\frac{1,000\ Nm}{1\ kJ}\right)$$
$$\times\left(\frac{1\ kg\ m/s^2}{1\ N}\right)/4.17\ kg/s$$

Simplifying

$$W_s = \frac{-\eta W_p}{\dot{m}} = -268.4\ m^2/s^2$$

Pump pressure rise, ΔP_p

$$\Delta P_p = \rho W_s = \left(\frac{1,000\ kg}{m^3}\right)\left(\frac{268.4\ m^2}{s^2}\right)\left(\frac{1\ kPa}{1,000\ N/m^2}\right)\left(\frac{1\ N}{1\ kg\ m/s^2}\right) = 268.4\ kPa$$

The negative sign is just because the calculated pressure drop was inlet minus exit. However, across the pump, pressure consistently increased.

UniSim Simulation

1. Start UniSim and create a new case.
2. Add water as the pure component, and select ASME steam for the fluid package.
3. Construct the piping process flowsheet and specify the feed stream conditions.
4. Double click on Pipe Segment on the process flowsheet, click on the Rating tab and then click on Append Segment to add the pipe specification (length, nominal size, elevation, and fittings) as shown in Figures 2.45.

Length - Elevation Profile

Segment	1	2	3
Fitting/Pipe	Pipe	User Elbow: 90	Pipe
Number of Fittings/Length [m]	2.000	1	5.000
Elevation Change [m]	-2.000	0.0000	0.0000
Outer Diameter [mm]	114.3	<empty>	114.3
Inner Diameter [mm]	102.3	102.3	102.3
Wall Thickness [mm]	6.020	<empty>	6.020
Material	Mild Steel	User Specified	Mild Steel
Increments	5	1	5

Segment	1	2	3	4	5	6
Fitting/Pipe	Pipe	Elbow: 45 Std	Pipe	Elbow: 90 Std	Coupling/Unic	Pipe
Number of Fittings/Length [m]	45.00	1	20.00	1	1	115.0
Elevation Change [m]	0.0000	0.0000	18.00	0.0000	0.0000	0.0000
Outer Diameter [mm]	114.3	<empty>	114.3	<empty>	<empty>	60.32
Inner Diameter [mm]	102.3	102.3	102.3	102.3	102.3	52.50
Wall Thickness [mm]	6.020	<empty>	6.020	<empty>	<empty>	3.912
Material	Mild Steel	Mild Steel	Mild Steel	Mild Steel	Mild Steel	Mild Steel
Increments	2	1	2	1	1	2

FIGURE 2.45 Pipe specifications for the line after the pump build in UniSim for the case described in Example 2.6.

While on the design page, click on Parameters to enter the pressure drop in the line before the pump (pressure drop = −19.56 kPa). The negative sign indicates the pressure rise due to the first pipe segment's negative elevation (Figure 2.46).

Add a new pipe segment to specify the line after the pump, while on the rating page, click on Append Segment and enter the required data. Double click on the pump segment in the flowsheet area, and while the Design page is open, click on Parameters, and specify the pressure drop as shown in Figure 2.47. While on the design page, click on Parameter and type 2.0 hp in the Duty cell. The pressure rise across the pump is 225.4 kPa (Figure 2.47). The final flowsheet should appear like that shown in Figure 2.48.

SIMULATION WITH PRO/II

In PRO/II, there should be one pipe segment for each change in diameter, length, height, and fitting K factors, as shown in Figure 2.49. The figure depicts the PFD and stream conditions.

ASPEN SIMULATION

Using Aspen, a procedure similar to that used in PRO/II is applied, as shown in Figure 2.50. In the first pipe segment, specify the pipe length as 2 m, the pipe rise as −2 m. In the second pipe segment, the size is 5 m, the elevation is zero, and

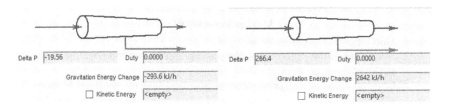

Delta P -19.56   Duty 0.0000   Delta P 266.4   Duty 0.0000

Gravitation Energy Change -293.6 kJ/h   Gravitation Energy Change 2642 kJ/h

☐ Kinetic Energy <empty>   ☐ Kinetic Energy <empty>

FIGURE 2.46 UniSim calculates the pressure drop across the first pipe segment (19.56 kPa) and the second pipe segment (266.4 kPa) for the case presented in Example 2.6.

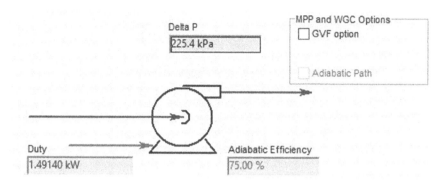

FIGURE 2.47 UniSim depicts the process flowsheet and stream summary for the case described in Example 2.6.

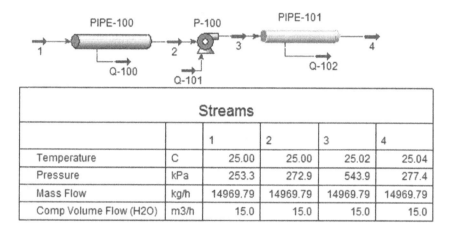

		1	2	3	4
Temperature	C	25.00	25.00	25.02	25.04
Pressure	kPa	253.3	272.9	543.9	277.4
Mass Flow	kg/h	14969.79	14969.79	14969.79	14969.79
Comp Volume Flow (H2O)	m3/h	15.0	15.0	15.0	15.0

FIGURE 2.48 The pump's duty is 2.0 hp (1.49 kW) at 75% adiabatic efficiency for the case described in Example 2.6.

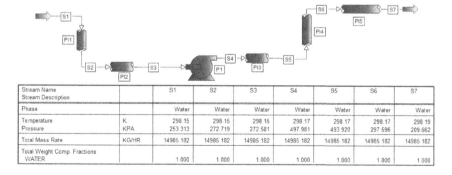

Stream Name Stream Description		S1	S2	S3	S4	S5	S6	S7
Phase		Water	Water	Water	Water	Water	Water	Water
Temperature	K	298.15	298.15	298.15	298.17	298.17	298.17	298.19
Pressure	KPA	253.313	272.719	272.581	497.981	493.920	297.596	209.662
Total Mass Rate	KG/HR	14985.182	14985.182	14985.182	14985.182	14985.182	14985.182	14985.182
Total Weight Comp Fractions WATER		1.000	1.000	1.000	1.000	1.000	1.000	1.000

FIGURE 2.49 PRO/II depicts the process flowsheet and stream summary for the case described in Example 2.6.

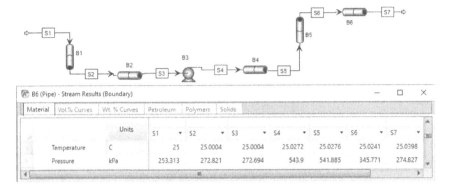

FIGURE 2.50 Aspen Plus generates the process flow diagram and stream summary for the case described in Example 2.6.

the fitting is one 90° elbow. For the pump, set the power required to 2 hp. The third pipe length is 45 m, and that of the fourth pipe is 20 m. The pipe rise is also 20 m, and the number of 90° elbows is 2. For the fifth pipe, the length is 115 m, the elevation is zero, and the NPS is 2 inches Schedule 40. Click on the Thermal Specification tab and select Adiabatically. Figure 2.50 shows the Aspen process flowsheet and the stream table.

AVEVA PROCESS SIMULATION

1. Start Aveva Process Simulation, create a new simulation, and rename it (e.g., Example 2.6).
2. Copy the Cooling water icon as the suitable method from the Fluids model library to the Example 2.6 model library.
3. Drag the pipe segment, pump, pipe, sudden contraction, and pipe to the canvas.
4. Drag Source and Sink to canvas and connect them to the first pipe and exit streams, respectively.
5. Fully specify the feed stream; T, P, and W.
6. Click on the pipes and type in the pipe's length and diameter. Click on the pump and specify the pump efficiency and its horsepower.
7. Figure 2.51 depicts the PFD and stream summary.

CONCLUSIONS

The pressure rise across the pump was calculated manually (268.4 kPa) and using four software packages, UniSim: 268.2 kPa, PRO/II: 268.1 kPa, Aspen Plus: 268.46 kPa, and Aveva Process Simulation: 268.45 kPa, all values were close to each other.

2.2   FLUID FLOW IN PUMPS

Pump, a device that spends energy to raise, transport, or compress fluids, moves liquids in a closed conduit or pipe. The pump increases the pressure of the liquid.

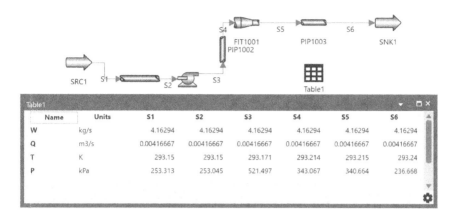

Name	Units	S1	S2	S3	S4	S5	S6
W	kg/s	4.16294	4.16294	4.16294	4.16294	4.16294	4.16294
Q	m3/s	0.00416667	0.00416667	0.00416667	0.00416667	0.00416667	0.00416667
T	K	293.15	293.15	293.171	293.214	293.215	293.24
P	kPa	253.313	253.045	521.497	343.067	340.664	236.668

FIGURE 2.51 Aveva Process Simulation generates a process flowsheet and stream summary for the piping system for the case presented in Example 2.6.

2.2.1 POWER AND WORK REQUIRED

Using the mechanical energy balance equation around the pump system, the pump's actual or theoretical energy W_s (kJ/kg) added to the fluid can be calculated. If η is the fractional efficiency and W_p the shaft work delivered to the pump,

$$W_p = -\frac{W_s}{\eta} \tag{2.16}$$

$$W_s = -\frac{P_b - P_a}{\rho} \tag{2.17}$$

$$\text{Brake power } kW = mW_p = -\frac{mW_s}{\eta} \tag{2.18}$$

The mechanical energy W_s in kJ/kg added to the fluid expressed as the developed head H of the pump in meters of the pumped liquid,

$$-W_s = H \, g \tag{2.19}$$

the head is

$$H = \frac{\Delta P_L}{\rho g} = \frac{32 \mu L V}{\rho g D^2} \tag{2.20}$$

2.3   FLUID FLOW IN COMPRESSORS

A compressor is a mechanical device that increases the pressure of a gas by reducing its volume. In compressors and blowers, pressure changes are significant, and also compressible flow occurs. In compression of gases, the density changes, and so the mechanical energy balance equation must be written in differential form and then integrated to obtain the work.

$$dW = \frac{dp}{\rho} \tag{2.21}$$

Integration between the suction pressure P_1 and discharge pressure P_2 gives the work of compression.

$$W = \int_{P_1}^{P_2} \frac{dp}{\rho} \tag{2.22}$$

For adiabatic compression, the fluid follows an isentropic path and

$$\frac{P_1}{P} = \left(\frac{\rho_1}{\rho}\right)^{\gamma} \tag{2.23}$$

where the ratio of heat capacities, $\gamma = C_p/C_v$,

$$-W_s = \frac{\gamma}{\gamma-1} \frac{RT_1}{M} \left[\left(\frac{P_2}{P_1}\right)^{(\gamma-1)/\gamma} - 1\right] \tag{2.24}$$

The adiabatic temperatures are related by

$$\frac{T_2}{T_1} = \left(\frac{P_2}{P_1}\right)^{(\gamma-1)/\gamma} \tag{2.25}$$

The brake power

$$\text{Brake kW} = \frac{-W_s \, m}{\eta} \tag{2.26}$$

Example 2.7: Flow-through Pump

Pure water is fed at a rate of 0.0126 kg/s into a pump at 121.11°C, 308.2 kPa. The exit pressure is 8,375 kPa. Use the pump module in UNISIM, PRO/II, Aspen Plus, and Aveva Process Simulation programs to model the pumping process. The adiabatic pump efficiency is 10%. Find the energy required.

SOLUTION

HAND CALCULATION

$$-W_s = \frac{(P_1 - P_2)}{\rho} = \frac{(8,375 - 308.2) \times 10^3 \text{ Pa}}{943.5 \frac{kg}{m^3}} = 8.55 \frac{kJ}{kg}$$

The density of water at 121.11°C is 943.5 kg/m³, the horsepower for 10% efficiency pump:

$$W_p = \frac{-W_s \times \dot{m}}{\eta} = \frac{1}{0.1}\left(8.55 \times 10^3 \frac{J}{kg}\right)\left(0.0126 \frac{kg}{s}\right) = 1,077.3 \text{ W} = 1.44 \text{ hp}$$

SIMULATION WITH UNISIM

Open a blank case in UniSim and then perform a process flowsheet of the pump with inlet, exit, and the red color energy stream connected to the pump. Specify the feed stream conditions and exit stream pressure is 8,375 kPa. Click on the Design tab, and then click on Parameters. In the Adiabatic Efficiency box on the parameter page, enter 10. Click on the Worksheet tab to view the results, as in Figure 2.52.

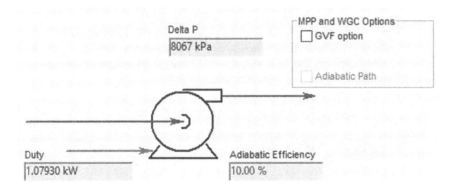

FIGURE 2.52 UniSim calculated adiabatic pump duty 1.08 kW (1.45 hp) for the case described in Example 2.7.

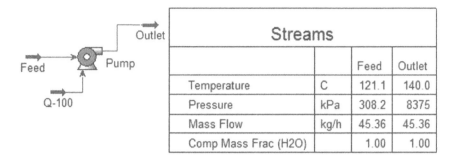

Streams			
		Feed	Outlet
Temperature	C	121.1	140.0
Pressure	kPa	308.2	8375
Mass Flow	kg/h	45.36	45.36
Comp Mass Frac (H2O)		1.00	1.00

FIGURE 2.53 UniSim generated pump process flow diagram and stream summary for the case described in Example 2.7.

Click the Workbook icon in the toolbar. From the Workbook, the menu clicks on setup. Once the setup appears, click on Add in the Workbook Tabs group and select the variable that should appear in the table. Right click on the PFD area below the PFD and click on Add Workbook Table. Figure 2.53 depicts the stream conditions for the pump. The new outlet temperature of the water is 140°C for the 10% efficient pump. Hand calculations and UniSim results are not separate from each other.

SIMULATION WITH PRO/II

Using PRO/II, add water as the pure component. The relevant fluid package is Peng–Robinson–Stryjek–Vera (PRSV) or Peng–Robinson (PR) equations of state. Perform the pumping process flowsheet. Specify the feed stream conditions. Set the pump exit pressure to 8,375 kPa (Figure 2.54). Run the system and generate the output report. The calculated pump work is 1.0787 kW (1.447 hp).

ASPEN SIMULATION

1. Start Aspen Plus, click New, and then create a blank simulation. Add pure water as the only component. Click on Methods/Specifications, and for Method name, select Stream-TA as the base method.
2. Click on the Simulation tab and then click on the Pressure Changers tab and select pump and click somewhere in the simulation area.
3. Double click on the feed stream and specify inlet stream conditions (temperature, pressure, total flow rate, and composition).

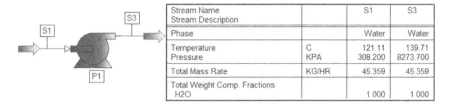

Stream Name Stream Description		S1	S3
Phase		Water	Water
Temperature	C	121.11	139.71
Pressure	KPA	308.200	8273.700
Total Mass Rate	KG/HR	45.359	45.359
Total Weight Comp. Fractions H2O		1.000	1.000

FIGURE 2.54 PRO/II generated pump process flow diagram and stream summary for the water pumping case described in Example 2.7.

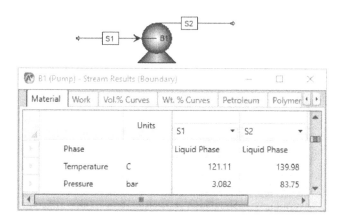

FIGURE 2.55 Aspen Plus generates a process flow diagram and a stream summary for the case described in Example 2.7.

4. Double click on the pump icon and set the Discharge Pressure to 82.66 atm (8,375 kPa) and the pump efficiency to 0.1. Note that the message "Required Input Completed" means that the system is ready to run in the status bar.
5. Click Run, and after the message Run is completed, click on results and click on block results. Figure 2.55 depicts the process flowsheet and the stream table. Figure 2.56 shows the brake power of the pump (1.447 hp).

AVEVA PROCESS SIMULATION

1. Start Aveva Process Simulation and create a new simulation and rename it to Example 2.7.
2. Copy the Cooling water icon as the suitable method from the Fluids model library to the Example model library.

Summary	Balance	Performance Curve	Utility Usage	✓ Status
Fluid power		0.1079	kW	▼
Brake power		1.44697	hp	▼
Electricity		10.79	kW	▼
Volumetric flow rate		0.802551	l/min	▼
Pressure change		80.668	bar	▼
NPSH available		11.0965	m-kgf/kg	▼
NPSH required				–
Head developed		873.235	m-kgf/kg	▼
Pump efficiency used		0.1		

FIGURE 2.56 Aspen Plus generates the pump performance for the case described in Example 2.7.

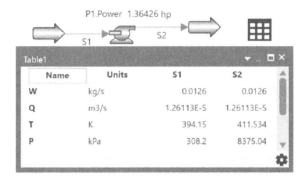

FIGURE 2.57 Aveva Process Simulation generates a process flow diagram and a stream summary for the case described in Example 2.7.

3. Drag the pump icon from the Process model library to the canvas.
4. Connect inlet and exit stream the pump.
5. Fully specifies the feed stream (T, P, W).
6. Double click the pump icon and specify P_2 = 8,375 kPa, eta = 0.1
7. The Aveva Process Simulation calculated power is 1.36 hp (Figure 2.57).

Conclusions

This example shows that pumping liquid can increase its temperature and pressure. In this case, the pump was only 10% efficient, causing an increase of 17°C in the water temperature. The less efficient a pump is, the more significant the pumped fluid's increase, attributed to the reason that more energy is needed to pump the liquid to get the same outlet pressure of a more efficient pump in a low-efficiency pump. Therefore, extra energy is transferred to the fluid. The brake power calculated by the manual calculation, UniSim, PRO/II, Aspen Plus, and Aveva Process Simulation is 1.44, 1.45, 1.447, 1.447, and 1.36 hp, respectively. Results were close to each other.

Example 2.8: Compression of Natural Gas

A compressor compresses 100 kg/h of natural gas consisting of 80 mol% methane, 10% ethane, 5% carbon dioxide, and nitrogen from 3 bars and 30°C to 10 bars. Find the compressor duty (brake kW) for 75% and 10% efficiency.

SOLUTION

Hand Calculation

The compressor shaft work,

$$-W_s = \frac{\gamma}{\gamma - 1} \frac{RT_1}{M} \left[\left(\frac{P_2}{P_1} \right)^{(\gamma-1)/\gamma} - 1 \right]$$

Natural gas average molecular weight (M)

$$M = 0.8(16) + 0.1(30) + 0.05(44) + 0.05(28) = 19.4 \text{ kg/kgmol}$$

Assuming that gas is incompressible,

$$-W_s = \left(\frac{1.31}{1.31-1}\right)\frac{\left(8.3143\dfrac{kJ}{kgmolK}\right)(30+273.15)K}{\left(19.4\dfrac{kg}{kgmol}\right)}\left[\left(\frac{10}{3}\right)^{(1.31-1)/1.31}-1\right]$$

The shaft work in kJ/kg,

$$-Ws = 181 \text{ kJ/kg}$$

The compressor brake kW,

$$\text{Brake kW} = \frac{-W_s \ m}{\eta}$$

The brake power for a compressor, 75% adiabatic efficiency

$$\text{Brake kW} = \frac{\left(181\dfrac{kJ}{kg}\right)\left(100\dfrac{kg}{h}\dfrac{h}{3,600s}\right)}{0.75} = 6.69 \text{ kW}$$

For 10% adiabatic efficiency,

$$\text{Brake kW} = \frac{(181 \text{ kJ/kg})\left(100 \text{ kg/h}\dfrac{h}{3,600 \text{ s}}\right)}{0.10} = 50.175 \text{ kW}$$

SIMULATION WITH UNISIM

The pump simulation is done by opening a recent UniSim case and selecting the methane, ethane, CO_2, and N_2. PRSV is the suitable thermodynamic fluid package for the property estimation method. The basis of the calculation is 100 kg/h of the natural gas stream. Select a compressor from the object palette, specify the stream conditions, keep the UniSim default compressor efficiency, and determine the outlet temperature and compressor duty. Neglect the heat loss or gain from the environment. Enter the pressure for the outlet stream, which is 10 bars.

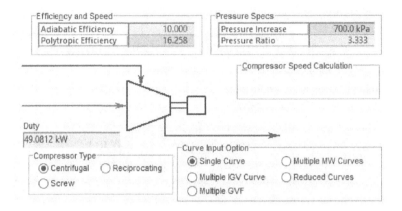

FIGURE 2.58　Specify compressor adiabatic efficiency in UniSim for the case described in Example 2.8.

Click on the Design tab, click on Parameters and see the UniSim default adiabatic efficiency and calculated Polytropic efficiency. In the Adiabatic Efficiency box on the parameter page, enter 10 (Figure 2.58). Click on the Worksheet tab to view the results. Double click on the pump PFD and click on the Worksheet tab. The pump's stream conditions are in Figure 2.59 for the 10% adiabatic efficiency, respectively. The new outlet temperature is 652.4°C for the 10% adiabatic efficiency and 144.4°C for an adiabatic efficiency of 75%.

Simulation with PRO/II

Following the same procedure as done previously with PRO/II, construct the process flowsheet shown in Figure 2.60, enter the inlet temperature, pressure, total inlet flow rate, and molar compositions. The figure shows the results of the 10% efficiency. From the output/generate text report, the actual compressor duty is 49.11 kW, and the exit temperature is 652.56°C, and the shaft power is 6.5 kW and an exit temperature of 144.89°C for an adiabatic efficiency of 75%.

Aspen Plus Simulation

1. Start Aspen Plus, click new, and create a blank simulation. Type components under Component ID (methane, ethane, CO_2, and N_2).

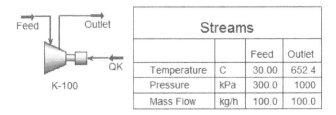

Streams			Feed	Outlet
Temperature	C		30.00	652.4
Pressure	kPa		300.0	1000
Mass Flow	kg/h		100.0	100.0

FIGURE 2.59　UniSim generates the simulated compressor conditions (10% adiabatic efficiency) for the case described in Example 2.8.

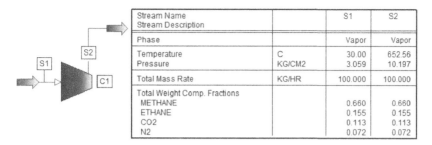

Stream Name Stream Description			S1	S2
Phase			Vapor	Vapor
Temperature	C		30.00	652.56
Pressure	KG/CM2		3.059	10.197
Total Mass Rate	KG/HR		100.000	100.000
Total Weight Comp. Fractions				
METHANE			0.660	0.660
ETHANE			0.155	0.155
CO2			0.113	0.113
N2			0.072	0.072

FIGURE 2.60   PRO/II generated the simulated compressor conditions (10% adiabatic efficiency) for the case described in Example 2.8.

2. Click on Methods/Specifications and select Peng–Robinson from the pull-down menu below Method name. Click on Parameters/Binary Interaction and click PRKBV-1.
3. Click on the Simulation button and click on Pressure Changers and select the compressor; click in the Main Flowsheet area. Click on Material stream and connect inlet and exit streams (start with the red lines).
4. Double click on the feed stream and specify inlet stream conditions (temperature, pressure, total flow rate, and composition).
5. Double click on the Compressor icon, for the type select Isentropic, and set the Discharge pressure to 10 bars and the compressor Isentropic and Mechanical efficiency to 0.10 for the 10% efficiency.
6. Note that the displayed message Required Input Completed in the status bar means that the system is ready to run.
7. Click on Run or F5; after completing the run successfully, click on results and then click on the block icon in the results menu. Figure 2.61 shows the process flowsheet and the stream table. Figure 2.62 shows the generated brake power of the pump with 10% efficiency.

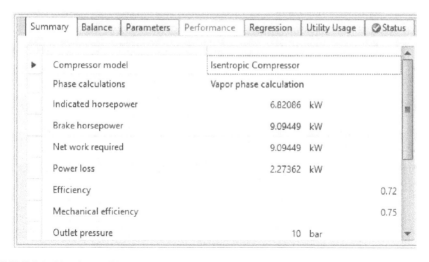

Summary	Balance	Parameters	Performance	Regression	Utility Usage	☑ Status

▶	Compressor model	Isentropic Compressor	
	Phase calculations	Vapor phase calculation	
	Indicated horsepower	6.82086 kW	
	Brake horsepower	9.09449 kW	
	Net work required	9.09449 kW	
	Power loss	2.27362 kW	
	Efficiency		0.72
	Mechanical efficiency		0.75
	Outlet pressure	10 bar	

FIGURE 2.61   Aspen Plus generates the compressor bakes power (75% efficiency) for the case described in Example 2.8.

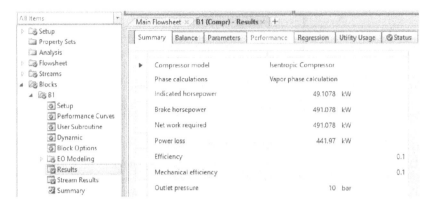

FIGURE 2.62 Aspen Plus generates the compressor bakes power (10% efficiency) for the case described in Example 2.8.

AVEVA PROCESS SIMULATION

1. Start Aveva Process Simulation and create a new simulation and rename it to Example 2.8.
2. Copy the Cooling water icon as the suitable method from the Fluids model library to the Example model library.
3. Drag the pump icon from the Process model library to the canvas.
4. Connect inlet and exit stream the pump.
5. Fully specifies the feed stream (T, P, W).
6. Double click the pump icon and specify the $P_2 = 10$ bars and eta 0.1.
7. The calculated compressor power is 49.23 kW.
8. Figure 2.63 shows the compressor flow diagram and stream summary.

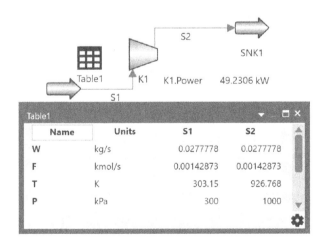

FIGURE 2.63 Aveva Process Simulation simulated the process flowsheet and stream summary for the case described in Example 2.8.

The pump's actual work with 10% efficiency with hand calculations is 50.175 kW; on the contrary, results obtained by UniSim, PRO/II, Aspen Plus, and Aveva Process Simulation were in good agreement and around 49 kW. The discrepancy in hand calculations is due to the ideal gas assumption.

PROBLEMS

2.1 PRESSURE DROP THROUGH A SMOOTH PIPE

Water is flowing in a 15-m smooth horizontal pipe at 8.0 m³/h and 35°C. The density of water is 998 kg/m³, and the viscosity of water is 0.8 cp. The tube is Schedule 40, 1-inch nominal diameter (2.66 cm ID). Water inlet pressure is 2 atm. Calculate the pressure drop using UniSim/Hysys (for water, the appropriate fluid package is ASME Steam) or any other available software packages (e.g., PRO/II, Aspen Plus, SuperPro, and Aveva Process Simulation).

2.2 PRESSURE DROP IN A HORIZONTAL PIPE

Calculate the pressure drop of water through a 50-m long smooth horizontal pipe. The inlet pressure is 100 kPa, the average fluid velocity is 1.0 m/s, the pipe diameter is 10 cm, and the pipe relative roughness is zero. Fluid density is 1.0 kg/L, and viscosity is 1.0 cp. Use UniSim/Hysys to calculate the pressure drop (the suitable fluid package is ASME Steam).

2.3 PRESSURE DROP IN A PIPE WITH ELEVATION

Calculate the pressure drop of water through a 50-m long pipe (relative roughness is 0.01 m/m). The inlet pressure is 100 kPa; the average fluid velocity is 1.0 m/s. The pipe diameter is 10 cm. Fluid density is 1.0 kg/L, and viscosity is 1.0 cp. The discharged water is at an elevation, 2 m higher than the water entrance. Use UniSim/Hysys (ASME Steam fluid package) to calculate the pressure drop.

2.4 PUMPING OF NATURAL GAS IN A PIPELINE

Natural gas contains 85 mol% methane, and 15 mol% ethane is pumped through a horizontal Schedule 40, 6-inch-diameter cast iron pipe at a mass flow rate of 363 kg/h. If the pipe inlet's pressure is 3.5 bars and 25°C and the pipe length is 20 km downstream, assume incompressible flow. Calculate the pressure drop across the pipe using UniSim, Aspen Plus, PRO/II, and Aveva Process Simulation (the suitable fluid package is Peng–Robinson).

2.5 COMPRESSION OF GAS MIXTURE

The mass flow rate of a gas stream 100 kg/h of feed contains 60 wt% methane and 40% ethane at 20 bars, and 35°C compressed to 30 bars. Determine the temperature of the exit stream in °C using UniSim/Hysys (the suitable fluid package is Peng–Robinson) or any alterative software packages mentioned in problem 2.1.

2.6 COMPRESSION OF NITROGEN

Find the compressor horsepower required to compress 100 kgmol/h of nitrogen from 1 atm and 25°C to 5 atm using UniSim/Hysys (the suitable fluid package is Peng–Robinson) or any other substitute software package.

2.7 PUMPING OF PURE WATER

Pure water is fed at 45.36 kg/h into a pump at 121°C, 308 kPa. The exit pressure is 8,375 kPa. Plot the adiabatic pump efficiency versus the energy required using UniSim/Hysys (the suitable fluid package is ASME Steam) or any other software package.

2.8 PUMPING OF WATER TO TOP OF BUILDING

Calculate the pump's size required to pump 100 kgmol/min of pure water at 1 atm and 25°C to the top of a building 12 m high (the suitable fluid package is ASME Steam).

2.9 POWER GENERATED FROM THE TURBINE

High-pressure compressed gas (200°C and 20 bar) enters a gas turbine at 100 kgmol/h (20% CO_2, 50% N_2, and the balance is water vapor). The exit gas pressure is 10 bar. Calculate the heat duty of the turbine and the exit gas temperature using UniSim/Hysys (Peng–Robinson) or any other software package.

2.10 PRESSURE DROP IN A THROTTLING VALVE

Pressurized natural gas (50°C and 20 bar) enters a throttling valve at 100 kgmol/h (80% CH_4, 20% C_2H_6). The exit gas pressure is 10 bar. Calculate the exit gas temperature using UniSim/Hysys or any other software (Peng–Robinson is a suitable fluid package for light hydrocarbon gases) package.

REFERENCES

1. Pritchard, F. and A. T. McDonald, 2015. Introduction to fluid mechanics, 9th edn, John Wiley & Sons, New York, NY.
2. McCabe, W. L., J. C. Smith and P. Harriott, 2005. Unit operations of chemical engineering, 7th edn, McGraw-Hill, New York, NY.
3. Geankoplis, C. J., D. Lepek and A. Hersel, 2018. Transport process and separation processes, 5th edn, Prentice-Hall, Englewood Cliffs, Upper Saddle River, NJ.
4. Young, D. F., B. R. Munson, T. H. Okiishi and W. W. Huebsch, 2010. A brief introduction to fluid mechanics, 5th edn, Wiley, New York, NY.

5. Finnemore, E. J. and J. B. Franzini, 2011. Fluid mechanics with engineering applications, 10th edn, McGraw-Hill, New York, NY.
6. Reuben, M. O. and S. J. Wright, 2012. Essentials of engineering fluid mechanics, 6th edn, Harpercollins College Div, New York, NY.
7. Colebrook, C. F. 1939. Turbulent flow in pipes with particular reference to the transition between the smooth and rough pipe laws. Journal of the Institute of Civil Engineering London, 11, 133–156.

3 Material and Energy Balance

At the end of this chapter, students should be able to:

1. Solve material balance problems for physical and chemical processes.
2. Perform energy balance on reactive and non-reactive processes.
3. Verify manual calculation results with the following software packages: UniSim/Hysys, PRO/II, Aspen Plus, SuperPro Designer, and Aveva Process Simulation.

3.1 INTRODUCTION

The fundamental law of conservation of mass is the basis of material balances. In particular, chemical engineers are concerned with doing mass balances around chemical processes. Chemical engineers do a mass balance to account for what happens to each of the chemicals used in a chemical process. By accounting for material entering and leaving a system, mass flows can be identified, which might have been unknown or difficult to measure without this technique. Heat transfer is the transition of thermal energy from a hotter mass to a colder one. When an object is at a different temperature from its surroundings or another entity, the transfer of thermal energy (also known as heat flow or heat exchange) occurs to transfer until the object and the surroundings reach thermal equilibrium.

3.2 MATERIAL BALANCE WITHOUT A REACTION

To apply a material balance, one needs to define the system and the quantities of interest. The system is a region of space defined by a real or imaginary closed envelope (envelope = system boundary); it can be a single process unit, a collection of process units, or an entire process. The general material balance equation is

Accumulation within the system = Input − output + generation - consumption

At steady state, accumulation = zero. If there is no reaction, generation and consumption are zero. Accordingly, the general material balance is input through system boundary = Output through system boundary.

Example 3.1: Material Balance on the Physical Mixing Process

A storage tank is employed to mix two streams at 5 atm pressure and 25°C. The first stream contains 20 kmol/h ethanol and 80 kmol/h water; the second stream consists of 40 kmol/h ethanol and 60 kmol/h water. Find the compositions and

DOI: 10.1201/9781003167365-3

molar flow rates of the product stream. Compare manual calculation with the predictions of the available software package (e.g., UniSim/Hysys, PRO/II, Aspen Plus, SuperPro, and Aveva Process Simulation).

SOLUTION

MANUAL CALCULATIONS

The process is just physical mixing, and no reaction is involved. Figure 3.1 shows the process flow diagram (PFD).

Basis: 100 kmol/h of stream 1

Total mass balance:

$$S1 + S2 = S3$$

$$100 + 100 = 200 \text{ kmol/h}$$

Component balance (ethanol):

$$0.2(100) + 0.4(100) = X(200); \text{ hence, } X = 0.3 \text{ (mole fraction of ethyl alcohol)}$$

The exit stream is 0.3 mole fraction ethanol, and the balance is water.

HYSYS/UNISIM SIMULATION

Start UniSim and select a New case. For Components, choose ethanol and water; for Fluid Package, choose the non-random two liquids (NRTL) activity coefficient model, and then enter the simulation environment. From the object palette, select Mixer and place it in the simulation area. Connect two inlet streams and one exit stream. Click on stream 1.0 and enter 25°C for temperature, 5 atm for pressure, and 100 kmol/h for molar flow rate. In the composition page, enter mole fraction 0.2 for ethanol and 0.8 for water. Click on stream S2 and enter 25°C for temperature and 5 atm for pressure to ensure that both the ethanol and water are in the liquid phase and 100 kmol/h for molar flow rate. On the composition page, enter 0.4 for ethanol and 0.6 mole fraction for moisture. To display the result below the process flowsheet, right click on each stream, select the show table, double click on each

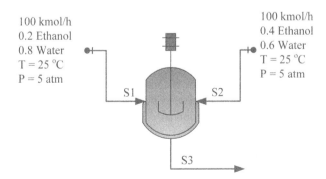

FIGURE 3.1 Process flow diagram of the physical mixing of the two streams, the case described in Example 3.1.

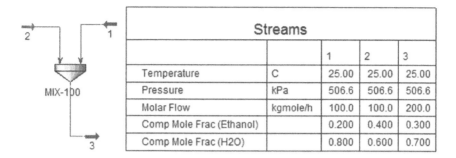

FIGURE 3.2 Process flow diagram of the mixing process and the stream summary generated by UniSim for the case described in Example 3.1.

table, click on Add Variable, select the component mole fraction, and click on Add Variable for both ethanol and water. Remove units, label for stream 2, and remove labels for stream 3. Results should appear like that shown in Figure 3.2.

PRO/II Simulation

The PRO/II process simulation program performs rigorous mass and energy balances for various chemical processes. Implement the following procedure to build the mixing process with PRO/II: open a New case in PRO/II; click on File and then New, which brings the user to the primary simulation environment. It is called the PFD screen. Then click on the Component Selection button in the top toolbar. This button appears like a benzene ring. Next, click on Thermodynamic Data. Once entered, click on Liquid Activity and then on NRTL. Then click on Add and then OK to return to the flow sheeting screen. Now it is ready to insert units and streams. The operation unit that we want to put in the simulation is the Mixer. Scroll down the toolbar until an icon looks like a mixer. Next, click on Streams, and create a feed stream, S1, entering the Mixer side. Next, create stream S2 from the other side of the Mixer. The third stream, S3, should be moved from the bottom side of the Mixer, as this will be the product stream.

The next task is entering the known data into the simulation. We will begin with the feed stream data. Double click on SI and the Stream Data screen will appear. Click on Flowrate and Composition, then Individual Component Flowrates, and then enter all the given feed flow rates. Once done, click on OK, and the Stream Data screen should reappear with Flowrate and Composition now outlined in blue. The two streams are composed of ethanol and water at 5 atm pressure and 25°C.

Stream S1: 20 kmol/h ethanol (ethyl alcohol)
80 kmol/h water
Stream S2: 40 kmol/h ethanol (ethyl alcohol)
60 kmol/h water

Now it is time to run the simulation. Click on Run on the toolbar, and the simulation should turn blue. If it does not, one can double click on the controller and then increase iterations until it converges. Otherwise, one should retrace steps to find the error and fix it. The next thing to do is to view the results. For this, click on Output and then Stream Property Table. Double click on the generated table and

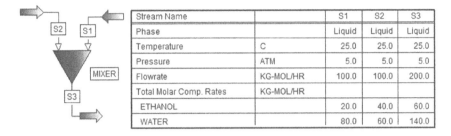

Stream Name		S1	S2	S3
Phase		Liquid	Liquid	Liquid
Temperature	C	25.0	25.0	25.0
Pressure	ATM	5.0	5.0	5.0
Flowrate	KG-MOL/HR	100.0	100.0	200.0
Total Molar Comp. Rates	KG-MOL/HR			
ETHANOL		20.0	40.0	60.0
WATER		80.0	60.0	140.0

FIGURE 3.3 PRO/II generated the mixer flowsheet and the stream summary for the case described in Example 3.1.

then select Material Balance List under the property list. Click on Add All to add available streams. Figure 3.3 shows the PFD and stream summary.

ASPEN PLUS SIMULATION

Start Aspen Plus, click on New, and create a blank simulation, add the components involved in the mixing process. Select the method and then enter the simulation environment. Select the Mixer from the Mixer/Splitter submenu and place it in the PFD area. Create the inlet streams by first clicking on Material Streams at the bottom-left corner of the window and move the cursor over the Mixer. Red and blue arrows appear around the Mixer. A red line signifies that a stream is required for the flow simulation, whereas a blue line icon indicates optional.

Click on Next to begin entering data. While in the Property Method, select the NRTL option from the list. Enter the data for the inlet stream labeled 1, enter 25°C for temperature and 5 atm for pressure. In the box labeled water, enter the value 80 and enter 20 in the box labeled EtOH. Note that the units of the values you just entered are displayed in the box labeled Compositions. These units can be changed by first clicking on the arrow by the box and selecting the appropriate measurement unit from the options. Click next. Enter 25°C for temperature and 5 atm for pressure to ensure that both the ethanol and water are in the liquid phase. Enter the flow rate values for both water and ethanol as instructed above. Enter the values of 60 and 40 kmol/h for water and ethanol, respectively.

Stream summary results are displayed on the process flowsheet by clicking on Report Options under the Setup folder. Then click on the Stream folder tab. For detailed data, make sure that both the boxes beside Mole and Mass contain a checkmark. Also, make sure that both the boxes are checked on a fraction basis. Click on Next. For summary results, just check the box below the mole fraction basis (Figure 3.4).

SUPERPRO DESIGNER

Start SuperPro and open a new case, selecting all components involved in the mixing process: Tasks ≫ Enter Pure Components.

Ethyl alcohol is selected. Water, oxygen, and nitrogen exist as default components. Select a mixing process: Unit Procedures ≫ Mixing ≫ Bulk flow ≫ 2 streams.

Connecting two inlet streams and one exit stream by double clicking on the inlet streams and enter the stream's temperature, pressure, and molar flow rate. To run the system, click on Solve ME Balance (the calculator icon) or press F9. Click on Toggle Stream Summary Table in the toolbar. Right click on the empty area in the

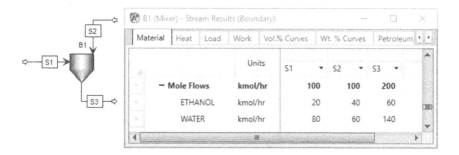

FIGURE 3.4 Aspen Plus generates a process flow diagram and streams summary for the case described in Example 3.1.

simulation area and click on update data. Select all streams. Results should appear like that shown in Figure 3.5.

AVEVA PROCESS SIMULATION

Start Aveva Process Simulation software and click on the plus sign to create a new simulation and rename it (e.g., Example 3.1). Drag the Mix object from the Process library to the simulation area (Canvas). Connect two Source streams and one exit stream to the Sink. Copy DefFluid from the process library to Example 3.1 model library. Edit DefFluid and select SRK from the pull-down menu as the suitable fluid thermodynamic property measurements method. Then click on the Component list and add water and ethanol. Double click on the feed stream, and for Fluid type, select Example 3.1 Model/DefFluid from the pull-down menu. To build a stream summary table, drag and drop the table from the Tools library to the canvas. Right click on feed stream and select Table/Add all Process. Stream. Double click on the table icon to display the table. Click on Expand the configuration section to add and delete unwanted items. Figure 3.6 shows the Aveva Process Simulation predicted results.

3.3 MATERIAL BALANCE ON REACTIVE PROCESSES

The extent of reaction ξ (or $\dot{\xi}$) is the number of moles (or molar flow rate) that are converted in a given reaction. The extent of reaction is a quantity that characterizes the reaction and simplifies our calculations. For a continuous process at a steady state,

$$\dot{n}_i = \dot{n}_i^o + \upsilon_i \dot{\xi} \qquad (3.1)$$

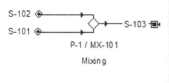

Time Ref: h			S-101	S-102	S-103
Total Mass Flow		kmol	100.0000	100.0000	200.0000
Temperature		°C	25.0	25.0	25.0
Pressure	atm	∨	1.000	1.000	1.000
Total Contents		mole frac	1.0000	1.0000	1.0000
Ethyl Alcohol			0.2000	0.4000	0.3000
Water			0.8000	0.6000	0.7000

FIGURE 3.5 SuperPro generates the process flowsheet and streams summary for the case described in Example 3.1.

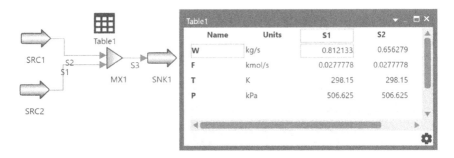

FIGURE 3.6 Aveva Process Simulation generated the mixing process and the stream summary of the case presented in Example 3.1.

where $\dot{n}_i^o$ and $\dot{n}_i$ are the molar flow rates of species i in the feed and outlet streams, respectively. For a batch process,

$$\dot{n}_i = \dot{n}_i^o + \upsilon_i \xi \tag{3.2}$$

where n_i^o and n_i are the initial and final molar amounts of species i, respectively. The extent of reaction ξ (or $\dot{\xi}$) has the same units as n (or $\dot{n}$). Generally, the syntheses of chemical products do not involve a single reaction but rather multiple reactions. The goal is to maximize the production of the desired product and minimize the production of unwanted by-products. For example, dehydrogenation of ethane produces ethylene [1]:

$$C_2H_6 \rightarrow C_2H_4 + H_2$$
$$C_2H_6 + H_2 \rightarrow 2CH_4$$
$$C_2H_4 + C_2H_6 \rightarrow C_3H_6 + CH_4$$

The definition of the yield is:

$$Yield = \frac{Moles\ of\ desired\ product\ formed}{Moles\ formed\ if\ there\ were\ no\ side\ reactions\ and\ limiting\ reactant\ reacts\ completely}$$

The selectivity is the moles of desired product formed divided by moles of undesired product in the product stream

$$Selectivity = \frac{Moles\ of\ desired\ product\ formed}{Moles\ of\ the\ undesired\ component\ in\ the\ product\ stream}$$

The multiple reactions, each reaction has its extent. Assuming a steady-state reactor, we can write the mole balance as a function of the extent of reaction:

$$\dot{n}_i = \dot{n}_i^o + \sum_j \upsilon_{ij} \dot{\xi}_j \tag{3.3}$$

where v_{ij} is the stoichiometric coefficient of substance i in reaction j; ξ; is the extent of reaction j. For a single reaction, the above equation reduces to the equation reported in a previous section.

Example 3.2: Conversion Reactor, Single Reaction

Ammonia is oxidized to produce nitric oxide. The fractional conversion of the limiting reactant is 0.5. The inlet molar flow rates of NH_3 and O_2 to the oxidizer are 5.0 kmol/h each. The operating temperature and pressure are 25°C and 1.0 atm, respectively. Calculate the exit components' molar flow rates. Assume that the reactor is operating adiabatically. Compare manual calculation with the predictions of the available software package (e.g., UniSim/Hysys, PRO/II, Aspen Plus, SuperPro, and Aveva Process Simulation).

SOLUTION

MANUAL CALCULATION

Figure 3.7 shows the PFD of the ammonia oxidation process.
The reaction stoichiometric coefficients:

$$v_{NH_3} = -4, v_{O_2} = -5, v_{NO} = 4, v_{H_2O} = 6$$

The extent of the reaction method (ξ) finds the product stream's molar flow rates. The material balance using the extent of the reaction method is as follows:

$$\dot{n}_i = \dot{n}_i^\circ + v_i \dot{\xi}$$

Ammonia exit molar flow rate, $\dot{n}_{NH_3}$

$$\dot{n}_{NH_3} = n_{NH_3}^\circ - 4\dot{\xi}$$

Oxygen exit molar flow rate, $\dot{n}_{O_2}$:

$$\dot{n}_{O_2} = n_{O_2}^\circ - 5\dot{\xi}$$

Nitrogen oxide exit molar flow rate, $\dot{n}_{NO}$

$$\dot{n}_{NO} = n_{NO}^\circ + 4\dot{\xi}$$

$$4NH_3 + 5O_2 \rightarrow 4NO + 6O_2$$

NH3 = 5 kmol/h
O2 = 5 kmol/h
T = 25 °C
P= 1 atm

NH3
O2
NO
H2O

Conversion reactor

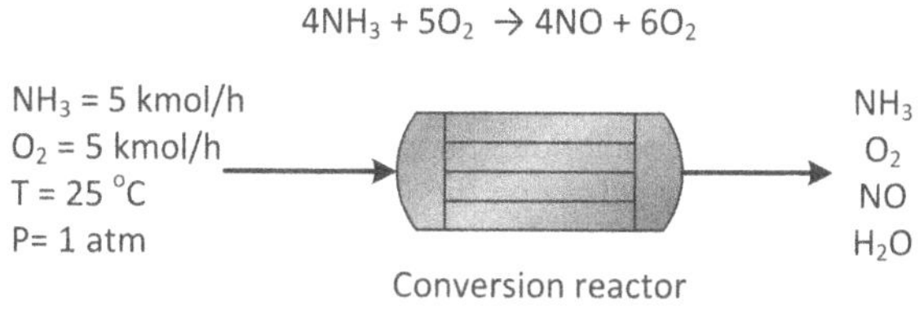

FIGURE 3.7 Process flowsheet of the ammonia oxidation process for the case described in Example 3.2.

Water exit molar flow rate, $\dot{n}_{H_2O}$

$$\dot{n}_{H_2O} = n^o_{H_2O} + 6\xi$$

The total number of moles at the outlet of the reactor:

$$n = n^o + (-4 - 5 + 4 + 6)\xi = n^o + \xi$$

Rearranging,

$$n = n^o + \xi$$

The inlet molar feed rates:

$$n^o_{NH_3} = 5 \text{ kmol/h}, \ n^o_{O_2} = 5 \text{ kmol/h}, \ n^o_{NO} = 0, \ n^o_{H_2O} = 0$$

Since the limiting reactant is oxygen, the fractional conversion of oxygen:

$$x = \frac{n^o_{O_2} - n_{O_2}}{n^o_{O_2}}$$

Substitution,

$$0.5 = \frac{5 - n_{O_2}}{5}, \ n_{O_2} = 2.5 \text{ kmol/h}$$

Substituting in the oxygen material balance equation,

$$2.5 = 5 - 5\xi, \ \Rightarrow \xi = 0.5$$

Solving the set of the material balance equations gives the following results:

$$\xi = 0.5, \ n_{NH_3} = 3 \text{ kmol/h}, \ n_{O_2} = 2.5 \text{ kmol/h}$$

The product stream component's molar flow rates:

$$n_{NO} = 2 \text{ kmol/h}, \ n_{H_2O} = 3 \text{ kmol/h}$$

UniSim/Hysys Solution

Five types of reactors are available in UniSim/Hysys; Conversion, Equilibrium, Gibbs, Plug Flow Reactor (PFR), and Continuously Stirred Tank Reactor (CSTR). Conversion, Equilibrium, and Gibbs reactors do not need reaction rates. On the contrary, CSTR and PFR reactors require reaction rate constants and order of the reaction. Five different reaction rates are available in UniSim/Hysys software packages: Conversion reaction, Equilibrium, Simple rate, Kinetic, and Catalytic. Gibbs reactors are established on minimizing the Gibbs free energy of all components involved in the chemical reactions. UniSim/Hysys reaction rates are in units of moles per volume of gas-phase per time.

$$r_{HYSYS} \ [=] \frac{mol}{m^3_{gas} \cdot s}$$

For catalytic reactions, reaction rates are in moles per mass of catalyst per time. For Example, the following reaction rate is in the units of a mole per kilogram catalyst per second.

$$r[=]\frac{mol}{kg_{cat} \cdot s}$$

The following equation converts catalytic reaction rate to the unit of moles per volume of gas:

$$r_{HYSYS} = r \times \rho_c \frac{(1-\phi)}{\phi}$$

Where ϕ is the reactor void fraction and ρ_c is the catalyst density. For this example, a conversion reactor will be sufficient to perform the material balance. Select a New case in Hysys, and add all components involved in the reaction and produced (NH_3, O_2, NO, and H_2O). Select Peng–Robinson as the fluid package. From the object palette, click on General Reactors, select Conversion Reactor, and then click anywhere in the simulation area. Double click on the reactor and attach the feed and the product streams. Double click on the feed stream; enter feed conditions (1 atm, 25°C, and total molar flow rate 10 kmol/h). On the compositions page, set a 0.5 mole fraction for ammonia and oxygen. The Flowsheet in the toolbar menu selects Reaction Package and then selects Conversion Reaction; fill in the Stoichiometry page as in Figure 3.8 and the Basis page as in Figure 3.9. The percent conversion is constant and is not a function of temperature, so CO = 50, and the C1 and C2 are zero. Figure 3.10 shows the process flowsheet and stream conditions. The results are the same as those obtained by hand calculation. Adding the involved reactions:

1. Open the Simulation Basis Manager by clicking the Home View icon (beaker) in the Simulation Basis Manager toolbar.
2. Select the Reactions tab.

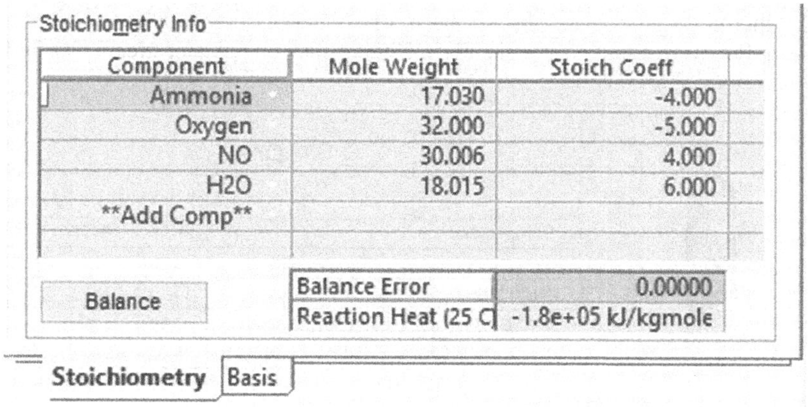

Stoichiometry Info

Component	Mole Weight	Stoich Coeff
Ammonia	17.030	-4.000
Oxygen	32.000	-5.000
NO	30.006	4.000
H2O	18.015	6.000
Add Comp		

Balance		
	Balance Error	0.00000
	Reaction Heat (25 C)	-1.8e+05 kJ/kgmole

Stoichiometry | Basis

FIGURE 3.8 UniSim stoichiometry info for setting the stoichiometric coefficient of the ammonia oxidation reaction (+ for the product, − for reactant) of the case described in Example 3.2.

Basis

Base Component	Oxygen
Rxn Phase	VapourPhase
C0	50.00
C1	0.0000
C2	0.0000

Conversion (%) = C0 + C1*T + C2*T^2

(T in Kelvin)

Stoichiometry   **Basis**

FIGURE 3.9   The reaction in the vapor phase; 50% conversion; O_2 is the base component.

3. Click on the Add Rxn button, opens the Reactions view.
4. From the available reaction list, types select the reaction type to use and click on Add Reaction.
5. The reaction menu opens the Reaction Property view. After adding the two reactions, using the default set name (i.e., Global Rxn Set), add the set to Current Reactions Sets. The reaction set should appear under Current Reaction Sets with the Associated Reactions; otherwise, it will not appear when returning to the simulation environment.

PRO/II Simulation

The PRO/II process simulation program performs rigorous mass and energy balances for various chemical processes. The conversion reactor in PRO/II is used as follows: open a New case in PRO/II, select the conversion reactor from the object palette, then connect inlet and exit streams. Click on the Component tab (benzene ring in the toolbar) and choose the components: NH_3, O_2, NO, and H_2O. From the thermodynamic data, choose the most commonly used fluid package, the Peng–Robinson equation of state (EOS). Double click on the inlet stream (S1), and

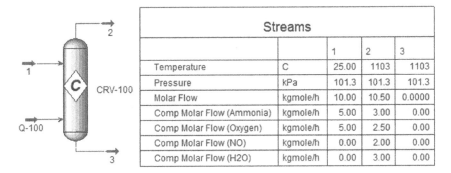

Streams			1	2	3
Temperature		C	25.00	1103	1103
Pressure		kPa	101.3	101.3	101.3
Molar Flow		kgmole/h	10.00	10.50	0.0000
Comp Molar Flow (Ammonia)		kgmole/h	5.00	3.00	0.00
Comp Molar Flow (Oxygen)		kgmole/h	5.00	2.50	0.00
Comp Molar Flow (NO)		kgmole/h	0.00	2.00	0.00
Comp Molar Flow (H2O)		kgmole/h	0.00	3.00	0.00

FIGURE 3.10   UniSim generated process flowsheet and stream summary for the simple conversion reaction case described in Example 3.2.

Reaction Name: CONVERSON

Reactant Stoichiometry

Copy NH3	NH3		4
Paste O2	O2		5
NO	NO		
H2O	H2O		

Product Stoichiometry

Copy NH3	NH3		
Paste O2	O2		
NO	NO		4
H2O	H2O		6

FIGURE 3.11 PRO/II reaction component and stoichiometry menu for the reaction described in Example 3.2.

the Stream Data screen should appear. Click on Flowrate and Composition, then Individual Component Flowrates, enter all component flowrates; 5 kmol/h for NH_3 and 5 kmol/h for O_2. For thermal conditions, set the pressure to 1 atm and temperature to 25°C. From Input in the toolbar menu, click on the Reaction Data tab, and enter the reaction name and description, enter the reactant and the product, as shown in Figure 3.11. Double click on the reactor R1, and from the pull-down menu of Reaction Set Name, select the conversion or the name chosen by the user. Save the File and click on Run. Generate the results from the Output menu, and choose Text Generate Report. Figure 3.12 shows the conversion reactor PFD and stream summary.

ASPEN PLUS SIMULATION

Aspen Plus chemical reactions models handle single and multiple chemical reactions and perform the material balance in the stoichiometric reactor available from the model library (Rstoic from Reactors). Click on Material Streams, and connect the inlet and product streams. Click on Components and choose the components involved (NH_3, O_2, NO, H_2O). Peng–Robinson EOS is selected as the thermodynamic fluid package. Double click on the conversion reaction block. Click on the Specification tab; enter pressure as 1 atm and temperature as 25°C. Then click on the Reactions tab, click on New and enter the components involved in the reaction, stoichiometric coefficient, and fractional conversion as shown in Figure 3.13. Close the stoichiometric windows and then double click on the inlet stream, specify temperature (25°C), pressure (1.0 atm), flow rate (10.0 kgmol/h), and composition (0.5 NH_3, 0.5 O_2). Click on Run and then generate the stream table as shown in Figure 3.14.

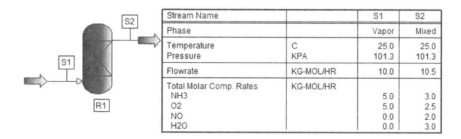

Stream Name		S1	S2
Phase		Vapor	Mixed
Temperature	C	25.0	25.0
Pressure	KPA	101.3	101.3
Flowrate	KG-MOL/HR	10.0	10.5
Total Molar Comp. Rates	KG-MOL/HR		
NH3		5.0	3.0
O2		5.0	2.5
NO		0.0	2.0
H2O		0.0	3.0

FIGURE 3.12 PRO/II generated a process flowsheet and streams summary for the simple conversion reaction case described in Example 3.2.

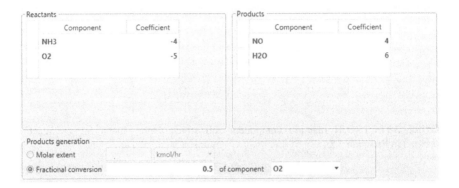

FIGURE 3.13   Aspen Plus reaction stoichiometry and fractional conversion for the ammonia oxidation reaction (coefficients: – for reactant, + product) for the chemical reaction described in Example 3.2.

SuperPro Designer Simulation

It is possible to simulate a conversion reaction in continuous stoichiometric CSTR or PFR reactors in SuperPro Designer software. Start New case, click on New case, and select continuous operation. Under the Tasks menu, select Pure Components, then select Register, Edit/view properties. Enter all components involved in the system (Ammonia and Nitric oxide); water and oxygen are available by default. Click on the Unit Procedure in the toolbar, and then pick the Continuous reactions/stoichiometric/in a CSTR. Click the Connect Mode in the toolbar, then connect a feed and a product stream. Double click on the feed streamline and set the feed stream conditions. A small amount of water is added to the feed stream to bypass the CSTR reactor's error with no liquid or solid in the feed stream. Insert the reaction data (double click on the unit, click on Reactions, and then click on Edit Stoichiometry). Add the reactant and product molar coefficient as shown in Figure 3.15. Assume a 50% completion for the limiting reactant.

Solve the mass and energy balances and generate the stream table. Under View, select Stream Summary Table, right click on the empty area, choose Edit Content, and choose inlet and exit streams. The PFR can also be used to generate the same results. On the operating conditions page. The reaction occurs in the

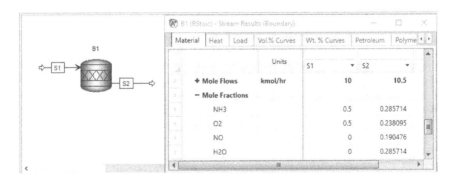

FIGURE 3.14   Aspen Plus generated a process flowsheet and stream summary for the simple conversion reaction case described in Example 3.2.

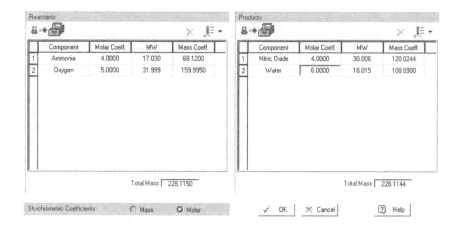

FIGURE 3.15 SuperPro interface for the components reactants and products molar coefficients, for the case described in Example 3.2.

vapor phase; otherwise, an error message: no material is available for chemical reaction. The generated table appears like that shown in Figure 3.16.

AVEVA PROCESS SIMULATION

The conversion reactor (CNVR) icon is available in the Process Library, the CNVR represents a conversion reactor, and it converts feed stream composition based on component reaction rates expressed through stoichiometric relations. It supports multiple reactions. Fully specify the feed stream and the following important variables: reactor length (L), reactor diameter (D), pressure drop (DP), and conversion of the base component in the individual reactions (Xrxn).

1. Start Aveva Process simulation. Click on the plus sign for creating a recent simulation case, right click and select Rename Simulation (e.g., Example 3.2).
2. Drag the CNVR object from the Process library to the simulation area (Canvas). Connect one Source streams and one Sink stream.

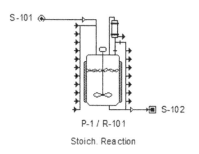

S-101

P-1 / R-101

Stoich. Reaction

Time Ref: h		S-101	S-102
Total Mass Flow	kmol	10.0007	10.5007
Temperature	°C	25.0	17.4
Pressure	bar	1.013	1.013
Total Contents	kmol	10.0007	10.5007
Ammonia		5.0000	2.9999
Nitric Oxide		0.0000	2.0001
Oxygen		5.0002	2.5001
Water		0.0006	3.0006

FIGURE 3.16 SuperPro process flow diagram and stream summary using conversion reactor, the results are generated for the case described in Example 3.2.

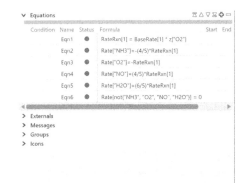 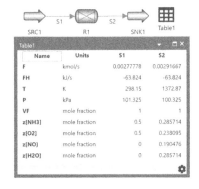

FIGURE 3.17 Aveva Process Simulation generated a process flowsheet and stream summary for the simple conversion reaction case described in Example 3.2.

3. Copy DefFluid from the process library to the example folder. Edit DefFluid and select SRK from the pull-down menu as the suitable fluid thermodynamic property measurements method. Then click on the components: NH_3, O_2, NO, H_2O. Copy DefRate from the Process library to the problem library. Edit the DefRate and add the equation as shown in Figure 3.17.
4. Double click on the feed stream, and for Fluid type, select Example 3.2 Model/DefFluid from the pull-down menu.
5. To build a stream summary table, drag and drop the table from the Tools library to the canvas. Right click on feed stream and select Table/Add all Process. Stream. Double click on the table icon to display the table. Click on Expand the configuration section to add and delete unwanted items.

CONCLUSIONS

For conversion reactors, manual calculations and simulation results using UniSim, PRO/II, Aspen Plus, SuperPro Designer, and Aveva Process Simulation were approximately identical.

Example 3.3: Multiple Reactions

Feed enters an isothermal reactor at 350°C, 30 atm, and 2110 mol/s with mole fractions, 0.098 CO, 0.307 H_2O, 0.04 CO_2, 0.305 hydrogens, 0.1 methane, and 0.15 nitrogen. The reactions take place simultaneously. Assume 100% conversion of methane and carbon monoxide (CO), neglecting the pressure drop across the reactor. Calculate the product components' molar flow rates, the following two reactions taking place in an isothermal reactor.

$$CH_4 + H_2O \rightarrow 3H_2 + CO$$
$$CO + H_2O \rightarrow CO_2 + H_2$$

Compare manual calculation with the predictions of the available software package (e.g., UniSim/Hysys, PRO/II, Aspen Plus, SuperPro, and Aveva Process Simulation).

SOLUTION

MANUAL CALCULATIONS

Figure 3.18 shows the labeled process flowsheet.
Material balance using the extent of the reaction method:

$$CH_4 + H_2O \rightarrow 3H_2 + CO \quad \xi_1$$
$$CO + H_2O \rightarrow CO_2 + H_2 \quad \xi_2$$

MATERIAL BALANCE

Mole balance on methane (CH_4):

$$n_{CH4} = 0.1\,(2110\text{ mol/s}) - \xi_1$$

Mole balance on the water (H_2O):

$$n_{H2O} = 0.307\,(2110\text{ mol/s}) - \xi_1 - \xi_2$$

Mole balance on hydrogen (H_2):

$$n_{H2} = 0.305\,(2110\text{ mol/s}) + 3\xi_1 + \xi_2$$

Mole balance on carbon monoxide (CO):

$$n_{CO} = 0.098\,(2110\text{ mol/s}) + \xi_1 - \xi_2$$

Mole balance on carbon dioxide (CO_2):

$$n_{CO_2} = 0.04\,(2110\text{ mol/s}) + \xi_2$$

Mole balance on nitrogen (N_2):

$$n_{N2} = 0.15\left(2110\frac{\text{mol}}{\text{s}}\right)$$

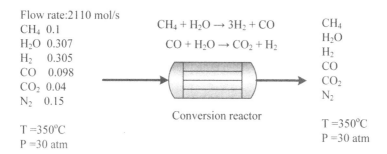

FIGURE 3.18 Process flowsheet of the conversion reactor defined in Example 3.3.

Auxiliary relations. Complete conversion of methane (CH_4):

$$X = 1 = \frac{(n_{CH_4})_{feed} - (n_{CH_4})_{product}}{(n_{CH_4})_{feed}} = \frac{0.1(2110 \text{ mol/s}) - n_{CH_4}}{0.1(2110 \text{ mol/s})}$$

Complete conversion of CO:

$$X = 1 = \frac{0.098 \ (2110 \text{ mol/s}) - n_{CO}}{0.098 \ (2110 \text{ mol/s})}$$

From the first relation, the exit molar flow rate of methane is zero ($n_{CH_4} = 0$), and substituting this value in methane material balance equations,

$$0 = 0.1(2110 \text{ mol/s}) - \xi_1$$

Hence, $\xi_1 = 211$ mol/s

The first and the second reactions take place simultaneously, and the percent conversions of both reactions are 100%.

$$1 = \frac{0.098(2110 \text{ mol/s}) - n_{CO}}{0.098(2110 \text{ mol/s})_1} = \frac{\xi_2}{0.098(2110 \text{ mol/s})} = \frac{\xi_2}{0.098(2110 \text{ mol/s})}$$

The second extent of reaction: $\xi_2 = 0.098(2110) = 206.78$ mol/s. Substituting calculated values of ξ_1 and ξ_2 in the material balance equations gives the following exit molar flow rates:

The molar flow rate of CH_4 in the product stream (complete conversion, $X = 1$), $n_{CH4} = 0.0$

The molar flow rate of H_2O in the product stream:

$$n_{H2O} = 0.307 \left(2110 \frac{\text{mol}}{\text{s}} \right) - 211 - 206.78 = 230 \frac{\text{mol}}{\text{s}} = 827.96 \frac{\text{kmol}}{\text{h}}$$

The molar flow rate of H_2 in the product stream,

$$n_{H2} = 0.305 \times \left(2110 \frac{\text{mol}}{\text{s}} \right) + 3 \times 211 + 206.78 = 1483.33 \frac{\text{mol}}{\text{s}} = 5340 \frac{\text{kmol}}{\text{h}}$$

The molar flow rate of CO in the product stream,

$$n_{co} = 0.098 \times \left(2110 \frac{\text{mol}}{\text{s}} \right) + 211 - 206.78 = 211 \frac{\text{mol}}{\text{s}} = 759.6 \frac{\text{kmol}}{\text{h}}$$

The molar flow rate of CO_2 in the product stream,

$$n_{CO2} = 0.04 \times \left(2110 \frac{\text{mol}}{\text{s}} \right) + 206.78 = 291.18 \frac{\text{mol}}{\text{s}} = 1048.25 \frac{\text{kmol}}{\text{h}}$$

The molar flow rate of N_2 in the product stream,

$$n_{N2} = 0.15 \times \left(2110 \frac{\text{mol}}{\text{s}} \right) = 316.5 \frac{\text{mol}}{\text{s}} = 1139.4 \frac{\text{kmol}}{\text{h}}$$

UNISIM/HYSYS SIMULATION

In a recent case in Hysys, click on the Components tab and add all components (reactants and products). Click on the Fluid PKgs tab and select Peng–Robinson as the fluid package. Select Reactions packages under Flowsheet, and on the Simulation Basis Manager's Reactions page, click on Add Rxn and select conversion, then click on Add Reaction. Select the components: methane, H_2O, H_2, and CO, under stoichiometric coefficient, enters –1 for methane and water as the reactant, 3 for hydrogen, and 1 for CO as the product components. The balance error should be 0.00. Click on the Basis tab, select CH_4, the base component, and set the conversion to 100% (Figure 3.19).

The second reaction (Rxn-2) entered the same way as the first reaction's coefficients. Click on basis; select CO as the base component, for Rxn Phase: overall, for CO: 100, and 0.0 for C1 and C2. Click on Add to FP and then on Add set to Fluid Package; note that "Basis-1" should appear under Assoc. Fluid Pkgs. Press Return to Simulation Environment to return to the simulation environment. Notice a button that was never there before when we looked at the reaction sets. Click on Ranking and change the default setting to 1 so that those reactions coincide in the parallel reactions, adding the involved as follows:

1. Open the Simulation Basis Manager by clicking the Home View icon located in the Simulation Basis Manager toolbar.
2. Select the Reactions tab.
3. Click on the Add Rxn button, opens the Reactions view.
4. From the available reaction list, types select the reaction type to use and click on Add Reaction.
5. After adding the two reactions, using the default set name (i.e., Global Rxn Set), add the set to Current Reactions Sets; the reaction package page appears like that shown in Figure 3.20. The reaction set should appear under Current Reaction Sets with the Associated Reactions; otherwise, it will not appear when returning to the simulation environment.

Enter the simulation environment by clicking on Return to Simulation Environment. To set up the conversion reactor, select a conversion reactor from the object palette and place it on the PFD. Connect feed and product streams and then double click on stream one and enter the temperature, 350°C; pressure, 30.4 bar; and molar flow, 7596 kgmol/h. Click on composition, enter the molar fraction, close the inlet stream window, and double click on the conversion reactor; then press the Reaction tab from the pull-down menu in front of Reaction Set and choose the reaction set.

Stoichiometry Info			
Component	Mole Weight	Stoich Coeff	
Methane	16.043	-1.000	
H2O	18.015	-1.000	
Hydrogen	2.016	3.000	
CO	28.011	1.000	
Add Comp			
Balance	Balance Error	0.00000	
	Reaction Heat (25 C)	2.1e+05 kJ/kgmole	

Stoichiometry | Basis

Basis	
Base Component	Methane
Rxn Phase	Overall
CO	100.0
C1	0.0000
C2	0.0000

Conversion (%) = C0 + C1*T + C2*T^2

(T in Kelvin)

Stoichiometry **Basis**

FIGURE 3.19 UniSim/Hysys set the reaction stoichiometry (left) and the reaction basis (right) for the case described in Example 3.3.

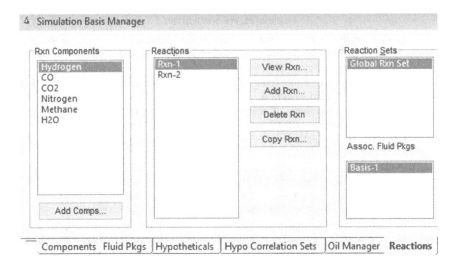

FIGURE 3.20   UniSim simulation basis managers set the two reactions involved in the case described in Example 3.3.

When the heat stream is not attached, it is assumed adiabatic; to make it iso-thermal, connect a heat stream and set the temperature on the product stream the same as the inlet stream temperature. On the Parameters page, a pressure drop across the reactor is zero. UniSim/Hysys requires attaching another product stream. Go ahead and connect a liquid product stream and call it 3. Figure 3.21 shows the PFD of the conversion reactor.

PRO/II SIMULATION

PRO/II is capable of handling multiple reactions. The same procedure of Example 3.2, first, perform the process flowsheet of the conversion reactor and specify the feed

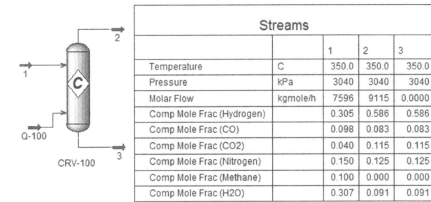

Streams				
		1	2	3
Temperature	C	350.0	350.0	350.0
Pressure	kPa	3040	3040	3040
Molar Flow	kgmole/h	7596	9115	0.0000
Comp Mole Frac (Hydrogen)		0.305	0.586	0.586
Comp Mole Frac (CO)		0.098	0.083	0.083
Comp Mole Frac (CO2)		0.040	0.115	0.115
Comp Mole Frac (Nitrogen)		0.150	0.125	0.125
Comp Mole Frac (Methane)		0.100	0.000	0.000
Comp Mole Frac (H2O)		0.307	0.091	0.091

FIGURE 3.21   UniSim/Hysys generates the process flowsheet and the stream summary for the reaction case presented in Example 3.3.

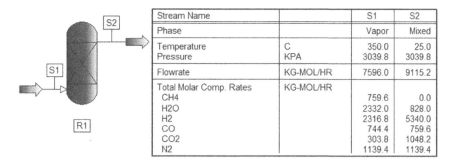

FIGURE 3.22 PRO/II kinetic rate calculation menu of the power law defined the two reactions presented in Example 3.3.

stream with the given total flow rate and compositions, the temperature is 350°C, and the pressure is 30 atm. Under the Reactions input menu, select Reaction Data and enter the two reactions shown in Figure 3.22. The user to choose R1 from the set name of the pull-down menu. Click on Extent of Reaction and specify 100% conversion (fractional conversion is 1) for both reactions. After running the system, the process flowsheet and the stream property table appear as shown in Figure 3.23.

ASPEN PLUS SIMULATION

Aspen can handle multiple reactions. Add all components, reactants, and products (CH_4, H_2O, H_2, CO, CO_2, and N_2). Select Peng–Robinson as the fluid package. Double click on the reactors block diagram from the Model Palette. For material balance on multiple reactions, select a stoichiometric reactor (Rstoic) from the reactors' subdirectory and place it in the primary process flowsheet area. Click on Material Stream, and connect feed and product streams. In the opened Data Browser window, click on the Specification tab; enter 1 atm for pressure and 25°C for temperature. Then click on the Reactions tab. Press New and enter components involved in the first reaction, enter the stoichiometric coefficient, and then the fractional conversion shown in Figure 3.24. Repeat the same procedure for adding the second reaction. The limiting component is CO, and the fractional conversion is (1). Close the stoichiometry page. Double click on stream one and enter the temperature, pressure, flow rate, and composition. Repeat the same for the second reaction, then click OK to close. Click Run to generate the stream table as shown in Figure 3.25.

Stream Name		S1	S2
Phase		Vapor	Mixed
Temperature	C	350.0	25.0
Pressure	KPA	3039.8	3039.8
Flowrate	KG-MOL/HR	7596.0	9115.2
Total Molar Comp. Rates	KG-MOL/HR		
CH4		759.6	0.0
H2O		2332.0	828.0
H2		2316.8	5340.0
CO		744.4	759.6
CO2		303.8	1048.2
N2		1139.4	1139.4

FIGURE 3.23 PRO/II generated the process flowsheet and the stream summary for the case described in Example 3.3.

Reaction No. ☑1   ▾

Reactants

Component	Coefficient
CH4 ▾	-1
H2O ▾	-1
▶	

Products

Component	Coefficient
H2	3
▶ CO	1

Products generation

○ Molar extent   kmol/hr   ▾

◉ Fractional conversion   1 of component CH4   ▾

[▶▶]   [Close]

FIGURE 3.24 Aspen Plus required reactants and products stoichiometric coefficient of the first reaction of the case described in the Example of 3.3.

SUPERPRO DESIGNER SIMULATION

Multiple reactions, following the same procedure for a single chemical reaction in the previous Example. The present case added another chemical reaction. The two reactions take place in parallel reactions. The SuperPro material balance menu is shown in Figure 3.26. The SuperPro defines the limiting component; however, the user needs to select the limiting component's radio button for each reaction. Figure 3.27 shows the stream summary.

AVEVA PROCESS SIMULATION

The CNVR in Process library represented a conversion reactor, and it converts feed stream composition based on component reaction rates expressed through stoichiometric relations. It supports multiple reactions. Fully specify the feed stream and the following important variables: reactor length (L), reactor

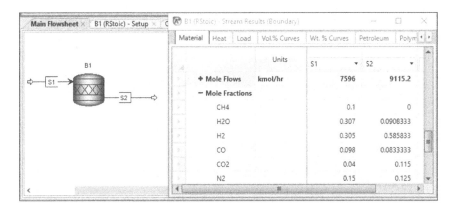

	Units	S1	S2
+ Mole Flows	kmol/hr	7596	9115.2
– Mole Fractions			
CH4		0.1	0
H2O		0.307	0.0908333
H2		0.305	0.585833
CO		0.098	0.0833333
CO2		0.04	0.115
N2		0.15	0.125

FIGURE 3.25 Aspen Plus generated the process flowsheet and the stream summary, the case described in Example 3.3.

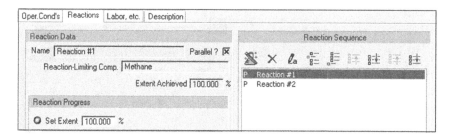

FIGURE 3.26 SuperPro required data for the two simultaneous reactions for the case described in Example 3.3.

diameter (D), pressure drop (DP), and conversion of the base component in the individual reactions (Xrxn).

Start Aveva Process Simulation. Click on the plus sign for a recent simulation case, right click and select Rename Simulation (here, Example 3.3). Drag the CNVR object from the Process library into the simulation area (Canvas). Connect one Source streams and one Sink stream. Copy DefFluid from the process library to the example folder. Edit DefFluid and select SRK from the pull-down menu as the suitable fluid thermodynamic property measurements method. Then click on the components: CH_4, CO, CO_2, H_2O, H_2, and N_2. Copy DefRate from the Process library to the problem library. Edit the DefRate and add the equation as shown in Figure 3.28.

Double click on the feed stream, and for Fluid type, select Example 3.3 Model/DefFluid from the pull-down menu. To build a stream summary table, drag and drop the table from the Tools library to the canvas. Right-click on feed stream and select Table/Add all Process. Stream. Double click on the table icon to display the table. Click on Expand the configuration section to add and delete unwanted items (Figure 3.28).

CONCLUSION

The results of material balance calculations obtained with hand calculations, UniSim/Hysys, PRO/II, Aspen, SuperPro Designer, and Aveva Process Simulation commercial software packages are the same.

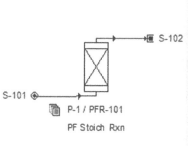

Time Ref: s			S-101	S-102
Total Mass Flow		mol	2110.000	2532.000
Temperature		°C	25.0	100.0
Pressure		atm	1.000	1.000
Total Contents		mole frac	1.000	1.000
Carb. Dioxide			0.040	0.115
Carbon Monoxide			0.098	0.083
Hydrogen			0.305	0.586
Methane			0.100	0.000
Nitrogen			0.150	0.125
Oxygen			0.000	0.000
Water			0.307	0.091

FIGURE 3.27 SuperPro generated the process flow diagram and streams summary for the case described in Example 3.3.

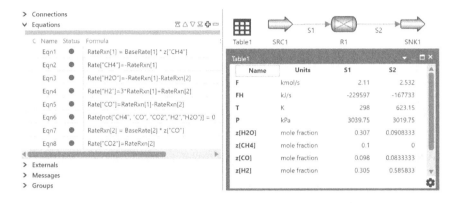

FIGURE 3.28 Aveva Process Simulation generates the process flow diagram and streams summary for the case described in Example 3.3.

3.4   ENERGY BALANCE WITHOUT REACTION

The general balance equation for a continuous open system at a steady state in the absence of generation/consumption term is: Energy input = Energy output

$$\text{Energy input} = \dot{U}_{in} + \dot{E}_{k,in} + \dot{E}_{p,in} + P_{in}\dot{V}_{in}$$

$$\text{Energy output} = \dot{U}_{out} + \dot{E}_{k,out} + \dot{E}_{p,out} + P_{out}\dot{V}_{out}$$

$$\text{Energy transferred} = \dot{Q} - \dot{W}_{s}$$

Energy transferred = Energy out–Energy in

$$\dot{Q} - \dot{W}_{s} = \left(\dot{U}_{out} + \dot{E}_{k,out} + \dot{E}_{p,out} + P_{out}\dot{V}_{out} \right)$$

Rearrange

$$\dot{Q} - W_{s} = \Delta\dot{U}_{in} + \Delta\dot{E}_{k} + \Delta\dot{E}_{p} + \Delta\left(P\dot{V} \right)$$

The enthalpy is defined as

$$\dot{H} = \dot{U} + P\dot{V}$$

The change in enthalpy

$$\Delta\dot{H} = \Delta\dot{U} + \left(P\dot{V} \right)$$

Rearranging the above equations leads to the First Law of Thermodynamics for an open system at a steady state.

$$\dot{Q} - \dot{W}_s = \Delta \dot{H} + \Delta \dot{E}_k + \Delta \dot{E}_p, \text{where } (\Delta = \text{Output} - \text{input})$$

where $\dot{Q}$ is the heat transferred to or from the system, $\dot{W}_s$ is the shaft work, $\Delta \dot{H}$ is the rate of change in enthalpy of the system, $\Delta \dot{E}_k$ is the rate of change of the kinetic energy, $\Delta \dot{E}_p$ is the rate of change in potential energy, and ΔU, is the rate of change in the internal energy of the system.

Example 3.4: Energy Balance on a Heat Exchanger

A shell and tube heat exchanger is used to cool hot water with cold water. Hot water enters through the tube side of a heat exchanger with a mass flow rate of 0.03 kg/s and 80°C cooled to 30°C. Coldwater in the shell side flows with a mass flow rate of 0.06 kg/s at a temperature of 20°C. Determine the outlet temperatures of the cold water. Compare manual calculation with the predictions of the available software package (e.g., UniSim/Hysys, PRO/II, Aspen Plus, SuperPro, and Aveva Process Simulation).

SOLUTION

MANUAL CALCULATIONS

The following energy balance equation calculates the heat transfer rate: $\dot{Q} = \dot{m} C_p \Delta T$
   For this case, the total heat transfer may be obtained from the energy balance equation considering the adiabatic heat exchanger.

$$\dot{m}_c C_{p_c} \Delta T_c = \dot{m}_h C p_h \Delta T_h$$

Calculate the rate of heat transfer from the following equation:

$$Q_h = \dot{m}_h C p_h \Delta T_h$$

The mass flow rate of the hot stream, $\dot{m}_h$

$$\dot{m}_h = 0.03 kg/s$$

The temperature difference of the hot stream inlet and exit

$$\Delta T_h = 80 - 30 = 50°C$$

Cp_h is the average specific heat
   For water at 80°C, $Cp_h = 4.198$ kJ/kg°C
   For water at 30°C, $Cp_h = 4.179$ kJ/kg°C
   The average heat capacity, C_{Pavg}

$$Cp_{avg} = \frac{(4.198 + 4.179)}{2} = 4.1885 \ kJ/kg°C$$

The heat transfer rate from hot stream to cold stream:

$$\dot{Q}_h = \dot{m}_h C p_h \Delta T_h = 0.33 \frac{kg}{s} \left(4.188 \frac{kJ}{kg^\circ C} \right) (80 - 30)^\circ C = 6.28 \; kJ/s$$

Calculate the water temperature at the outlet of the shell side ($T_{c,\,out}$, °C).

$$Q_c = \dot{m}_c \, C_{pc} \Delta T_c, \text{ and } \dot{m}_c = 0.06 \, kg/s$$

The specific heat of the cold water at 20°C and atmospheric pressure is

$$C_p \text{ at } 20^\circ C = 4.183 \; kJ/kg^\circ C$$

Rearranging for calculating the temperature difference of the cold stream:

$$\Delta T_c = \frac{\dot{Q}_h}{\dot{m}_c \times C p_c}$$

Substituting values,

$$\Delta T_c = \frac{6.28 \, (kJ/s)}{0.06 \, (kg/s) \times 4.183 \; (kJ/kJ^\circ C)} = 25^\circ C$$

Hence, the exit temperature of the cold water:

$$T_{c,out} = T_{c,in} + \Delta T_c = 25 + 20 = 45^\circ C$$

UniSim/Hysys Simulation

Start UniSim, and in a UniSim recent case, select all the components involved in the problem; for this case, it is only water. For fluid package, ASME steam is the most suitable for water, then Enter Simulation Environment. The situation required using the shell and tube heat exchanger from the object palette, followed by specifying the heat exchanger's required parameters. Information of streams, such as temperature, pressure, flow rate, and compositions, should be determined to make the software run properly. After identifying all the necessary information to the cold and hot inlet streams and the hot exit temperature, pressure drop in the tube and Shell sides set to zero. Hysys calculates automatically once enough information is received. Figure 3.29 shows the solution for this problem. To avoid the warning message, Ft Correction Factor is Low, while in the Design/Parameters page, set the cell below Shells in Series at the bottom of the screen to 2 instead of 1.

PRO/II Simulation

Open a new case in PRO/II; click on File and then New. Then click on Component Selection on the toolbar. Select water. Next, click on Thermodynamic Data. Once inside, click on Most Commonly Used and then select Peng–Robinson. Then click

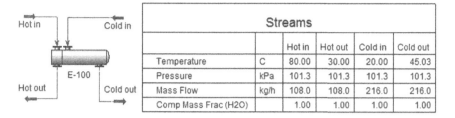

Streams					
		Hot in	Hot out	Cold in	Cold out
Temperature	C	80.00	30.00	20.00	45.03
Pressure	kPa	101.3	101.3	101.3	101.3
Mass Flow	kg/h	108.0	108.0	216.0	216.0
Comp Mass Frac (H2O)		1.00	1.00	1.00	1.00

FIGURE 3.29 UniSim/Hysys generated the process flow diagram of the shell and tube heat exchanger and stream summary for the case described in Example 3.4.

Add and then OK to return to the PFD screen. Now it is ready to insert units and streams. The system that we want to put in the simulation is the heat exchanger. Scroll down the toolbar until you see the shell and tube heat exchanger, Rigorous HX. Next, click on Streams. First, create a feed stream, S1 (Hot in), entering into the tube side. Next, create stream S2 (Hot out) from the other side of the heat exchanger, the tube side's exit. The next stream, S3 (Cold in), should go into the shell side, and S4 (Cold out) should leave the shell side.

Double click on S1 and specify the tube mass flow rate (0.03 kg/s). The inlet temperature is 80°C, and the pressure is 1 atm. Double click on stream S3 and specify the cold stream's mass flow rate as 0.06 kg/s; temperature, 20°C; and pressure 1 atm. Double click on the heat exchanger icon and specify the calculation type such that the tube outlet temperature is 30°C. Set the Area/Shell to an arbitrary value, 2 m². Now it is time to run the simulation. Click on Run on the toolbar. The simulation should turn blue. The next is to view the result. To view the results from each stream, right click on the stream and then choose View Results. Figure 3.30 is displayed by clicking Output in the toolbar and then selecting Stream Property Table. Note that a Simple HEX exchanger generates the same results.

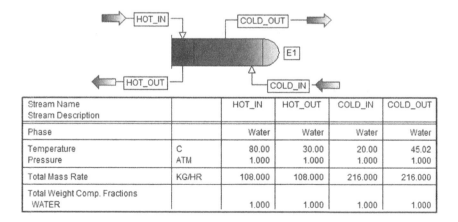

Stream Name Stream Description		HOT_IN	HOT_OUT	COLD_IN	COLD_OUT
Phase		Water	Water	Water	Water
Temperature	C	80.00	30.00	20.00	45.02
Pressure	ATM	1.000	1.000	1.000	1.000
Total Mass Rate	KG/HR	108.000	108.000	216.000	216.000
Total Weight Comp. Fractions WATER		1.000	1.000	1.000	1.000

FIGURE 3.30 PRO/II generates the process flowsheet and the stream summary of the shell and tube heat exchanger for the case described in Example 3.4.

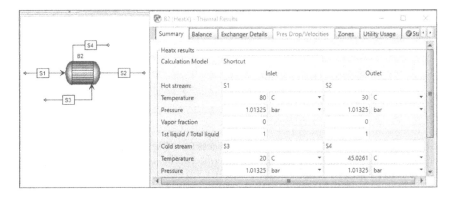

FIGURE 3.31 Aspen Plus generates the process flow diagram of the shell and tube exchanger and stream summary for the case described in Example 3.4.

ASPEN PLUS SIMULATION

Start Aspen Plus, create a blank simulation, and select the system's components (here only water). Steam-TA is a suitable property method. The case required working on a shell and tube heat exchanger. Click on the Simulation tab at the bottom left corner, and then click on the Exchangers from the submenu, select the shell and tube heat exchanger icon (HEATX) and click somewhere in the simulation area. Clicking on the arrow near the icon allows the user to select the desired heat exchange shape. The hot and cold inlet streams are fully specified. In the heat exchanger specifications, the hot stream outlet temperature is set and defined as 30°C. Specify the two feed streams and the cold stream's outlet temperature, and the system is ready to run. The results should appear like that shown in Figure 3.31. The cold stream outlet temperature is 45°C.

SUPERPRO DESIGNER

Water is a default component in SuperPro; consequently, there is no need for component selection. The heat exchanger is selected as follows: Unit Procedures ≫ Heat Exchanger ≫ Heat exchanging. The two inlets are connected and fully specified (temperature, pressure, flow rates, and compositions). Two exit streams are combined. Double click on the exchanger block in the PFD area, and under Performance Options, select the hot stream outlet temperature button and enter the hot stream exit temperature as 30°C. Figure 3.32 shows the results.

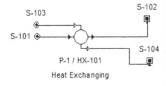

Time Ref: h			S-101	S-102	S-103	S-104
Total Mass Flow		kg	108.00	108.00	216.00	216
Temperature		°C	80.00	30.00	20.00	45.00
Pressure		bar	1.01	1.01	1.01	1.013
Total Contents	mass frac		1.00	1.00	1.00	1
Water			1.00	1.00	1.00	1

FIGURE 3.32 SuperPro generates the process flow diagram of the shell and tube exchanger and stream summary for the case described in Example 3.4.

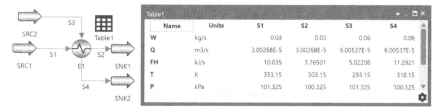

FIGURE 3.33   Aveva Process Simulation generates the process flow diagram of the shell and tube exchanger and stream summary for the case described in Example 3.4.

AVEVA PROCESS SIMULATION

If we plan to use cooling water for this case, it is easy to use the built-in water-cooled heat exchanger. Drag the heat exchanger icon to the simulation area (Canvas), then connect two inlets and two exit streams. Change the inlet temperatures, pressure, and flow rate. Adjust the overall heat transfer coefficient if needed in only a few steps. Aveva Process Simulation has calculated the area for the exchanger and the general utility requirement. This heat exchanger is a great and easy option for using default utilities like cooling water and steam to cool or heat stream.

Select the HX heat exchanger from the Process library to the simulation area (Canvas). Select the DefFluid and add water. Connect the two feed streams (hot fluid into the tube and cold to the shell). Specify the pressure drop and the S2 temperature at 30°C. Figure 3.33 shows the generated exchanger PFD and stream summary.

CONCLUSIONS ·

According to manual calculation, the cold water outlet temperature was 45°C. The manual calculation results matched the values obtained from the software packages: Hysys, PRO/II, Aspen Plus, SuperPro, and the Aveva Process Simulation. The selection of a suitable fluid package is essential to get the correct results. Also, providing the software with the proper temperature, pressure, flow rate, and composition will lead to obtaining the right solution.

3.5   ENERGY BALANCE ON REACTIVE PROCESSES

Material balances are performed by writing compound balances (which required the extent of reaction) or elements (which only need the material balance without generation terms for each component); we can also do energy balances using either compounds or elements. We have two methods for solving these problems [2, 3]: The heat of reaction method and the heat formation method.

These two methods differ in the choice of the reference state. The heat of reaction method is ideal when there is a single reaction for which $\Delta \hat{H}r°$ is known. This method requires calculating the extent of reaction ($\dot{\xi}$). The reference state is such that all reactant and product species are at 25°C, 1 atm in the states for which the reaction heat is known. $\hat{H}_i$ accounts for the change in enthalpy with T and phase (if necessary). Hence, the rate of change in enthalpy of a single reactive process is

$$\Delta \dot{H} = \dot{\xi}\Delta H_r^o + \sum_{out} \dot{n}_i \hat{H}_i - \sum_{in} \dot{n}_i \hat{H}_i \tag{3.4}$$

For multiple reactions,

$$\Delta \dot{H} = \sum_{reactions} \dot{\xi}_i \Delta H_{rj}^o + \sum_{out} \dot{n}_i \hat{H}_i - \sum_{in} \dot{n}_i \hat{H}_i \qquad (3.5)$$

The reference state is such that the reactants and products are at 25°C and 1 atm. If considering a reference temperature other than 25°C, calculate the reaction's heat at the new reference state.

$$\Delta \dot{H} = \sum_{reactions} \dot{\xi}_i \Delta H_{rj(Tref)} + \sum_{out} \dot{n}_i \hat{H}_i - \sum_{in} \dot{n}_i \hat{H}_i \qquad (3.6)$$

where the change in the heat of reaction at any temperature ($\Delta \dot{H}_r$):

$$\Delta \dot{H}_r (T) = \Delta \dot{H}_{rxn}^o (T = 25°C) + \int_{25°C}^{T} \Delta Cp \, dT \qquad (3.7)$$

The change in the specific heat, ΔC_p

$$\Delta Cp = \sum v_i Cp_i \qquad (3.8)$$

where v_i, the stoichiometric coefficient of component i involved in the reaction, is positive for products (+) and negative (-) for reactants.

Example 3.5: Oxidation of Ammonia

In the ammonia oxidation reaction, 100 kmol/h ammonia (NH_3) and 200 kmol/h oxygen (O_2) at 25°C and 1.0 atm enters into a conversion reactor, where the reaction processed to completion consuming all the ammonia in the feed stream. The product gas emerges at 300°C. Calculate the rate at which heat must be transferred to or from the reactor [4]. The standard heat of reaction for the oxidation of ammonia:

$$4NH_3 + 5O_2 \rightarrow 4NO + 6H_2O$$

Compare manual calculation with the predictions of the available software package (e.g., UniSim/Hysys, PRO/II, Aspen Plus, SuperPro, and Aveva Process Simulation).

SOLUTION

MANUAL CALCULATIONS

Basis: 100 kmol/s of NH_3
    Dividing by the stoichiometric coefficient of NH_3

$$NH_3 + \frac{5}{4}O_2 \rightarrow NO + \frac{6}{4}H_2O$$

Neglecting change in kinetic and potential energy, the difference in the enthalpy of a single reaction taking place in a stoichiometric reactor is $Q - W = \Delta \dot{H}$

$$\Delta \dot{H} = \dot{\xi} \Delta \hat{H}_r^\circ + \sum_{out} \dot{n}_i \hat{H}_i - \sum_{in} \dot{n}_i \hat{H}_i$$

Material balance (extent of reaction). The mole balance on ammonia, NH_3

$$n_{NH3} = 100 - \xi$$

Mole balance on oxygen, O_2

$$n_{O2} = 200 - \frac{5}{4}\xi$$

Mole balance on nitric oxide, NO_2

$$n_{NO_2} = 0 + \xi$$

Mole balance on the water, H_2O

$$n_{H2O} = 0 + \frac{6}{4}\xi$$

Complete conversion of ammonia ($x_c = 1$):

$$\frac{\xi}{100} = 1 \Rightarrow \xi = 100$$

$n_{NH3} = 0$ kmol/s, $n_{O2} = 75$ kmol/s, $n_{NO} = 1\,00$ kmol/s, and $n_{H_2O} = 150$ kmol/s.
Energy balance (reference temperature: 25°C):

$$\Delta \dot{H} = \dot{\xi} \Delta H_r^\circ + \sum_{out} \dot{n}_i \hat{H}_i - \sum_{in} \dot{n}_i \hat{H}_i$$

Substitution

$$\Delta \hat{H}_r^\circ = \frac{6}{4}(-241,800) + (90,370) - (-46,190) = -226,140 \text{ kJ/kmol}$$

The overall change in the system sensible heat of all components:

$$\Delta \dot{H}_s = \left[75 \int_{25}^{300} Cp_{O_2}\, dT + 100 \int_{25}^{300} Cp_{NO}\, dT + 150 \int_{25}^{100} Cp_{H_2O,g}\, dT \right]_{out} - 0$$

Substitution,

$$\Delta \dot{H}_s = \left[(75 \times 8.47 + 100 \times 8.453 + 150 \times 9.57) \right] = 2.92 \times 10^6 \text{ kJ/h}$$

The change in the process enthalpy equals the difference in the sensible heat plus the heat of reaction:

$$\Delta\dot{H} = 100 \frac{kmol}{h}\left(-226{,}140 \frac{kJ}{kmol}\right) + 2.92\times10^6 \frac{kJ}{h} = -1.97\times10^6 \frac{kJ}{h}$$

UniSim/Hysys Method

From the UniSim object palette, select the conversion reactor. The percent conversion of ammonia is set to 100%, as stated in the problem statements. Figure 3.34 shows the heat flow from the conversion reactor. Peng–Robinson EOS used for property measurement.

PRO/II Simulation

In a new case in PRO/II, the components involved are ammonia, oxygen, nitric oxide, and water. Select Peng–Robinson EOS for the Thermodynamic Data in the toolbar and the Most Commonly Used Property. Select the Conversion Reactor from the object palettes to the PFD. A feed stream S1 is connected and fully specified, and an exit stream S2 is connected. Once completed the process flowsheet, it is time to enter the reaction.

Click on Reaction Data in the pop-up menu, and name the reaction set and reaction description. Click on Enter Data, rename the reaction, and then click on the Reactants = Products. Set the stoichiometry of the chemical reaction based on the balanced chemical reaction. Click OK three times to return to the PFD. Click on the reactor block diagram. Select the chemical reaction set from the down arrow in front of the Reaction Set Name. Click on Reaction Extent, select ammonia as the base component, and set fraction conversion to 1. (A = 1, B, and C = 0). The system is ready to run. Click on Run, the block color changes to blue. To generate results and place them on the process flowsheet as shown in Figure 3.35, click on Output in the toolbar and then select Stream Property Table. Double click on the table block. Under the Property, list to be used, select Comp. Molar Rates. Click Add All to add the Property of all streams. The calculated heat duty is –19.8 M kJ/h, extracted from the generated text report. The value is close to that obtained by hand calculations.

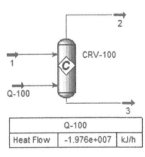

Streams				
		1	2	3
Temperature	C	25.00	300.0	300.0
Pressure	kPa	101.3	101.3	101.3
Molar Flow	kgmole/h	300.0	325.0	0.0000
Comp Molar Flow (Ammonia)	kgmole/h	100	0	0
Comp Molar Flow (Oxygen)	kgmole/h	200	75	0
Comp Molar Flow (NO)	kgmole/h	0	100	0
Comp Molar Flow (H2O)	kgmole/h	0	150	0

Q-100		
Heat Flow	-1.976e+007	kJ/h

FIGURE 3.34 UniSim/Hysys generates the heat released from the conversion reaction for the reaction described in Example 3.5.

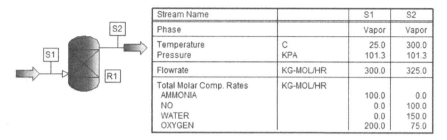

Stream Name		S1	S2
Phase		Vapor	Vapor
Temperature	C	25.0	300.0
Pressure	KPA	101.3	101.3
Flowrate	KG-MOL/HR	300.0	325.0
Total Molar Comp. Rates	KG-MOL/HR		
AMMONIA		100.0	0.0
NO		0.0	100.0
WATER		0.0	150.0
OXYGEN		200.0	75.0

FIGURE 3.35 PRO/II generates the process flowsheet and the stream summary for the reaction described in Example 3.5.

ASPEN PLUS SIMULATION

Start the Aspen Plus software, click on the New tab and create a blank simulation. Enter the compounds contributed to the reaction process in the component's ID: NH_3, O_2, NO, H_2O.

Click on Next and select the Peng–Robinson as the appropriate thermodynamic method from the pull-down menu located under Methods. The stoichiometric reactor in Aspen Plus estimates the heat transfer rate from the conversion reactor. Figure 3.36 shows the components' molar flow rates. The heat released is shown in Figure 3.37. The results are within the range of hand calculations.

SUPERPRO DESIGNER

Selecting the conversion reactor through: Unit procedures ≫ Continuous reactions ≫ Stoichiometric ≫ in PFR.

The limiting reactant is ammonia with complete conversion. The reaction takes place in the vapor phase, and the heat of the reaction is needed. Figure 3.38 shows the PFD and stream summary. Figure 3.39 shows and the SuperPro calculated heat duty.

AVEVA PROCESS SIMULATION

The Aveva conversion reactor (CNVR) icon available in the Process library represented a conversion reactor, and it converts feed stream composition based on

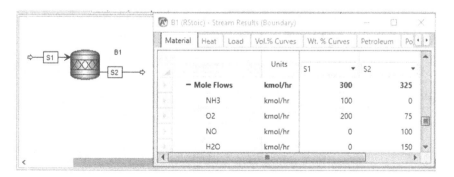

FIGURE 3.36 Aspen Plus generated the process flowsheet of the conversion reactor and the stream summary for the case described in Example 3.5.

Summary	Balance	Phase Equilibrium	Reactions	Selectivity	Utility Usage

Outlet temperature	300	C	▼
Outlet pressure	1.01325	bar	▼
Heat duty	-1.97683e+07	kJ/hr	▼
Net heat duty	-1.31155e+06	cal/sec	▼
Vapor fraction	1		
1st liquid / Total liquid			

FIGURE 3.37 Heat duty of a stoichiometric reaction calculated by Aspen Plus for the case presented in Example 3.5.

component reaction rates expressed through stoichiometric relations. It supports multiple reactions. Fully specify the feed stream and enter the following essential parameters: reactor length (L), reactor diameter (D), pressure drop (DP), and conversion of the base component in the individual reactions (Xrxn).

Start Aveva Process Simulation. Click on the plus sign for a recent simulation case, right click and select Rename Simulation (e.g., Example 3.5). Drag the CNVR object from the Process library into the simulation area (Canvas). Connect one Source streams and one Sink stream. Copy DefFluid from the process library to the example folder. Edit DefFluid and select SRK from the pull-down menu as the suitable fluid thermodynamic property measurements method. Then click on the components: NH_3, H_2O, NH_3, and NO. Copy DefRate from the Process library to the problem library. Double click on the feed stream and for Fluid type, select Example 3.5 Model/DefFluid from the pull-down menu or drag and drop the DefFluid to the feed stream. To build a stream summary table, drag and drop

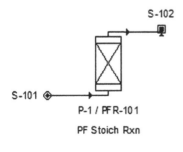

Time Ref: h		S-101	S-102
Total Mass Flow	kmol	300.0	325.0
Temperature	°C	25.0	300.0
Pressure	bar	1.013	1.013
Total Contents	mole frac	1.000	1.000
Ammonia		0.333	0.000
Nitric Oxide		0.000	0.308
Oxygen		0.667	0.231
Water		0.000	0.462

FIGURE 3.38 SuperPro generates the process flowsheet of the conversion reactor and the stream summary for the reaction case described in Example 3.5.

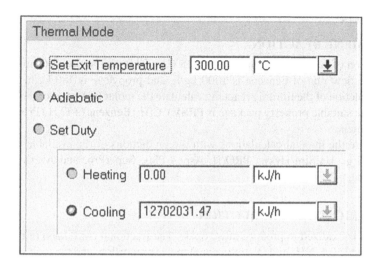

FIGURE 3.39 SuperPro calculates the heat duty using the isothermal conversion reactor for the ammonia oxidation case described in Example 3.5.

the table from the Tools library to the canvas. Right click on feed stream and select Table/Add all Process. Stream. Double click on the table icon to display the table. Click on Expand the configuration section to add and delete unwanted items (Figure 3.40).

Conclusions

UniSim/Hysys, Aspen Plus, SuperPro Designer, and Aveva Process Simulation results precisely lead to approximately the same results. There is a discrepancy between manual calculations and those obtained by SuperPro due to the difference in the values of physical properties such as specific heat.

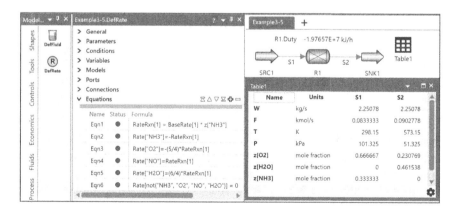

FIGURE 3.40 Aveva Process Simulation generated process flowsheet of the conversion reactor and the stream summary for the case described in Example 3.5.

PROBLEMS

3.1 CUMENE REACTION

The reaction of benzene and propylene at 25°C and 1.0 atm produced Cumene. The inlet mass flow rate of benzene is 1000 kg/h, and propylene is 180 kg/h. Assume 45% completion of the limiting reactant, calculate the molar flow rates of the product stream (the suitable property package is PRSV). C_6H_6(Benzene) + C_3H_6(Propene) $\rightarrow$ C_9H_{12}(Cumene)

Compare the manual calculations with the predictions of the available software package (e.g., UniSim/Hysys, PRO/II, Aspen Plus, SuperPro, and Aveva Process Simulation).

3.2 NITRIC OXIDE PRODUCTION

Ammonia is oxidized to produce nitric oxide. The fractional conversion of the limiting reactant is 0.6. NH_3 and O_2 are in equal molar proportion, and the total inlet flow rate is 100 kmol/h. The operating temperature and pressure are 25°C and 1.0 atm, respectively. Calculate the exit molar flow rates of all components. Assume that the reactor is operating adiabatically. Compare manual calculation results with any available software package (e.g., UniSim/Hysys, PRO/II, Aspen Plus, SuperPro, and Aveva Process Simulation). The recommended fluid package is PRSV.

3.3 MULTIPLE REACTIONS

A feed enters a conversion reactor at 350°C, and 30 atm, a molar flow rate of 7600 kmol/h. The feed molar fractions are 0.098 CO, 0.307 H_2O, 0.04 CO_2, 0.305 hydrogens, and 0.25 methane. The reactions take place in series. Assume 100% conversion of methane and CO, and neglect the pressure drop across the reactor. Calculate the molar flow rates of the product components. The reactions take place in an isothermal conversion reactor.

$$CH_4 + H_2O \rightarrow 3H_2 + CO$$
$$CO + H_2O \rightarrow CO_2 + H_2$$

Compare the results of manual calculation with any available software package (e.g., UniSim/Hysys, PRO/II, Aspen Plus, SuperPro, and Aveva Process Simulation). The appropriate fluid package for hydrocarbon gases is Peng–Robinson.

3.4 GAS-PHASE REACTION

The gas-phase reaction proceeds with 80% conversion. Estimate the heat that must be provided or removed if the gases enter at 400°C and leave at 500°C; the proper fluid package for this case is Peng–Robinson. The following reaction takes place: $CO, + 4H_2 \rightarrow 2H_2O + CH_4$. Compare the results of manual calculation with any available software package (e.g., UniSim/Hysys, PRO/II, Aspen Plus, SuperPro, and Aveva Process Simulation).

3.5 BURNING OF CO

CO at 10.0°C is burned completely at 1.0 atm pressure with 50% excess air fed to a burner at a temperature of 540°C. The combustion products leave the burner chamber at a temperature of 425°C. Calculate the heat evolved, Q, from the burner. Compare manual calculation results with those obtained from any available software package (e.g., UniSim/Hysys, PRO/II, Aspen Plus, SuperPro, and Aveva Process Simulation). The proper fluid package for this case is Peng–Robinson (PR).

3.6 PRODUCTION OF KETENE FROM ACETONE

Pure acetone reacts isothermally to form ketene and methane at 2.0 atm and 650°C, the percent conversion of acetone is 70%. Calculate the reactor exit flow rate of acetone, ketone, and methane. Calculate the heat added or removed from the reactor. As a basis, assume 100 kmol/h of pure acetone. Note that 1.0 mole of acetone reacts to form 1.0 mole of ketene and 1.0 mole of methane. Compare manual calculation results with those predicted from any available software package (e.g., UniSim/Hysys, PRO/II, Aspen Plus, SuperPro, and Aveva Process Simulation). The proper fluid package for this case is PRSV.

3.7 PRODUCTION OF DIMETHYL ETHER (DME)

The production of DME is via the catalytic dehydration of methanol. A feed stream of 134.2 kgmol/h at 250°C and 1470 kPa contains 0.97 methanol, 0.02 ethanol, and 0.01 steam to an adiabatic conversion reactor. If the single-pass conversion is 90%, calculate the molar flow rate of the exit stream. Use UniSim/Hysys software and the PRSV fluid package and compare the software predictions with manually calculated results.

3.8 TOLUENE PRODUCTION FROM HEPTANE

The catalytic dehydrogenation of n-heptane produces toluene. The reaction is taking place isothermally in a conversion reactor. The feed stream is at 430°C, 1 atm, and 100 kgmol/h (pure n-heptane). The single-pass conversion of n-heptane is 20%. Calculate the molar flow rate of the exit stream. The catalytic dehydrogenation reaction:

$$C_7H_{16} \rightarrow C_6H_5CH_3 + 4H_2$$

Use UniSim/Hysys software (Peng–Robinson) and the PRSV fluid package and compare the software predictions with manually calculated results.

3.9 KETENE PRODUCTION FROM ACETIC ACID

Catalytic cracking of acetic acid at 700°C and 1.0 bar pressure produces ketene as an intermediate product via the following primary reaction with low conversion (6%):

$$CH_3COOH \rightarrow CH_2CO + H_2O$$

The side reaction is with higher conversion (74%):

$$CH_3COOH \rightarrow CH_4 + CO_2$$

The feed stream is pure acetic acid fed to a conversion reactor at a rate of 100 kgmol/h, 1 bar, and 300°C. Calculate the required heating rate using UniSim/Hysys (PRSV fluid package) software or any other available software packages (e.g., PRO/II, Aspen Plus, SuperPro, and Aveva Process Simulation).

3.10   METHANE PRODUCTION

Methane produced from the reaction of carbon monoxide and hydrogen via the following reactions:

$$CO + 3H_2 \rightarrow CH_4 + H_2O$$

The feed flow rate is 100 kmol/h carbon monoxide and 300 kmol/h hydrogens at 550°C and 1,000 kPa. The single-pass conversion of CO is 80%, the reaction is taking place in an adiabatic conversion reactor. Determine the reactant exit temperature. Compare the manually calculated result with predictions of software packages such as UniSim/Hysys (Peng–Robinson) software or any other available software packages (e.g., PRO/II, Aspen Plus, SuperPro, and Aveva Process Simulation).

REFERENCES

1. Ghasem, N. M. and R. Hend, 2015. Principles of Chemical Engineering Processes Material and Energy Balances, 2nd edn, CRC Press, New York, NY.
2. Felder, R. M., R. W. Rousseau and L. G. Bullard, 2015. Elementary Principles of Chemical Processes, 4th edn, John Wiley, New York, NY.
3. Reklaitis, G. V., 1983. Introduction to Material and Energy Balances, John Wiley & Sons, New York, NY.
4. Himmelblau, D. M. and J. B. Riggs, 2012. Basic Principles and Calculations in Chemical Engineering, 8th edn, Prentice-Hall, Englewood Cliffs, NJ.

4 Shell and Tube Heat Exchangers

At the end of this chapter, students should be able to:

1. Identify the main types of shell and tube heat exchange equipment.
2. Estimate the overall heat transfer coefficients for a shell-and-tube heat exchanger.
3. Calculate the pressure drops on shell and tube sides.
4. Perform mechanical design of the most appropriate shell-and-tube heat exchanger to meet desired heat duty and pressure drops.
5. Verify the manually designed heat exchanger with the five software packages: Hysys/UniSim, PRO/II, Aspen Plus, SuperPro Designer, and the Aveva Process Simulation.

4.1 INTRODUCTION

The heat exchange process between two fluids at different temperatures, separated by an actual wall, occurs in many engineering applications. The device used to implement this exchange is called a heat exchanger, and specific applications are in heating and air conditioning, power production, waste heat recovery, and chemical processing. Flow arrangement and type of construction classified the types of heat exchangers. In the first classification, flow can be countercurrent or co-current. There are different heat exchangers; double pipe heat exchanger, shell and tube heat exchanger, plate heat exchanger, and phase change heat exchangers (boilers and condensers). Shell and tube heat exchanger is the most common heat exchanger in oil refineries and other extensive chemical processes [1, 2]. The tube side is for corrosive, fouling, scaling, hazardous, high temperature, high-pressure, and more expensive fluids. The shell side is for a more viscous, cleaner, lower flow-rate, evaporating, and condensing fluids. Gas or vapor heat exchange fluids, typically introduced on the shell side, high-viscosity liquids, for which the pressure drop for flow through the tubes might be prohibitively large, can be introduced on the shell side (Figure 4.1).

4.2 DESIGN OF SHELL AND TUBE HEAT EXCHANGER

When two fluids of different temperatures flow through the heat exchanger, one fluid flows in the tube side (the tubes are called a tube bundle), and the other fluid flows in the shell side outside the tubes. Heat is transferred from hot fluid to the cold fluid through the tube walls, either from tube side to shell side or vice versa. The fluids can be either liquids or gases on either the shell or the tube side. Heat

DOI: 10.1201/9781003167365-4

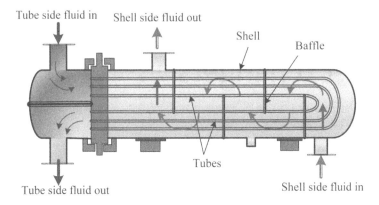

Tube side fluid in   Shell side fluid out

Shell

Baffle

Tubes

Tube side fluid out   Shell side fluid in

FIGURE 4.1 A typical most commonly used labeled schematic diagram for 1–2 type shell-and-tube heat exchanger.

transfer coefficients, pressure drops, and heat transfer area depend on the design's geometric configuration of the heat exchanger, which needs to be determined. Computation of shell-and-tube heat exchangers involves iteration. The geometric design to be selected includes shell diameter, tube diameter, tube length, tube configuration, and the number of tubes and shell passes [3–5].

4.2.1 REQUIRED HEAT DUTY

The starting point of any heat transfer calculation is the overall energy balance and the rate equation. Assuming only the transfer of sensible heat, the required heat duty, Q_{req}

$$Q_{req} = m_h Cp_h \left(T_{h,in} - T_{h,out} \right) = m_c Cp_c \left(T_{c,out} - T_{c,in} \right) \tag{4.1}$$

The heat exchanger has to meet or exceed this requirement, the basic design equation:

$$Q_{req} = U_i A_i F \left(\Delta T_{lm} \right) \tag{4.2}$$

where U_i is the overall heat transfer coefficient (Table 4.1), A_i is the inside heat transfer area, and the symbol F stands for a correction factor used with the log mean temperature difference for a countercurrent heat exchanger, ΔT_{lm}:

$$\Delta T_{lm} = \frac{\left(T_1 - t_2 \right) - \left(T_2 - t_2 \right)}{\ln \left(T_1 - t_2 / T_2 - t_2 \right)} \tag{4.3}$$

T_1 and t_1 are the hot and cold side inlet temperatures, respectively, and T_2 and t_2 are the corresponding outlet temperatures. The value of F depends upon the exact

TABLE 4.1

Selected Overall Heat Transfer Coefficient for Shell and Tube Heat Exchangers [6]

Shell Side	Tube Side	Design (W/m²K)	Included Total Dirt
Ethanolamine solutions	Water or amine solutions	195–1,136	0.003
Organic solvents	Water	284–852	0.003
Demineralized water	Water	1,704–2,840	0.001
Water	Water	795–1,476	0.003
Low boiling hydrocarbon	Water	454–1,136	0.003
Air, N2, compressed	Water	227–454	0.005
Propane, butane	Steam condensing	1,136–1,704	0.0015
Water	Steam condensing	1,420–2,270	0.0015

arrangement of the streams within the exchangers. The range of hot side fluid to the cold side fluid temperatures, R,

$$R = \frac{T_1 - T_2}{t_2 - t_1} = \frac{T_{hot,in} - T_{hot,out}}{t_{cold,out} - t_{cold,in}} \tag{4.4}$$

The range of cold fluid to maximum temperature differences, S,

$$S = \frac{t_2 - t_1}{T_1 - t_1} = \frac{t_{cold,out} - t_{cold,in}}{T_{hot,in} - t_{cold,in}} \tag{4.5}$$

The mathematical relationship between F, R, and S can be found from graphical representation or calculated from 4.6 and 4.7 [4, 5]. The formula for one shell pass and two tube passes (F_{1-2}):

$$F_{1-2} = \frac{\left[\sqrt{R^2+1}/(R-1)\right]\ln\left[(1-S)/(1-SR)\right]}{\ln\left[A+\sqrt{R^2+1}/A-\sqrt{R^2+1}\right]}, \quad A = \frac{2}{S} - 1 - R \tag{4.6}$$

The formula for two-shell pass and four, eight, tube passes or any multiple of four, F_{2-4}

$$F_{2-4} = \frac{\left[\sqrt{R^2+1}/2(R-1)\right]\ln\left[(1-S)/(1-SR)\right]}{\ln\left[\left(A+B+\sqrt{R^2+1}\right)/\left(A+B-\sqrt{R^2+1}\right)\right]} \tag{4.7}$$

where

$$B = \frac{2}{S}\sqrt{(1-S)(1-SR)}$$

F's value should be greater than 0.8 because F's low values mean that substantial additional area must be supplied in the heat exchanger to overcome the inefficient thermal profile. The approximate heat transfer area, A, can be calculated using a reasonable guess for the overall heat transfer coefficient; selected values are available in Table 4.1 [2]. The next step is to determine the approximate number of tubes (N_t) needed to do the job.

4.2.2 TUBE SELECTION

Select the suitable tube length such as m (ft): 2.4384(8), 3.048(10), 3.6576(12), 4.8768(16), and 6.096(20). Likewise, and the most common tube outside diameter m (inch): 0.00635(0.25), 0.009525(0.375), 0.0127(0.5), 0.015875(0.625), 0.01905(0.75), 0.0254(1), and 0.03175(1.25). Define the tube wall thickness is by the Birmingham Wire Gauge (BWG). The tubes are typically specified to be 14 BWG. The most common tube lengths m(ft): 4.8768(16) and 6.096(20); and the most common tube outside diameter values are 0.01905 m (0.75 inch) and 0.0254 m (1 inch). The velocity through a single tube should be between m/s(ft/s): 0.9144(3) and 3.048(10); to keep the pressure under reasonable constraints and maintain turbulent flow and minimize fouling. The number of tube passes can adjust the velocity to fall in this range. Table 4.2 lists the most commonly used heat exchanger tube data. Further information can be found elsewhere [3, 6–10].

TABLE 4.2
Heat Exchanger Tube Data

Outside Diameter (OD) (inch)	Birmingham Wire Gauge (BWG)	Inside Diameter (ID) (inch)	Inside Diameter (ID) (cm)
$\frac{1}{2}$	12	0.282	0.716
	14	0.334	0.848
	16	0.370	0.940
	18	0.402	1.020
$\frac{3}{4}$	10	0.482	1.224
	11	0.510	1.295
	12	0.532	1.351
	13	0.560	1.422
1	8	0.670	1.702
	9	0.704	1.788
	10	0.732	1.859
	11	0.760	1.930

4.2.3 SHELL INSIDE DIAMETER (ID), D_s

Once the number of tubes is known, pitch (square or triangle), and the number of tube passes is known, the shell size can be defined. A square pitch is chosen for convenience in cleaning the outside of the tubes because cleaning is relatively easy when the tubes are in-line. Pipes on a triangular pitch are difficult to clean with tools but washed by passing a chemical solution over the side of the shell. The standard choice is the square pitch, 0.03175 m (1.25 inch) for 0.0254 m (1 inch) outside tube diameter, and 0.0254 m (1 inch) square pitch for 0.01905 m (0.75 inch) outside tube diameter. Knowing the number of tubes and number of passes, the required ID of the shell can be used from standard tables. Shells of heat exchangers are from commercial steel pipes. Table 4.3 shows the selected tube sheet layouts. More information is available in Kern [3].

4.2.4 NUMBER OF BAFFLES

Typically, baffles are equally spaced. The minimum baffle spacing is one-fifth of the shell diameter, but not less than 2 inches, and maximum baffle spacing is $74D_o^{0.75}$, where D_o is the outside tube diameter in inches [3]. The number of baffles, N_{bf}

$$N_{bf} = \frac{L}{b} - 1$$

where L is the tube length, b is the baffle spacing.

4.2.5 HEAT TRANSFER COEFFICIENTS

Overall, the heat transfer coefficient checks the designed heat exchanger's thermal performance; it requires calculating the tube side and shell heat transfer

---

TABLE 4.3
Shell Inside Diameter (Shell ID)

$\frac{3}{4}$ Inch OD Tubes on 1 inch Square Pitch				1 inch OD Tubes on $1\frac{1}{4}$ inch Square Pitch			
Shell ID (inch)	1-Pass	2-Pass	4-Pass	Shell ID (inch)	1-Pass	2-Pass	4-Pass
8	32	26	20	8	21	16	14
10	52	52	40	10	32	32	26
12	81	76	68	12	48	45	40
$13\frac{1}{4}$	97	90	82	$13\frac{1}{4}$	61	56	52
$15\frac{1}{4}$	137	124	116	$15\frac{1}{4}$	81	76	68
$19\frac{1}{4}$	224	220	204	$19\frac{1}{4}$	138	132	128
$21\frac{1}{4}$	277	270	246	$21\frac{1}{4}$	177	166	158
25	413	394	370	25	260	252	238
31	657	640	600	31	406	398	380

coefficients, the tube wall contribution to the resistance, and the appropriate fouling resistance. The overall heat transfer coefficient, U_o, based on the outside surface area of the tubes is,

$$\frac{1}{U_o} = \frac{1}{h_o} + \frac{\Delta x}{k_W}\left(\frac{A_o}{A_{Lm}}\right) + \frac{1}{h_i}\left(\frac{A_o}{A_i}\right) + R_{fi}\left(\frac{A_o}{A_i}\right) + R_{fo} \tag{4.8}$$

where h_o is the outside film heat transfer coefficient, h_i the inside film heat transfer coefficient, Δx the tube wall thickness, k_w is the tube metal conductivity, R_{fi} and R_{fo} is the inside and outside fouling resistances, respectively. A_o is the tube outside the area, A_i is the tube inside the area, and A_{LM} is the log mean of A_i and A_o. The wall thickness, Δx,

$$\Delta x = \frac{D_o - D_i}{2}, \quad A_{LM} = \frac{A_o - A_i}{\ln\left(A_o/A_i\right)} \tag{4.9}$$

The overall heat transfer coefficient based on the inside area as a function of D_o and D_i is

$$\frac{1}{U_i} = \left(\frac{D_i}{D_o}\right)\frac{1}{h_o} + \frac{D_i\,\Delta x}{D_{LM}k_w} + \frac{1}{h_i} + R_{fi} + \left(\frac{D_i}{D_o}\right) + R_{fo} \tag{4.10}$$

where the log mean diameter, D_{LM}

$$D_{LM} = \frac{D_o - D_i}{\ln\left(D_o/D_i\right)} \tag{4.11}$$

4.2.5.1   Tube Side Heat Transfer Coefficient, h_i

Calculate the heat transfer coefficient for inside tubes (h_i) using Sieder–Tate equation for laminar flow [3]:

$$\mathrm{Nu}_i = \frac{h_i D_i}{k_i} = 1.86\left[\left(N_{Rei}\right)\left(N_{Pri}\right)\left(\frac{D_i}{L}\right)\right]^{1/3}\left(\frac{\mu_i}{\mu_w}\right)^{0.14} \tag{4.12}$$

Calculate for turbulent flow, the heat transfer coefficient inside tubes (h_i) using Sieder–Tate equation for the flow in a constant diameter pipe:

$$\mathrm{Nu}_i = \frac{h_i D_i}{k_i} = 0.027 N_{Rei}^{0.8} N_{Pri}^{1/3}\left(\frac{\mu_i}{\mu_w}\right)^{0.14} \tag{4.13}$$

where

$$N_{\text{Rei}} = \frac{\rho_i u_i D_i}{\mu_i} \quad \text{and} \quad N_{\text{Pri}} = \frac{C_{p_i} u_i}{k_i}$$

where ρ_i, u_i, D_i, μ_i, C_{p_i}, and k_i are the density, velocity, inside diameter, viscosity, specific heat, and thermal conductivity of tube side fluid, respectively [6].

4.2.5.2 Shell Side Heat Transfer Coefficient, h_o

For turbulent flow, the Kern method can be used [3]:

$$\text{Nu}_o = \frac{h_o D_e}{k} = 0.36 \, N_{\text{Reo}}^{0.55} N_{\text{Pro}}^{1/3} \left(\frac{\mu_o}{\mu_w} \right)^{0.14} \tag{4.14}$$

The Reynolds number, $N_{\text{Reo}} = D_e V_{\text{max}} \rho_o / \mu_o$, where D_e is the effective hydraulic diameter. V_{max} is the fluid's maximum velocity through the tube bank, which equals the shell side fluid volumetric flow rate divided by the shell side crossflow area. Among the physical properties (ρ, μ, k, and C_p), μ demonstrates the most substantial dependence on temperature and has the most considerable effect on the transfer process. The effective hydraulic diameter,

$$D_e = \frac{4 \times \text{free area}}{\text{Wetted perimeter}} = \frac{4 \left(P_t^2 - \pi D_o^2 / 4 \right)}{\pi D_o} \tag{4.15}$$

To calculate the mass velocity normal to tubes at the centerline of the exchanger, G_o, first, calculate the cross-sectional area between baffles and shell axis, A_{cf}

$$A_{\text{cf}} = \frac{D_s}{P_t} \times \text{clearance} \times b \tag{4.16}$$

where

$$\text{Clearance} = P_t - D_o \tag{4.17}$$

Since the flow entering the shell distributes itself into space between the tubes, the flow turns around each baffle. The alternative method is the Donohue equation [3], based on the weighted average of the mass velocity of the shell-side fluid flowing parallel to the tubes (G_b) and the mass flow rate of the shell-side fluid flowing across the tubes (G_c):

$$\left(\frac{h_o D_o}{k_o} \right) = 0.2 \left(\frac{D_o G_e}{\mu_o} \right)^{0.6} \left(\frac{C_{po} \mu_o}{k_o} \right)^{0.33} \left(\frac{\mu_o}{\mu_w} \right)^{0.14} \tag{4.18}$$

where

$$G_c = (G_b \ G_c)^{1/2}$$

The mass velocity of the shell-side flowing parallel to the tubes, G_b

$$G_b = \dot{m}/S_b, \ \ S_b = f_b \frac{\pi D_s^2}{4} - N_b \frac{\pi D_o^2}{4} \tag{4.19}$$

The mass velocity of the shell-side flowing across the tubes, G_c

$$G_c = \dot{m}/S_b, \ \ S_c = b \times D_s \left(1 - \frac{D_o}{P_t}\right) \tag{4.20}$$

where
f_b = fraction of the shell cross-section occupied by the baffle window, commonly 0.1995 for 25% baffle
N_b = number of tubes in baffle window = $f_b \times$ number of tubes
$\dot{m}$ = the mass flow rate of the shell-side fluid
D_o = Outside diameter of tubes
D_s = diameter of the shell
b = baffle spacing
P_t = tube pitch

4.2.6 PRESSURE DROP

Allowable pressure drop for both streams is an essential parameter for heat exchanger design. Generally, for liquids, a value of 48,263–68,948 Pa (7–10 psi) is permitted per shell. A higher pressure drop is frequently acceptable for viscous fluids, especially on the tube side. The allowed value is generally 4.826–20.684 kPa (0.7–3 psi), with 10.342 kPa (1.5 psi) being typical [8].

4.2.6.1 Pressure Drop in the Tube Side

The pressure drop for the flow of a liquid or gas without phase change through straight tubes:

$$-\Delta P_i = P_{in} - P_{out} = 1.2 \frac{N_P f_D G_i^2 L}{2 g_c \rho_i D_i \phi} \tag{4.21}$$

where N_p is the number of tube passes, and L is the tube length,

$$f_D = (1.82 \log_{10} N_{Rei} - 1.64)^{-2} \tag{4.22}$$

the tube side mass velocity, $G_i = \rho_i u_i$.

ϕ is the correction factor for the non-isothermal turbulent flow = 1.02 (μ/μ_w).

4.2.6.2 Pressure Drop in the Shell Side

The pressure drop for the flow of liquid without phase change across the tubes in the shell side is given by the following equations [7]:

$$-\Delta P_o = P_{in} - P_{out} = K_s \frac{2N_R f' G_o^2}{g_c \rho_c \phi} \tag{4.23}$$

$$N_{reversals} = \frac{L}{b} = \frac{\text{Length of tube}}{\text{Baffle spacing}} \tag{4.24}$$

The correction factor,

$$K_s = 1.1 \times N_{reversals} \tag{4.25}$$

The number of tubes at the centerline, N_{cl}

$$N_{cL} = D_s / P_t \tag{4.26}$$

The number of tube rows across which the shell fluid flows (N_R) equals the total number of tubes at the center plane minus the number of tube rows that pass through the baffles' cut portions. For 25% cut baffles, N_R is 50% of the tubes at the center plane.

$$N_R = 0.5 \times \frac{D_s}{P_t} \tag{4.27}$$

The modification friction factor, f'

$$f' = b_f \left(\frac{D_o G_o}{\mu_o} \right)^{-0.15} \tag{4.28}$$

where b for square pitch is

$$b_f = 0.044 + \frac{0.08 x_L}{(x_T - 1)^{0.43 + 1.13/x_L}} \tag{4.29}$$

The pitch ratio is transverse to the tube OD flow, x_T, pitch parallel to the tube, and OD is x_L. For square pitch $x_L = x_T$; hence,

$$x_L = x_T = \frac{P_t}{D_o} \tag{4.30}$$

4.2.7 ALTERNATIVE PRESSURE DROP METHOD

The following sections present an alternative method used to determine pressure drop in the tube side and the shell side of the shell and tube heat exchanger.

4.2.7.1 Pressure Drop in the Tube Side

Calculate the pressure drop in the tube side using the following equation:

$$\Delta P_i = f \frac{L}{D}\left(\frac{1}{2}\rho V^2\right) \times N_p \qquad (4.31)$$

L is the tubes' length, D is the tubes' ID, ρ is the fluid density in tube side, and V is the average flow velocity through a single tube, N_p is the number of tube passes. The Fanning friction factor, f, can be calculated from Darcy friction factor, f_D:

$$f_D = \left(1.82\log_{10}N_{Re} - 1.64\right)^{-2} \qquad (4.32)$$

The Darcy friction factor is related to the Fanning friction factor by $f_D = 4f$.

4.2.7.2 Pressure Drop in the Shell Side

Calculate the pressure drop in the shell side using the following equation:

$$\Delta P_{shell} = \frac{2 f G_s^2 D_s (N_B + 1)}{\rho D_e (\mu/\mu_s)} \qquad (4.33)$$

where f is the Fanning friction factor for flow on the shell side [5], G_s is the mass velocity on the shell side, D_s is the ID of the shell, N_B is the number of baffles, ρ is the density of the shell-side fluid, and D_e is an equivalent diameter. The mass velocity $G_s = m/S_m$, where m is the fluid's mass flow rate, and S_m is the crossflow area measured close to the shell's central symmetry plane containing its axis. The crossflow area, S_m

$$S_m = D_s \times L_B \times \frac{\text{clearance}}{\text{pitch}} \qquad (4.34)$$

where L_B is the baffle spacing, the equivalent diameter, D_e, is defined as follows:

$$D_e = \frac{4\left(C_p S_n^2 - \pi D_o^2/4\right)}{\pi D_o} \qquad (4.35)$$

The D_o is the outside diameter of the tubes, and S_n is the pitch, center-to-center distance of the tube assembly. The constant C_p is 1.0 for a square pitch and 0.86 for a triangular pitch [5]. Calculate the Fanning friction factor using Reynolds number based on equivalent diameter as

$$R_e = \frac{D_e G_s}{\mu_o} \qquad (4.36)$$

4.2.8 SUMMARY OF DESIGN STEPS

The ideal procedure to design a shell and tube heat exchanger:

1. The required heat duty (Q_{req}) is fixed by the required service, and the designed heat exchanger has to meet or exceed this requirement.
2. Select the streams on the tube side and shell side.
3. Calculate the heat transfer area required using a reasonable guessed overall heat transfer coefficient.
4. Select suitable tube specifications (OD/ID/Pt and length).
5. Calculate tube cross-sectional area by assuming appropriate velocity inside tubes.
6. Estimate the number of tubes and tube passes.
7. Estimate heat transfer coefficient of the inside and outside tubes.
8. Calculate the overall heat transfer coefficient and compare it with the assumed overall value.
9. If there is discrepancy between assumed overall heat transfer coefficients (U) and estimated U, change the baffle spacing and re-estimate the new U, repeat until they are close to each other.

4.3 CONDENSERS AND BOILERS

Condensers are typically multi-pass shells and tube exchangers with floating heads that removed heat by contacting vapor with a cold surface through the tube wall. The liquid then flows off the tube under the influence of gravity, collects, and flows out of the exchanger. In some cases, vapor flow rates may be high enough to sweep the liquid off the tubes. Boilers are closed vessels used to heat water or other fluid. The heated or vaporized fluid exits the boiler for use in various processes or heating applications. The correlation for predicting heat transfer coefficients is presented by Hewitt [10].

Example 4.1: Aqueous Diethanolamine (DEA) and Water Exchanger

Design a shell and tube heat exchanger to cool 6.3 kg/s of aqueous DEA solution (mass fractions 0.2 DEA/0.8 water) from 335.3K $(62°C)$ to 318.15K $(45°C)$. Using water at 298.15 K (25°C) heated to 311 K (38°C). Assume tube inside fouling resistance, $R_{fl} = 7.044 \times 10^{-4}$ m^{2}k/W, ignoring shell-side fouling resistance. Compare the manually calculated results with the five software packages, UniSim/Hysys, PRO/II, Aspen Plus, SuperPro, and Aveva Process Simulation.

SOLUTION

MANUAL CALCULATION

Figure 4.2 shows the schematic diagram of the 1–4 type shell and tube heat exchanger, the shell and tube side's physical properties at the average pass temperature are listed in Table 4.4.

The required heat duty, Q_{req}

$$Q_{req} = \dot{m}_h Cp_h \left(T_{h,in} - T_{h,out}\right) = \dot{m}_c Cp_c \left(T_{c,out} - T_{c,in}\right)$$

Substitute giving values,

$$Q_{req} = \left(6.3\frac{kg}{s}\right)\left(3.8578\frac{kJ}{kg.K}\right)(335.3 - 318.15)K = 416.8\frac{kJ}{s}$$

The mass flow rate of the cold stream, $\dot{m}_c$

$$\dot{m}_c = \frac{Q_{req}}{Cp_c\left(T_{c,out} - T_{c,in}\right)} = \frac{416.8\ (kJ/s)}{\left(4.1868\frac{kJ}{kg.\ K}\right)(311 - 298.15)K} = 7.75\ kg/s$$

Assume an appropriate value of the overall heat transfer coefficient (U), the suitable designed overall heat transfer coefficient for DEA solution–water system, the U value is between 795 and 1,135 $\frac{W}{m^2\ K}$ (Table 4.1). Assume $U_i = 852\ \frac{W}{m^2\ K}$

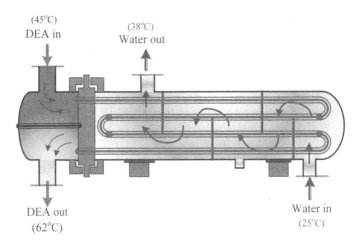

FIGURE 4.2 Process flowsheet for 1–4 type shell and tube heat exchanger for cooling aqueous DEA with water for the case described in Example 4.1.

TABLE 4.4

Physical Properties for the Case Described in Example 4.1

Parameters	Tube Side (Hot) DEA Solution (54°C)	Shell Side (Cold) Water (31°C)
ρ (kg/m³)	957	959
C_p (kJ/kg K)	3.85	4.187
μ (cP)	0.75	0.77
k (W/m K)	0.519	0.623

The overall heat transfer area based on the tube outside area (A_i):

$$A_i = \frac{Q}{U_i F \, \Delta T_{LM}}$$

The overall heat transfer area based on the tube outside area (A_o):

$$A_o = \frac{Q}{U_o F \, \Delta T_{LM}}$$

The log means temperature difference (ΔT_{LM}):

$$\Delta T_{LM} = \frac{(T_1 - t_2) - (T_2 - t_1)}{\ln\left[(T_1 - t_2)/(T_2 - t_1)\right]} = \frac{(335.3 - 311) - (318.15 - 298.15)}{\ln\left[(335.3 - 311)/(318.15 - 298.15)\right]} = 22K$$

The temperature range between hot and cold fluid, R,

$$R = \frac{T_1 - T_2}{t_2 - t_1} = \frac{335.3 - 318.15}{311 - 298.15} = 1.35$$

The range of cold fluid temperature to maximum temperature difference, S,

$$S = \frac{t_2 - t_1}{T_1 - t_1} = \frac{311 - 298.15}{335.3 - 318.15} = 0.343$$

The configuration factor, F can be found from figures in [4–8] or using the following equation:

$$F = \frac{\sqrt{R^2 + 1} \, \ln\{1 - S/1 - R.S\}}{(R - 1) \ln\left[2 - S\left(R + 1 - \sqrt{R^2 + 1}\right)\Big/2 - S\left(R + 1 + \sqrt{R^2 + 1}\right)\right]}$$

Substituting values in the configuration factor (F):

$$F = \frac{\sqrt{1.35^2 +1}\ln\left\{\dfrac{1-0.343}{1-1.35\times0.343}\right\}}{(1.35-1)\ln\left[\dfrac{2-0.343\left(1.35+1-\sqrt{1.35^2+1}\right)}{2-0.343\left(1.35+1+\sqrt{1.35^2+1}\right)}\right]} = \frac{1.68\times0.2}{0.362} = 0.93$$

The inside heat transfer area ($A_{i,}$):

$$A_i = \frac{Q}{U_i F\,\Delta T_{LM}} = \frac{\left(416800W\right)}{\left(852\dfrac{W}{m^2K}\right)(0.93)(22K)} = 23.9 \text{ m}^2$$

The velocity inside tubes should be assumed to maintain turbulent flow; an appropriate value is between 0.9144 and 3.048 m/s, let the velocity inside tubes be $u_i = 1.524$ (m/s). The total cross-sectional area per pass (DEA solution in the tube side) (A_{ci}):

$$A_{ci} = \frac{m_i}{\rho_i u_i} = \frac{\left(6.3\dfrac{kg}{s}\right)}{\left(957.26\dfrac{kg}{m^3}\right)\left(1.524\dfrac{m}{s}\right)} = 4.32\times10^{-3}\text{ m}^2$$

Selecting tube length, $L = 4.2672$ m, OD $= 0.75$ inch (0.01905 m), 11 BWG (ID $= 0.01224$ m.). The number of tubes per pass (N_t):

$$N_t = \frac{A_{ci}}{\pi D_i^2/4} = \frac{4.32\times10^3\text{ m}^2}{\pi(0.0122\text{ m})^2/4} = 36.68\frac{tube}{pass} \cong 37\frac{tube}{pass}$$

Heat transfer area per tube, $A_t = \pi D_i L$

$$A_t = \pi D_i L = \pi(0.0122\text{ m})(4.2672\text{ m}) = 0.1635\frac{m^2}{tube}$$

The number of tube passes (N_p):

$$N_p = \frac{A_i}{A_t N_t} = \frac{23.9\ m}{(0.1635\text{ m}^2/\text{tube})(37\text{ tube/pass})} = 3.94 \text{ passes}$$

Let $N_p = 4$, then $A_{i,}$ based on the new number of tube passes,

$$A_i = N_p N_t \left(\pi D_i\right)L = (4)(37)(\pi)(0.0122\ m)(4.2672\ m) = 24.2 \text{ m}^2$$

The corrected overall heat transfer coefficient based on the calculated internal heat transfer area (U_i):

$$U_i = \frac{(416800 \ W)}{(24.2 \ m^2)(0.93)(22 \ K)} = 841 \frac{W}{m^2.K}$$

The shell inside diameter (D_s) is estimated using tabulated data [7] or using the relation given by Coulson and Richardson [8]. The tube bundle diameter (D) = OD × $(N_t/k)^{1/n}$, where OD is the outside diameter of a single tube, N_t is the number of tubes, and k and n are constants depending on the number of tube pass (for two passes $k = 0.249$ and $n = 2.207$). The calculation gives the size of the tube bundle and not the shell diameter. The obtained shell diameter is by adding the tube bundle diameter and the clearance between the tube bundle and the shell. The clearance typically ranges from 10 to 90 mm [8]. Using Table 4.3 [7], in four tube passes exchanger and 148 tubes, the shell ID, D_s = 0.4385 m. An approximate number of baffles is twice the length of tubes in meters; if the tube length is 5 m, then the number of baffles is 10. Add more baffles to increase the heat transfer coefficients. By contrast, more baffles increase pressure drop, which must be within acceptable limits. The range of baffle spacing, b, is determined as follows:

$$\left(\frac{D_s}{5}\right) < b < D_s$$

So, b is in the range: (0.061 m) < b < 0.44 m, let $b = 0.254$ m. The shell heat transfer coefficient (h_o) is estimated using the heat transfer equation from Kern [3].

$$Nu_o = \frac{h_o D_e}{k} = 0.36 \ N_{Reo}^{0.55} \ N_{Pro}^{1/3} \left(\frac{\mu}{\mu_w}\right)^{0.14}$$

where $N_{Reo} = D_e G_o / \mu_o$ and $N_{Pro} = Cp_o \mu_o / k_o$.
   The effective hydraulic diameter (D_e):

$$D_e = \frac{4(P_t^2 - \pi D_o^2/4)}{\pi D_o} = \frac{4(0.025^2 - \pi(0.01905)^2/4)}{\pi \times 0.01905} = 0.023 \ m$$

The shell side mass velocity normal to tubes at the centerline of the heat exchanger is the shell mass flow rate to the shell cross-sectional area between baffle and shell axis. The cross-sectional area between baffles and shell axis, A_{cf},

$$A_{cf} = \frac{D_s}{P_t} \times clearance \times b$$

Substituting values to calculate the cross-sectional area between baffles and the shell axis, A_{cf}

$$A_{cf} = \frac{0.4385\ m}{0.025\ m} \times (0.00635\ m) \times (0.254\ m) = 0.0282\ m^2$$

The shell side mass velocity (G_o):

$$G_o = \frac{m_o}{A_{cf}} = \frac{28122\ (kg/h)}{0.02787\ m^2} = 1 \times 10^6\ \frac{kg}{m^2 h}$$

The shell side Reynolds number (N_{Reo}):

$$N_{Reo} = \frac{D_e G_o}{\mu_o} = \frac{0.02408\ m \times 1 \times 10^6}{0.77 \times 10^{-3}\ \frac{kg}{m.s}} = 8765$$

The Shell side Prandtl number (N_{Pro}):

$$N_{Pro} = \frac{C_{po} \mu_o}{k_o} = \frac{\left(4.1868 \times 10^3\ \frac{J}{kg.K}\right)\left(0.00077\ \frac{kg}{m.s}\right)}{0.623\left(\frac{W}{m.K}\right)} = 5.2$$

Substituting values of Reynolds number and Prandtl number in the heat transfer equations and neglecting the effect of change in viscosity,

$$Nu_o = \frac{h_o D_e}{k_o} = 0.36(8765)^{0.55}(5.2)^{1/3}(1)^{0.14} = 92$$

The shell heat transfer coefficient, h_o

$$h_o = Nu_o \frac{k_o}{D_e} = 92\frac{\left(0.623\ \frac{W}{m.K}\right)}{0.0241\ m} = 2378.25\ \frac{W}{m^2.K}$$

The tube side coefficient (h_i) is calculated using Sieder–Tate equation [7]:

$$Nu_i = \frac{h_i D_i}{k_i} = 0.027 N_{Rei}^{0.8}\ N_{Pri}^{1/3}\left(\frac{\mu_i}{\mu_w}\right)^{0.14}$$

The tube side Reynolds number (N_{Rei}):

$$N_{Rei} = \frac{D_i \rho_i \mu_i}{\mu_i} = \frac{(0.01224\ m)\left(957.263\ \frac{kg}{m^3}\right)\left(1.524\ \frac{m}{s}\right)}{\left(0.75 \times 10^{-3}\ \frac{kg}{m.s}\right)} = 23{,}808$$

The tube side Prandtl number (N_{Pri}):

$$N_{pri} = \frac{C_{p_i}\mu_i}{k_i} = \frac{\left(3.852\times10^3 \dfrac{J}{kg.K}\right)\left(0.75\times10^{-3}\dfrac{kg}{m.s}\right)}{\left(0.52\dfrac{W}{m.K}\right)} = 5.567$$

Substituting estimated values of Reynolds number and Prandtl number in tube heat transfer coefficient,

$$Nu_i = \frac{h_i D_i}{k_i} = 0.027(23{,}808)^{0.8}(5.567)^{1/3}\times1 = 152$$

The tube heat transfer coefficient (h_i):

$$h_i = Nu_i\frac{k}{D_i} = 152\frac{\left(0.51887\dfrac{W}{m.K}\right)}{0.01224\ m} = 44.48\frac{W}{m^2 K}$$

The overall heat transfer coefficient based on internal area (U_i):

$$\frac{1}{U_i} = \frac{D_i}{D_o h_o} + \frac{D_i\,\Delta x}{D_{LM} k_w} + \frac{1}{h_i} + R_{fi} + \left(\frac{D_i}{D_o}\right)R_{fo}$$

The tube thickness (Δx):

$$\Delta x = \frac{D_o - D_i}{2} = \frac{0.01905 - 0.0122}{2} = 3.405\times10^{-3}\ m$$

The log mean diameter (D_{LM}):

$$D_{LM} = \frac{D_o - D_i}{\ln(D_o/D_i)} = \frac{0.01905 - 0.0122}{\ln(0.01905 - 0.0122)} = 0.0154\ m$$

The thermal conductivity of carbon steel is 51.887 W/m. K. The overall heat transfer coefficient depends on the design specifications of the shell and tube heat exchanger.

$$\frac{1}{U_i} = \frac{0.0122\ m}{(0.01905\ m)\left(2379.19\dfrac{W}{m^2 K}\right)} + \frac{(0.0122\ m)(0.0034\ m)}{(0.0154\ m)\left(51.887\dfrac{W}{m.K}\right)}$$

$$+ \frac{1}{\left(6{,}433.47\dfrac{W}{m^2 K}\right)} + 0.0007\frac{m^2 K}{W}$$

Simplifying further, the internal overall heat transfer coefficient, U_i

$$\frac{1}{U_i} = 2.690 \times 10^{-4} + 5.191 \times 10^{-5} + 1.554 \times 10^{-4} + 7.044 \times 10^{-4} = 1.18 \times 10^{-3} \frac{m^2 K}{W}$$

The calculated U-value based on the exchanger-designed specifications is $U_i = 847.46$ W/m²K.

Consequently, the overall heat transfer coefficient based on the tube outside area (U_o)

$$U_o = U_i \frac{D_i}{D_o} = 847.46 \frac{W}{m^2 K} \times \frac{0.0122}{0.01905} = 542.73 \frac{W}{m^2 K}$$

The calculated overall heat transfer coefficient based on the tube inside area is 847.46 W/m²K, close to the assumed designed value (841 W/m² K). Consequently, the specifications of the designed exchanger are satisfactory. If the calculated heat transfer coefficient based on the exchanger specifications is lower than the initially assumed value, then the heat provided by the designed exchanger is less than the required heat. The estimated heat transfer coefficient based on the exchanger design qualifications should be close to the assumed design value; the simplest way is to calculate the shell transfer coefficient, h_o, based on the assumed corrected value:

$$h_o = \frac{1}{D_o/D_i \left[1/U_i - 1/h_i - \left(D_i \Delta x/D_{LM} k \right) - R_{fi} \right]}$$

Substitute required data

$$h_o = \frac{1}{\left(\dfrac{0.01905}{0.0122} \right) \left[\left(\dfrac{1}{847.46} \right) - \left(\dfrac{1}{6,433.47} \right) - 5.1911 \times 10^{-5} - 7.044 \times 10^{-4} \right]} = 2,380 \frac{W}{m^2 K}$$

The shell side Nusselt number (Nu_o):

$$Nu_o = \frac{h_o D_o}{k_o} = \frac{2380 \text{ W/m}^2 \text{ K} \times (0.024079 \text{ m})}{0.5188 \dfrac{J}{m.s.K}} = 115.04$$

The shell side Reynolds number (N_{Reo}):

$$N_{Reo} = \left[\frac{Nu_o}{0.36 \, N_{Pro}^{1/3}} \right]^{1/0.55} = \left[\frac{115.04}{0.36 (5.2)^{1/3}} \right]^{1/0.55} = 13175$$

The shell side mass velocity (G_o):

$$G_o = \frac{N_{Reo}\mu_o}{D_e} = \frac{13{,}175 \times 0.77 \times 10^{-3}\ \dfrac{kg}{m.s}}{0.024079\ m} = 421.3\ \frac{kg}{m^2.s}$$

The new shell side cross-sectional area (A_{cfo}):

$$A_{cfo} = \frac{m_o}{G_o} = \frac{7.8118\ kg/s}{421.3\ kg/m^2 s} = 0.0185\ m^2$$

The new baffle spacing (b):

$$b = \frac{A_{cfo}}{D_s/P_t \times \text{Clearance}} = \frac{\left(0.0185\ m^2\right)}{\left(\dfrac{0.438\ m}{0.0254\ m}\right)(0.00635\ m)} = 0.16895\ m$$

The new baffle spacing is 0.16895 m, which is higher than the previously assumed value ($b = 0.254$ m); the increase in baffle spacing decreases the shell side coefficient. The pressure drop for the flow of liquid without phase change through a circular tube is given by the following equations [7]:

$$-\Delta P_o = P_{in} - P_{out} = K_s \frac{2N_R f' G_o^2}{g_c \rho_c \phi}$$

The value of $N_{reversals}$

$$N_{reversals} = \frac{L}{b} = \frac{4.2672\ m}{0.16895\ m} = 25.25$$

The value of K_s,
$K_s = 1.1 \times N_{reversals} = 1.1 \times 25.25 = 27.78$
Number of tubes at centerline (N_{cl}):

$$N_{cl} \cong \frac{D_s}{P_t} = \frac{0.438\ m}{0.0254\ m} = 17.25$$

The number of tube rows across which the shell fluid flows (N_R) equals the total number of tubes at the center plane minus the number of tube rows that pass through the baffles' cut portions. For 25% cut baffles, N_R is 50% of the tubes at the center plane.
$N_R = 0.5 \times 17.25 = 8.625 \sim 9$
The modification friction factor, f'

$$f' = b\left(\frac{D_o G_o}{\mu_o}\right)^{-0.15}$$

The square pitch (b) is

$$b = 0.044 + \frac{0.08 x_L}{(x_T - 1)^{0.43 + 1.13/x_L}}$$

x_T is the transverse pitch ratio to flow to tube outside diameter (OD), and x_L is the ratio of pitch parallel to tube OD. For square pitch $x_L = x_T$

$$x_L = x_T = \frac{P_t}{D_o} = \frac{1}{0.75} = 1.33$$

Hence,

$$b = 0.044 + \frac{0.08 \times 1.333}{(1.333 - 1)^{0.43 + 1.13/1.333}} = 0.48$$

Substitute the calculated values of (b):

$$f' = b \left(\frac{D_o G_o}{\mu_o}\right)^{-0.15} = 0.48 \left[\frac{(0.01905 \ m)\left(421.3 \frac{kg}{m^2 s}\right)}{\left(0.325 \times 10^{-3} \frac{kg}{m.s}\right)}\right]^{-0.15} = 0.1053$$

Substituting in the pressure drop equation of the shell side (ΔP_s):

$$-\Delta P_s = K_s \frac{2 N_R f' G_o^2}{g_c \rho_c \phi} = 15.73 \frac{2(9)(0.1146)\left(421.3 \frac{kg}{m^2 s}\right)^2}{\left(1 \frac{m.kg}{N.s^2}\right)\left(959.025 \frac{kg}{m^3}\right)(1)} = 6177.6 \frac{N}{m^2} = 6.2 \ kPa$$

Estimating the pressure drop on the tube side is much easier than the calculated pressure drop on the shell side. Calculate the pressure drop in the tube side using the following equations [7]:

$$-\Delta P_t = 1.2 \frac{N_P f_D G_t^2 L}{2 g_c \rho_i D_i \phi}$$

where

$$f_D = (1.82 \log_{10} N_{Rei} - 1.64)^{-2}$$

Substituting the values

$$f_D = (1.82 \log_{10} 23812 - 1.64)^{-2} = 0.025$$

Tube side mass velocity

$$G_i = \rho_i \mu_i = \left(957.26 \frac{kg}{m^3}\right)\left(1.524 \frac{m}{s}\right) = 1458.86 \frac{kg}{m^2 s}$$

Substituting in pressure drop equation $(-\Delta P_t)$:

$$-\Delta P_t = 1.2 \frac{N_p f_D G_i^2 L}{2 g_c \rho_i D_i \phi}$$

Substitute required value:

$$-\Delta P_t = 1.2 \frac{4(0.025)\left(1458.86 \frac{kg}{m^2 s}\right)^2 (4.2672\ m)}{2 \times \left(1 \frac{m.kg}{N.s^2}\right)\left(957.26 \frac{kg}{m^3}\right)(0.0122\ m)(1)} = 46658.6\ N/m^2$$

The tube side calculated pressure drop (ΔP_t) is 46.66 kPa. Table 4.5 shows the summary of the resultant-designed heat exchanger specifications.

TABLE 4.5

Specifications of the Designed Shell and Tube Heat Exchanger of the Case Described in Example 4.1

	Shell Side	Tube Side
Components	Water (100%)	DEA/water (0.2/0.8)
Mass flow rate (kg/s)	12.6	6.3
Temperature (°C)	25/37.78	62.22/45°C
Pressure (kPa)	100	100
Pass	1	4
Shell ID (m)	0.43	
Tubes: OD/ID/P_t (m)		0.0019/0.0.012/0.0254
Tube configuration		Square pitch
Length (m)		4.27
Total number of tubes		148
Number of baffles		13
Baffle spacing (m.)	0.30	0.30
Fouling factor (m² K/W)		0.0007
Pressure drop (kPa)	6.2	46.66
Log mean temperature difference (LMTD) (°C)	4.44	
F factor	0.93	
U_o (W/m² K)	543	
Duty (kW)	418	

UniSim/Hysys Simulation

The difference in the heat duty between manual calculations and Hysys/UniSim is the value of specific heat calculated using both software. UniSim has no features to calculate heat transfer coefficients of shell and tube sides.

Start by opening a new case in UniSim; add the component; water and DEA amine. Select the non-random two liquids (NRTL) as the fluid package. On the connection page, S1 and S2 are assigned for tube side streams, and S3 and S4 are for inlet and exit shell side streams, respectively.

Double click on the exchanger icon in the process flow diagram area, while on *Design, Parameters* page, select the *steady-state rating* from the Heat Exchanger Model's pull-down menu. In the *rating,* then the *Sizing Data* menu (Figure 4.3). Click on the following radio buttons and enter the following pieces of information:

Overall: Tube passes per shell are 4, and the number of shell passes is 1.
Shell: Shell diameter is 0.43815 m, number of tubes per shell is 148 tubes, tube pitch is 0.0254 m, tube layout is square, baffle cut is 25%, and baffle spacing is 0.29845 m.
Tube: The OD/ID, tube length, and tube fouling are as follows: 0.01905/ 0.012243 m, 4.2672 m, and 0.0007 m^2K/W, respectively.

The results are on the Performance page, or a table can be generated by right clicking on the exchanger block and then select the show table. Double click on the table and then add or remove variables, as shown in Figure 4.4. The UniSim/Hysys calculated overall heat transfer coefficient is based on the area provided in the tube's menu and is found by dividing the overall heat transfer coefficient time the area (UA) to the overall area (A).

PRO/II Simulation

PRO/II has a rigorous model available for complex heat exchangers to understand how to design this piece of equipment by manipulating various parameters. The thorough model allows for more accurate simulations to the unit with a more detailed report of the findings. Many parameters were taken into consideration when designing the heat exchanger. These parameters include the number of tubes per shell, number of passes, tube configuration and length, heat transfer coefficients,

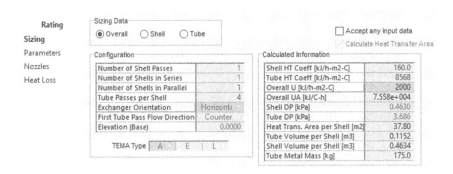

Rating	Sizing Data					
Sizing	⊙ Overall ○ Shell ○ Tube				☐ Accept any input data	
Parameters					☑ Calculate Heat Transfer Area	
Nozzles	Configuration			Calculated Information		
Heat Loss	Number of Shell Passes		1	Shell HT Coeff [kJ/h-m2-C]		160.0
	Number of Shells in Series		1	Tube HT Coeff [kJ/h-m2-C]		8568
	Number of Shells in Parallel		1	Overall U [kJ/h-m2-C]		2000
	Tube Passes per Shell		4	Overall UA [kJ/C-h]		7.558e+004
	Exchanger Orientation	Horizontal		Shell DP [kPa]		0.4630
	First Tube Pass Flow Direction	Counter		Tube DP [kPa]		3.686
	Elevation (Base)		0.0000	Heat Trans. Area per Shell [m2]		37.80
				Tube Volume per Shell [m3]		0.1152
	TEMA Type A E L			Shell Volume per Shell [m3]		0.4634
				Tube Metal Mass [kg]		175.0

FIGURE 4.3   UniSim required heat exchanger sizing data menu for the heat exchanger presented in Example 4.1.

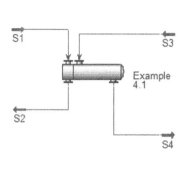

Example 4.1		
Duty	1.534e+006	kJ/h
Shell Side Feed Mass Flow	2.812e+004	kg/h
Tube Inlet Temperature	62.22	C
Tube Outlet Temperature	45.00	C
Shell Inlet Temperature	25.00	C
Shell Outlet Temperature	37.90	C
Control UA	7.558e+004	kJ/C-h
LMTD	20.30	C
Overall U	2000	kJ/h-m2-C
Uncorrected LMTD	22.09	C
Ft Factor	0.9187	
Heat Trans. Area Per Shell	37.80	m2
Tube Side Pressure Drop	3.686	kPa
Shell Side Pressure Drop	0.4630	kPa

FIGURE 4.4 UniSim generated a process flow diagram and stream summary for the heat exchanger design case described in Example 4.1. The selected property package is NRTL-Ideal.

area per shell, shell diameter, pitch, baffles, material type, and pressure drop. After approaching the heat exchanger design by hand, all specifications are entered into PRO/II to simulate the design. Compare the simulation results against those obtained by manual calculations to ensure a reasonable agreement between the two.

The process flowsheet is performed by selecting the Rigorous block *HX* and connecting the inlet and exit stream, as shown in Figure 4.5. If the inlet streams and heat exchanger label are in red, that means data is required. Click on the

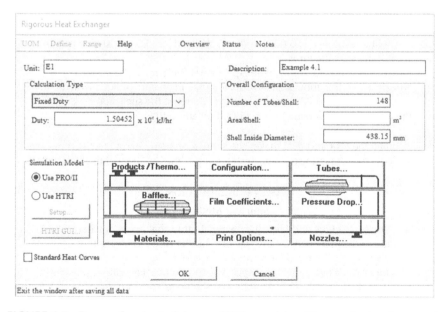

FIGURE 4.5 Process flowsheet of rigorous HX in provision. Rigorous heat exchanger calculation type and overall configuration.

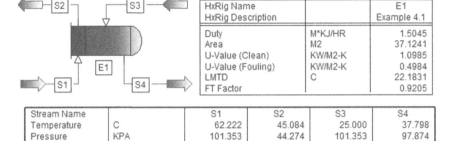

Stream Name		S1	S2	S3	S4
Temperature	C	62.222	45.084	25.000	37.798
Pressure	KPA	101.353	44.274	101.353	97.874
Flowrate	KG-MOL/HR	1050.270	1050.270	1561.047	1561.047

FIGURE 4.6 PRO/II generated the process flow diagram and stream summary for the heat exchanger design case described in Example 4.1.

component selection icon and select water and DEA. From the thermodynamic data, use NRTL for this example. Double click on tube side stream S1 (DEA solution) and shell side inlet stream S3 (water) and specify flow rate, composition, temperature, and pressure. Once these two streams are fully defined, the stream S1 and S2 color labels are changed to black.

Double click on the Rigorous HX block, a window will pop up, as shown in Figure 4.5. The data of the tubes, configuration, film coefficients, baffle, nozzles, and pressure drop should be provided (to fill in data click on the button of each item):

Tubes: ID 0.012243 m, length 4.27 m, OD 0.01905 m, pitch 0.0254 m, pattern square 90°.
Configuration: The number of tube passes/shell is 4.
Film coefficients: Overall U-value estimate is 851.74 W/m^2 K, tube side fouling resistance is 0.0007 m^{2}K/W, shell side fouling resistance is zero, and the scale factor is 1.
Baffle: Cut is 0.25 and the center spacing 0.29845 m.
Nozzles: Set the tube side nozzle internal inlet and outlet diameter to 0.1016 m. and 0.0762 m. for inlet and outlet of the shell side.
Pressure drop: Calculated using PRO/II default setting.

Filling in all the required data makes the simulation ready to run. Figure 4.6 shows the process flowsheet, the exchanger description, and the stream table parameters.

Aspen Plus Simulation

Shortcut Method

1. Start Aspen Plus and click the "New" button that appears when Aspen Plus starts, then click on the "Create" button or go to file -> new.
2. The first step of creating a simulation is defining the properties of the simulation. Start by telling Aspen which compounds will be present in our system, which should be the first screen to come up.
3. We will simply have water and DEA in our system. Because water is a joint compound, we can type it directly into the Component ID box, type water, and press enter. For DEA, use the find menu.

4. Notice that the folder named Methods on the navigation pane, the left side of the screen, has a small red half-filled circle icon on the folder icon. It indicates that this section needs more input data.

5. Click on the little white arrow to the left of this folder and then click on Specifications. Here is where we will have to tell Aspen which equations to use to estimate each of our components' properties. Click on the drop-down menu for Method Name and find NRTL. Aspen filled in the rest of the sheet, and now the Methods folder on the navigation pane has a blue checkmark on it, which means that we have provided enough information.

6. The Properties environment is the background information that we just put in that Aspen Plus needs to calculate all the things we will want it to figure later.

7. Now we are ready to start our simulation. Click on simulation right below Properties to go to the Simulation environment.

8. We now have a blank sheet to make a flowsheet of our process. To draw our process, we will need to use blocks and streams to tell Aspen Plus what components are going where and what is happening to them between the entrance and the exit of our process.

9. The picture below is the Model Palette, and it has everything we will need to draw the flowsheet. Click on the Exchangers tab, and Aspen Plus gives us various ways to model a heat exchanger and other unit operations. We will be using HeatX. Notice that the heat exchanger has red and blue arrows going in and out of it. The red arrow on the left is where the feed stream connects; the coming out are the product streams. Hover over the arrows with the mouse describes what each arrow is for. When connecting the streams, connect each stream to the exact place where Aspen wants it to be; otherwise, it will be confusing. The red arrows indicate required streams, and the blue arrows indicate optional streams.

10. Click on the red arrow going into the heat exchanger, and then click again somewhere to the left of the picture. Click on the other red arrows and also somewhere to the picture's right to get a product stream.

11. Our flowsheet is complete, but we still have to tell Aspen Plus the background information of each of the blocks and streams. Go back to the folders on the left side of the screen (navigation pane). Click the little white arrow next to Streams; click on 1; the inlet cold water stream (tube side), temperature, and pressure set to 25°C and 1 bar, respectively. Click on 3, the hot feed stream (shell side), and enter two of either temperature, pressure, and vapor fraction. Use temperature and pressure and set them to 62°C and 1 bar, respectively. To change the units on the pressure, use the drop-down menu and find a bar. Our total flow rate will be 6.3 kg/s. First, change the flow basis to mass instead of mole, and then find kg/s in the units box. To the right, there is a box for composition. To specify this, we want Mass-Flow in kg/s, with 0.8 of the mass being water and the other 0.2 being DEA. For stream two, it is pure water. The design computes the energy balance, the pressure drop, the exchanger area, and velocities in a shell and tube heat exchanger. The HeatX exchanger requires two process streams; a hot and cold stream.

12. Start preparing the simulation as follows: double click on the exchanger block in the Setup page and enter the value of manually

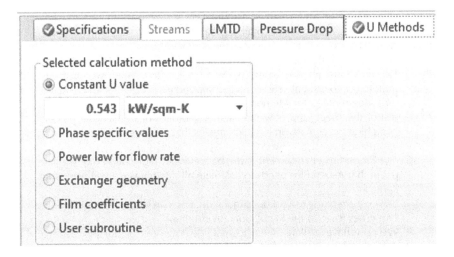

FIGURE 4.7 Aspen Plus selected calculation method menu, constant U value of the case described in Example 4.1.

calculated U (Figure 4.7); under Calculation mode, choose the design. For the exchanger, specifications select Hot stream outlet temperature from the drop-down menu and set the value to 45°C (Figure 4.8). Now our simulation is ready to run or calculate.

13. Click the Run button (under the home tab, there is always a run button that is just the triangle on the very top ribbon, to the right of the save icon) and wait a few seconds for Aspen Plus to calculate the simulation.
14. The calculated overall heat transfer coefficient, the LMTD, and Pressure Drop depend on the exchanger design geometry. It needs to specify how

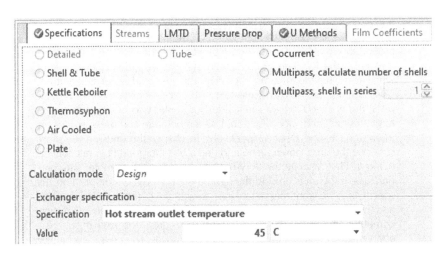

FIGURE 4.8 Aspen Plus specified the calculation mode of the heat exchanger presented in Example 4.1.

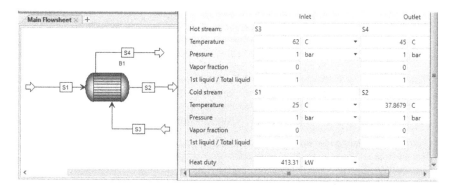

FIGURE 4.9 Aspen Plus generated the process flow diagram and stream summary of the heat exchanger design case described in Example 4.1.

Aspen will calculate the U-value for the exchanger. There are several options; for this example, the individual heat transfer coefficient (h_o, h_i) calculates the overall heat transfer coefficient (U).

15. The Film heat transfer coefficients depend on the exchanger geometry, for both sides (hot and cold) of the exchanger need to be specified. The fouling factors for each stream need to be provided in the space area. The next step is to set out the heat exchanger's geometry (Rating).

16. Detailed design (Rating mode) requires additional information supplied from literature or manually calculated, such as the overall heat transfer coefficient (U). User needs to provide the number of tube passes, the shell diameter, the tube number, and length; typical values are 8–20 ft (2.5–6 m), the IDs and ODs of the pips, the pitch, the material of pipes, the number of baffles, and the baffle spacing.

The simulation is now ready to run. Click on Next and run the simulation. Figure 4.9 shows if some of the required input is incomplete, then Aspen will point to the appropriate input sheet where the information is needed. Process flowsheet and stream table properties. Figure 4.10 shows the summary of the results page for the heat exchanger shown on the Exchanger Details page.

Detailed Design

1. While in Block Specifications, select a rating for the Calculation mode and choose the Shell and Tube type of method.

2. The tube passes, and the shell ID needs to be specified. The Tube menu provides the number of tubes, outside/inside tube diameter, tube layout, and pitch.

3. The Baffles menu provides baffle type, the number of baffles, baffle cut, and baffle spacing. For the baffle type, segmental baffles are typical. In the number of baffles, all passes specify the number of baffles in the exchanger. Add more baffles to increase the heat transfer coefficients; the pressure drop must be within acceptable limits. For the baffle cut, specify the fraction of cross-sectional area for the shell fluid flow. For example, 0.25 means that one baffle covers 75% of the shell cross-sectional area while 25% left for fluid flow. The baffle cut must be

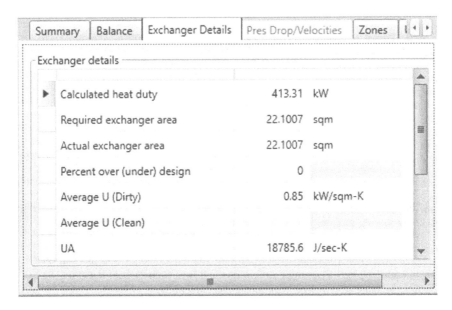

| Summary | Balance | Exchanger Details | Pres Drop/Velocities | Zones | ◄ ► |

Exchanger details

▶	Calculated heat duty	413.31	kW
	Required exchanger area	22.1007	sqm
	Actual exchanger area	22.1007	sqm
	Percent over (under) design	0	
	Average U (Dirty)	0.85	kW/sqm-K
	Average U (Clean)		
	UA	18785.6	J/sec-K

FIGURE 4.10 Aspen Plus shortcut method calculates the heat exchanger's detailed design specifications for the heat exchanger described in Example 4.1.

between 0 and 0.5. Baffle-to-baffle spacing, and if the baffle spacing is unknown at the start of the simulation, the best way is to choose the spacing between the tube sheet and the first/last baffle. Then Aspen will automatically calculate the inner baffle spacing [10].

4. On the Nozzle page, set the shell side inlet and outlet nozzle diameter to approximately one-fourth shell ID, and the tube side inlet and outlet nozzle diameter to roughly one-fifth of the shell ID. The data used in carrying out the detailed design are:
   - **Shell:** The number of tube passes is 4; inside the shell's diameter 0.43815 m (17.25 inches).
   - **Tubes:** Number of tubes: 148, square pattern, the length is 4.2672 m (14 ft), pitch 0.0254 m (1 inch), the outer diameter: 0.01905 (0.75 inches).
   - **Baffles:** Number of baffles all passes is 13, baffle cut 0.25, baffle-to-baffle spacing 3.6576 cm (1.44 inch).
   - **Nozzles:** Shell inlet and outlet nozzle diameter is 1.2192 m. Tube side inlet and outlet nozzle diameter is 0.9144 m.

SuperPro Designer Simulation

The heat exchange in the SuperPro designer calculates only simple energy balance. Calculate the product stream by providing the inlet stream conditions. From the Unit Procedure in the toolbar menu, select the Heat exchange and then Heat exchanging. While in the Connection mode, connect inlet and exit streams. The Tasks/Edit Pure component adds the component involved in the process, and then the exchanger components are registered. Specify feed stream temperature, pressure, flow rates, and compositions. Double click on the heat exchanger to display the operation conditions page. Enter the Flow type (countercurrent) and the Performance options; the cold stream outlet temperature is 37.8°C. Click on the

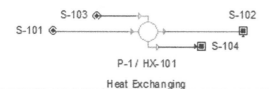

Heat Exchanging

Stream Summary Table (Example 4.1-v2)						
Time Ref: s			S-101	S-103	S-102	S-104
Total Mass Flow	kg		6.3000	7.7500	6.3000	7.7500
Temperature	°C		62.0	25.0	45.0	37.8
Pressure	bar		1.013	1.013	1.013	1.013
Total Contents	mass frac		1.0000	1.0000	1.0000	1.0000
Diethanolamine			0.2000	0.0000	0.2000	0.0000
Water			0.8000	1.0000	0.8000	1.0000

StreamSummary

FIGURE 4.11 SuperPro generated a process flow diagram and stream summary for the heat exchanger design case described in Example 4.1.

material balance icon to run the simulation. Figure 4.11 presented the results in a stream summary table.

AVEVA PROCESS SIMULATION

1. Start Aveva Process Simulation. Simply drag the heat exchanger onto the simulation.
2. Connect inlet stream to the source, specify temperature, pressure, flow rate, and composition. Connect the outlet of the exchanger to sink.
3. Change the inlet and outlet temperatures and adjust the overall heat transfer coefficient if needed in only a few steps; Aveva Process Simulation has calculated the area for the exchanger, as well as the general utility requirement (Figure 4.12).

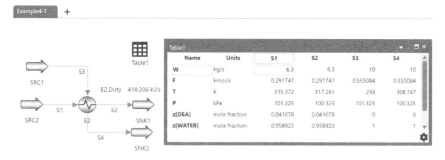

FIGURE 4.12 Aveva Process Simulation generated a process flow diagram and stream summary for the heat exchanger design case described in Example 4.1.

Example 4.2: Rigorous Design of Heat Exchanger

The saturated vapor of propanol flowing at 20.8 kg/s is to be condensed in an existing shell and tube heat exchanger at 2.3 bar. The exchanger contains 900 steel tubes (18 BWG, 3.6576 m long, tube OD/Pitch is 0.75/0.9375 inches, triangular tube pattern in 0.94 m shell ID). The exchanger is one shell, and two tube passes. The propanol flows in the shell side, and cooling water flows in the tube side. If the cooling water enters at 24°C and 2.3 bar and exits at 46°C, baffle spacing is 1.397 m (55 inches), and the number of baffles is 2. Will the exchanger perform the required duty successfully based on the below specifications?

Heat exchanger specifications:

- The boiling point of n-propanol, T_b at 2.3 bar = 391 K (118°C)
- Latent heat of vaporization, λ = 597.782 kJ/kg
- Thermal conductivity of the tubes, k = 44.97 W/mK
- Shell side heat transfer coefficient, h_o = 6,813.916 W/m²K (2.453 kJ/h.m².K)
- The tube heat transfer coefficient, h_i = 1,135.65 W/m²K (4,088 kJ/h.m².K)
- Fouling resistance for water, R_f = 0.00035 m²K/W

SOLUTION

MANUAL CALCULATIONS

The schematic diagram of the n-propanol–water exchanger is shown in Figure 4.13. The required heat duty to condense n-propanol at 2.3 bar:

$$Q_{required} = 20.8 \frac{kg}{s} \times 597.782 \frac{kJ}{kg} = 12430 \text{ kW}$$

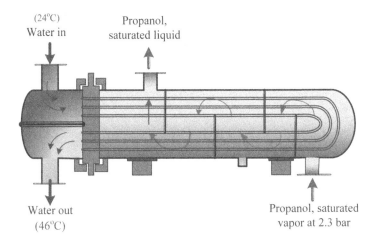

(24°C)
Water in

Propanol,
saturated liquid

Water out
(46°C)

Propanol, saturated
vapor at 2.3 bar

FIGURE 4.13 Schematic diagram of the condensation n-propanol using cold water in a heat exchanging process for the case described in Example 4.2.

where m is the mass flow rate of propanol and λ is the latent heat of vaporization at condenser operating pressure. The cold stream mass flow rate, m_c

$$\dot{m}_c = \frac{12430 \text{ kJ/s}}{4.1868 \dfrac{\text{kJ}}{\text{kg.K}} \times (46-24)\,^\circ\text{C}} = 133.37 \frac{\text{kg}}{\text{s}}$$

The designed heat within the existing heat exchanger depends on the heat exchanger configuration.

$$Q = U_o A_o F \Delta T_{LM}$$

The correction factor for condensation, $F = 1$

$$\Delta T_{LM} = \frac{(118-24)-(118-46)}{\ln(118-24)/(118-46)} = \frac{(94-72)}{\ln(94/72)} = 82.3\,^\circ\text{C}$$

The tubes outside total heat transfer surface area, A_o

$$A_o = N_t \left(\pi D_o L\right) = 900\left(\pi \times 0.01905 \text{ m} \times 3.6576 \text{ m}\right) = 197 \text{ m}^2$$

The tubes inside the heat transfer surface area, A_i

$$A_i = N_t \left(\pi D_i L\right) = 900\left(\pi \times 0.0165 \text{ m} \times 3.6576 \text{ m}\right) = 171 \text{ m}^2$$

The thickness of the tube wall, Δx

$$\Delta x = \frac{0.01905 - 0.01656}{2} = 1.245 \times 10^{-3}$$

The log mean area, A_{LM}

$$A_{LM} = \frac{A_o - A_i}{\ln(A_o/A_i)} = \frac{197-171}{\ln(197/171)} = 183.82 \text{ m}^2$$

The overall heat transfer coefficient based on the outside area, U_o

$$\frac{1}{U_o} = \frac{A_o}{A_i h_i} + \frac{A_o \Delta x}{A_{LM} k_w} + \frac{1}{h_o} + R_{fi} \frac{A_o}{A_i} + R_{fo}$$

Substituting required values in the above equations,

$$\frac{1}{U_o} = \frac{\left(197 \ m^2\right)}{\left(171 \ m^2\right)\left(1{,}135.63\frac{W}{m^2K}\right)} + \frac{\left(197 \ m^2\right)\left(0.00124 \ m\right)}{\left(183.83 \ m^2\right)\left(44.97\frac{W}{mK}\right)}$$

$$+ \frac{1}{\left(6{,}813.92\frac{W}{m^2K}\right)} + 0.00035\frac{m^2K}{W}$$

The overall heat transfer coefficient, U_o

$$\frac{1}{U_o} = 0.00101 + 2.96\times10^{-5} + 1.468\times10^{-4} + 0.00035 = 0.00154\frac{m^2K}{W}$$

The calculated overall heat transfer coefficients, U_o

$$U_o = 649\frac{W}{m^2 \ K}$$

The heat provided by the constructed heat exchanger, Q_{design}

$$Q_{design} = U_o A_o F \ \Delta T_{LM} = \left(0.649\frac{kW}{m^2K}\right)\left(197 \ m^2\right)\left(82.3 \ K\right) = 10522 \ kW$$

The required heat duty is 12,430 kW is higher than the heat duty provided by the designed heat exchanger (10,522 kW), and therefore, the existing heat exchanger is not acceptable and cannot fully condense the propanol saturated vapor stream. Table 4.6 shows the summary of the designed exchanger specifications.

UNISIM/HYSYS SIMULATION

Drag the Heat Exchanger icon in the object pallet into the simulation environment to construct the process flow diagram. Select NRTL as the appropriate fluid package. While on Design/Parameters page, select the Exchanger Design (weighted) from the pull-down menu. Pressure drop needs to be specified or set to zero for both shell and tube side. The Rating/Sizing menu sets the UA from the manual calculation section to 4.6 × 10⁵ kJ/h.°C. The data of the shell and tube side data are from the manual calculations. Filling in the required data leads to the results that should appear like Figure 4.14. The heat duty based on the exchanger specifications obtained by UniSim (Figure 4.15) is 3.788 × 10⁷ kJ/h (10,522 kW) which is less than the required heat duty for condensing the entire propanol saturated vapor to saturated liquid (12,430 kW). Accordingly, the heat exchanger with the specifications mentioned in Example 4.2 is not acceptable and will not condense the entire propanol saturated vapor saturated liquid at 2.3 bars.

TABLE 4.6
Design Specifications of the Manual Calculations
for the Case in Example 4.2

	Shell Side	Tube Side
Components	Propanol	Water
Mass flow rate (kg/s)	20.79	133.56
Temperature (°C)	118 (vapor)/118(liquid)	24/46°C
Pressure (bar)	2.3	2.3
Pass	1	2
Shell ID (m)	0.94	
Tubes: OD/ID/P_t (cm)		1.91/1.66/2.38
Tube configuration		Triangle pitch
Length (m)		3.66
Total number of tubes		900
Number of baffles		2
Baffle spacing (m)		1.397
Fouling factor (m²K/W)		0.00035
LMTD (°C)	82°C	
F factor	1.0	
Duty (kW)	10,522	

PRO/II SIMULATION

The heat exchange is designed with Rating mode under the calculation type and specifying tube side and shell side film coefficients. The PRO/II simulated the process successfully, and the results are in good agreement with manual calculations. Results show that the heat provided by the designed exchanger 2.922×10^7 kJ/h (8,117 kW) is less than the required heat duty (12,430 kW). Accordingly, the heat exchanger with the specifications provided in Example 4.2 is not satisfactory. Figure 4.16 shows the process flowsheet and stream table properties obtained using PRO/II software.

FIGURE 4.14 UniSim required sizing data of the rating mode (overall, tube and shell) for the heat exchanger with the specifications defined in Example 4.2.

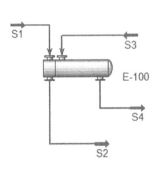

E-100		
Duty	3.788e+007	kJ/h
Tube Side Feed Mass Flow	4.053e+005	kg/h
Shell Side Feed Mass Flow	7.484e+004	kg/h
Tube Inlet Temperature	23.89	C
Tube Outlet Temperature	46.00	C
Shell Inlet Temperature	117.8	C
Shell Outlet Temperature	117.8	C
Control UA	4.600e+005	kJ/C-h
Overall U	2335	kJ/h-m2-C
LMTD	82.34	C

FIGURE 4.15 UniSim generated a process flow diagram and stream summary for the shell and exchanger designed case defined in Example 4.2. The selected property package is NRTL-SRK.

ASPEN SIMULATION

The example utilizes the rigorous heat exchanger method in Aspen Plus software. First, open the converged shortcut heat exchanger using the NRTL-RK method and purge the results (Figure 4.17) by clicking on the Next blue arrow at the top of the screen. On the menu tree, click on Setup under the HeatX block, change the calculation type from Design to Rating, and the model fidelity as shown in the dialog box will appear asking if you would like to convert to a rigorous exchanger. Click Convert and then click OK. Change the calculation mode again to the rating mode for this simulation specifying the outer tube diameters at 0.025 meters, the shell's inner diameter as 0.94 meters, the tube length as 3.66 meters, the baffle spacing as 1 meter, baffle cut to 0.25, and the number of baffles as two, and the number of tubes per pass as 900. Click on the size and then accept the design. Run the simulation, and click on thermal results in the menu tree. Notice how the actual exchanger area is only slightly larger than the required exchanger area, which means that the heat exchanger is only slightly overdesigned by about 0.45%.

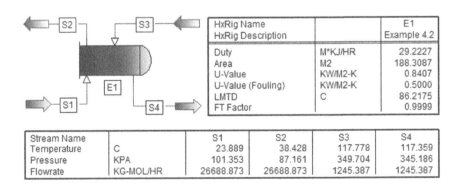

HxRig Name		E1
HxRig Description		Example 4.2
Duty	M*KJ/HR	29.2227
Area	M2	188.3087
U-Value	KW/M2-K	0.8407
U-Value (Fouling)	KW/M2-K	0.5000
LMTD	C	86.2175
FT Factor		0.9999

Stream Name		S1	S2	S3	S4
Temperature	C	23.889	38.428	117.778	117.359
Pressure	KPA	101.353	87.161	349.704	345.186
Flowrate	KG-MOL/HR	26688.873	26688.873	1245.387	1245.387

FIGURE 4.16 PRO/II generated a process flow diagram and stream summary for the heat exchanger design case described in Example 4.2.

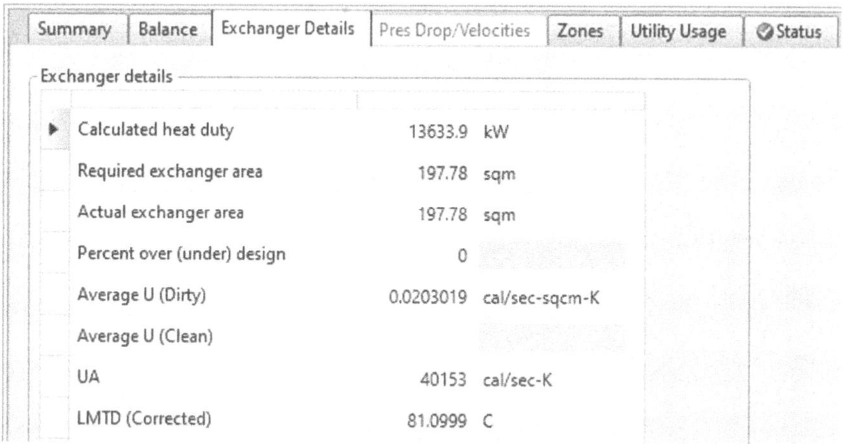

Summary	Balance	Exchanger Details	Pres Drop/Velocities	Zones	Utility Usage	⊘ Status

Exchanger details

▶ Calculated heat duty	13633.9	kW
Required exchanger area	197.78	sqm
Actual exchanger area	197.78	sqm
Percent over (under) design	0	
Average U (Dirty)	0.0203019	cal/sec-sqcm-K
Average U (Clean)		
UA	40153	cal/sec-K
LMTD (Corrected)	81.0999	C

FIGURE 4.17 Aspen Plus generated exchanger details of the heat exchanger design case described in Example 4.2.

Furthermore, Aspen also provides a UA value for the heat exchanger. Click on the Summary tab to view the streams summary and the heat duty for the exchanger. The heat exchanger specifications are summarized as follows:

Tubes: Tubes outside/ID 0.75 inch/0.652 (0.01905/0.0165608 m), triangle-pitch, number of tubes to 900 tubes, tube length 12 ft (3.66 m).
Shell: The ID (0.94 m).
Baffle: Set the values of the baffle cut to 0.25, baffle spacing to 1.016 m, and baffle number to 2.
Nozzle: Set the inlet nozzle value and outlet nuzzle diameter to 0.2286 m, tube side inlet, and outlet nuzzles to 0.1778 m.

Figure 4.18 shows that the designed heat exchange is not successful and will not meet the required heat.

SuperPro Simulation

A simple energy balance with SuperPro is possible. The process is carried out by providing temperature, pressure, flow rates, and compositions of the two inlet streams and providing one of the exit streams' temperature. Figure 4.19 shows the process flow diagram and stream summary.

Aveva Process Simulation

1. Start Aveva Process Simulation. Simply drag the heat exchanger onto the simulation.
2. Connect inlet stream to the source, specify temperature, pressure, flow rate, and composition. Connect the outlet of the exchanger to sink.
3. Change the inlet and outlet temperatures and adjust the overall heat transfer coefficient if needed in only a few steps; Aveva Process Simulation has calculated the area for the exchanger, as well as the overall utility requirement.

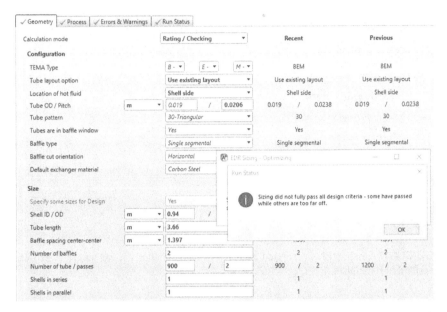

FIGURE 4.18 Aspen Plus generated the results using the rigorous heat exchanger design method (Rating) of the heat exchanger case described in Example 4.2.

Conclusions

Manual calculations and software simulations were in good agreement because the existing exchanger cannot provide the required heat duty to condense the shell side saturated vapor n-propanol. Exit streams show that the propanol is partially liquefied. Comparing software results with hand calculations shows that PRO/II results are the closest to the manual computations.

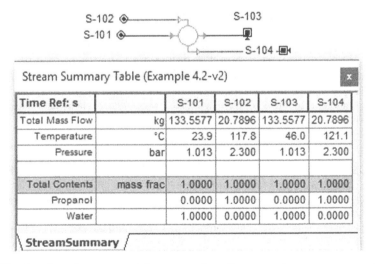

FIGURE 4.19 SuperPro generated the process flow diagram and stream summary for the designed exchanger of the case described in Example 4.2.

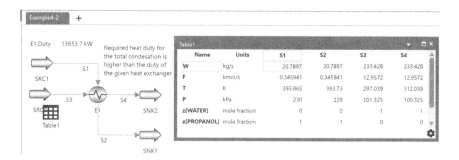

FIGURE 4.20 Aveva Process Simulation generated process flow diagram and stream summary for the designed case of the shell and tube heat exchanger described in Example 4.2.

Example 4.3: Ethylene Glycol–Water Heat Exchanger

Design a shell and tube heat exchanger that cools 12.6 kg/s of ethylene glycol (EG) at 121°C cooled to 54°C using cooling water heated from 32 to 49°C. Assume that the tube side fouling resistance is 7.04×10^{-4} (m²·K/W). Furthermore, neglecting shell-side fouling resistance. Compare manual calculations with the available software packages (e.g., UniSim, PRO/II, Aspen Plus, SuperPro, and Aveva Process Simulation).

SOLUTION

MANUAL CALCULATIONS

Figure 4.21 shows the schematic diagram of the exchanger process flowsheet. Table 4.7 shows the physical properties of the shell side and tube side fluids determined at average temperatures.

Assuming that ethylene glycol enters the shell side of the heat exchanger and cooling water is flowing inside the tubes, the required heat duty, Q_{req}

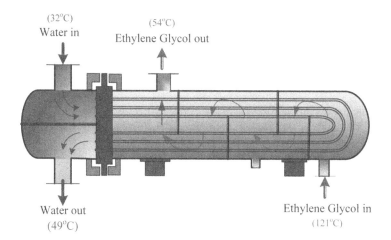

FIGURE 4.21 Schematic diagram of ethylene glycol–water countercurrent heat exchanger for the case presented in Example 4.3.

TABLE 4.7

Physical Properties at Average Temperatures for the Case Described in Example 4.3

	Shell Side	Tube Side
Parameters	Ethylene glycol (87.5°C)	Water (40.5°C)
ρ (kg/m3)	1,099	1,000
Cp (kJ/kg.K)	2.72	4.23
μ (cP)	3.50	0.67
k (W/m.K)	0.277	0.628

$$Q_{req} = m_o Cp_o \Delta T_o = \left(12.6\frac{kg}{s}\right)\left(2.721\frac{kJ}{kg.K}\right)(394.26 - 327.6)\,K = 2,285.4\;kW$$

The mass flow rate of the cooling stream (tube side), $\dot{m}_c$

$$\dot{m}_c = \frac{Q_{req}}{Cp_i \Delta T_i} = \frac{2285.4\;kW}{\left(4.228\frac{kJ}{kg.K}\right)(322 - 305)K} = 32.43\frac{kg}{s}$$

The log means temperature difference, ΔT_{LM}

$$\Delta T_{LM} = \frac{(394 - 322) - (327 - 305)}{\ln\left(\frac{394 - 322}{327 - 305}\right)} = 42\;K$$

The correction factor, F

$$F = \frac{\left[\sqrt{R^2 + 1}/R - 1\right]\ln[1 - S/1 - SR]}{\ln\left[A + \sqrt{R^2 + 1}/A - \sqrt{R^2 + 1}\right]}$$

Calculate the R-value as follows:

$$R = \frac{T_{h,in} - T_{h,out}}{T_{c,out} - T_{c,in}} = \frac{394 - 327}{322 - 305} = 4.0$$

Calculate the S-value as follows:

$$S = \frac{T_{c,out} - T_{c,in}}{T_{h,in} - T_{c,in}^a} = \frac{322 - 305}{394 - 305} = 0.19$$

Calculate the A-value as follows:

$$A = \frac{2}{S} - 1 - R = \frac{2}{0.19} - 1 - 4 = 5.53$$

Substituting values of R, S, and A to calculate F:

$$F = \frac{\left[\sqrt{4^2 + 1}/4 - 1\right] \ln\left[1 - 0.19/1 - 0.19 \times 4\right]}{\ln\left[5.64 + \sqrt{4^2 + 1}/5.64 - \sqrt{4^2 + 1}\right]} = 0.85$$

Assuming a reasonable guess for the overall heat transfer coefficient, let U_i = 567.83 W/m²K consequently, the inside overall heat transfer area, A_i is

$$A_i = \frac{Q_{req}}{U_i F \Delta T_{LM}} = \frac{(2285.4 \ kW)}{\left(567.83\frac{W}{m^2 K}\right)(0.85)(42 \ K)} = 113 \ m^2$$

Assuming that the tube side fluid velocity is 1.524 m/s, the total inside tubes cross-sectional area/pass, A_{ci}

$$A_{ci} = \frac{F_i}{u_i} = \frac{m_i/\rho_i}{u_i} = \frac{\left(32.76\frac{kg}{s}\right)/\left(1000\frac{kg}{m^3}\right)}{\left(1.524\frac{m}{s}\right)} = 0.0215 \ m^2/pass$$

Select the approximate number of tubes; for this purpose, select tube dimensions. For this example, the selected tubes have an OD of 0.01905 m (0.75 inches), 16 BWG tubing (ID = 0.0157 m) arranged on a 0.0254 m (1 inch) square pitch, and a length of tube 5.4864 m (18 ft). The inside cross-sectional area per tube (A_{tc})

$$A_{tc} = \frac{\pi D_i^2}{4} = \frac{\pi (0.0157 \ m)^2}{4} = 1.936 \times 10^{-4} \ m^2/tube$$

The total number of tubes per pass, N_t

$$N_i = \frac{A_{ci}}{A_{tc}} = \frac{0.0215 \ m^2/pass}{1.94 \times 10^{-4} \ m^2/tube} = 110 \ tube/pass$$

The inside surface area per tube, A_t

$$A_t = \pi D_i L = \pi (0.0157 \ m)(5.4864 \ m) = 0.27 \ m^2$$

The total number of tube passes, N_p

$$N_p = \frac{A_i}{N_t \, A_t} = \frac{111 \, m^2}{\left(110 \dfrac{tubes}{pass}\right)\left(0.2717 \dfrac{m^2}{tube}\right)} = 3.73$$

Since there are no 3.73 passes, the number of tube passes rounded to 4 ($N_p = 4$); consequently, the total heat transfer area, A_i, must be corrected,

$$A_i = N_p N_t \left(\pi D_i L\right) = 4 \times 110 \times \pi \times 0.0157 \, m \times 5.4864 \, m = 119 \, m^2$$

Since the inside heat transfer area is changed, the overall heat transfer coefficient must be corrected accordingly.

$$U_i = \frac{Q_{req}}{A_i F \, \Delta T_{lm}} = \frac{\left(2285.4 \, kW\right)}{\left(119 \, m^2\right)\left(0.85\right)\left(42 \, K\right)} = 0.572 \frac{kW}{m^2 K}$$

Compare the overall heat transfer coefficient of the designed heat exchanger with the guessed corrected value; for this purpose, determine the tube side (h_i) and shell side (h_o) film heat transfer coefficient. The tube heat transfer coefficient h_i is calculated using the following equation [7]:

$$Nu_i = \frac{h_i D_i}{k_i} = 0.027 \, N_{Rei}^{0.08} N_{Pri}^{1/3} \left(\frac{\mu}{\mu_w}\right)^{0.14}$$

where $N_{Rei} = D_i \rho_i u_i / \mu_i$ and $N_{Pri} = C_{pi} \, \mu_i / k_i$.
   Reynolds number, N_{Rei}

$$N_{Rei} = \frac{D_i \, \rho_i \, u_i}{\mu_i} = \frac{\left(0.0157 \, m\right)\left(1000 \dfrac{Kg}{m^3}\right)\left(1.524 \dfrac{m}{s}\right)}{\left(0.67 \times 10^{-3} \dfrac{kg}{m.s}\right)} = 35,800$$

Prandtl number, N_{Pri}

$$N_{Pri} = \frac{C_{pi} u_i}{k_i} = \frac{\left(4.1868 \times 10^3 \dfrac{J}{kg.K}\right)\left(0.67 \times 10^{-3} \dfrac{kg}{m.s}\right)}{\left(0.628 \dfrac{J}{m.K.s}\right)} = 4.47$$

Substituting calculated values of Reynolds number and Prandtl number in the film heat transfer equation. The Nusselt number, Nu_i,

$$Nu_i = \frac{h_i D_i}{k_i} = 0.027 N_{Rei}^{0.8} N_{Pri}^{1/3} \left(\frac{\mu}{\mu_w}\right)^{0.14} = 0.027 (35,800)^{0.8} (4.47)^{1/3} \times 1 = 195.5$$

The film side heat transfer confident, h_i

$$h_i = \frac{195.5 \times k_i}{D_i} = 195.54 \frac{\left(0.628 \frac{W}{m.K}\right)}{(0.0157 \ m)} = 7821.6 \frac{W}{m^2 K}$$

The shell transfer coefficient, h_o, is calculated using the heat transfer equation from Kern [3]:

$$Nu_o = \frac{h_o D_o}{k_o} = 0.36 N_{Reo}^{0.55} N_{Pro}^{1/3} \left(\frac{\mu_o}{\mu_w}\right)^{0.14}$$

where $N_{Reo} = D_e G_o / \mu_o$ and $N_{Pro} = C_{po} \mu_o / k_o$.

The effective hydraulic diameter $= D_e = \dfrac{4 \times \left(P_t^2 - \left(\pi D_o^2/4\right)\right)}{\pi D_o}$

$$D_e = \frac{4\left(0.0254^2 - \left(\pi (0.01905)^2/4\right)\right)}{\pi \times 0.01905} = 1.68 \ m$$

The mass velocity normal to tubes at the exchanger's centerline (G_o) depends on the cross-sectional area between baffles and shell axis (A_{cf}). The shell side diameter for 4.0 tube passes and 440 tubes, 0.01905 m OD, 0.0254 m square pitch, can be found from tube sheet layouts tables [7]. Since the number is between 432 and 480 tubes, choose the shell side packed with 480 tubes. Assuming that, the shell ID is 0.7366 m, and the baffle spacing is 0.2032 m.

$$A_{cf} = \frac{D_s}{P_t} \times clearance \times b$$

Substitute required parameters

$$A_{cf} = \frac{0.7366 \ m}{0.0254 \ m} \times (0.00635 \ m) \times (0.2032 \ m) = 0.0374 \ m^2$$

The shell side mass velocity, G_o

$$G_o = \frac{m_o}{A_{cf}} = \frac{12.6 \ (kg/s)}{0.0374 \ m^2} = 336.9 \frac{kg}{m^2 s}$$

The shell side Reynolds number (N_{Reo})

$$N_{Reo} = \frac{D_e G_o}{\mu_o} = \frac{(0.024 \text{ m})\left(336.9\frac{\text{kg}}{\text{m}^2\text{s}}\right)}{\left(3.5\times10^{-3}\frac{\text{kg}}{\text{m.s}}\right)} = 2310$$

Shell side Prandtl number, N_{Pro}

$$N_{pro} = \frac{C_{po}\mu_o}{k_o} = \frac{\left(2.72\times10^3\frac{\text{J}}{\text{kg.K}}\right)\left(3.5\times10^{-3}\frac{\text{kg}}{\text{m.s}}\right)}{\left(0.2767\frac{\text{W}}{\text{mK}}\right)} = 34.4$$

Substituting values of Reynolds number and Prandtl number in the shell heat transfer equations (h_o) and neglecting the effect of change in viscosity,

$$\text{Nu}_o = \frac{h_o D_e}{k_o} = 0.36(2310)^{0.55}(34.4)^{1/3}(1)^{0.14} = 83$$

The shell heat transfer coefficient, h_o

$$h_o = \text{Nu}_o\frac{k_o}{D_e} = 83\frac{\left(0.2767\frac{\text{W}}{\text{m.K}}\right)}{(0.02408 \text{ m})} = 953.74\frac{\text{W}}{\text{m}^2\text{K}}$$

The designed overall heat transfer coefficient (U_i) is based on the inside tube surface area (A_i).

$$\frac{1}{U_i} = \frac{D_i}{D_o h_o} + \frac{D_i \Delta x}{D_{LM} k_w} + \frac{1}{h_i} + R_{fi}$$

The tube thickness, Δx

$$\Delta x = \frac{D_o - D_i}{2} = \frac{0.01905 - 0.01575}{2} = 1.65\times10^{-3}\,m$$

The tube log means diameter, D_{LM}

$$D_{LM} = \frac{D_o - D_i}{\ln(D_o/D_i)} = \frac{0.01905 - 0.01575}{\ln(0.01905 - 0.01575)} = 0.01735\,m$$

Substituting values involved in the overall heat transfer equation,

$$\frac{1}{U_i} = \frac{0.01575\ m}{0.01905\ m\left(953.74\dfrac{W}{m^2K}\right)} + \frac{(0.0158\ m)(0.0017\ m)}{(0.01735\ m)\left(51.89\dfrac{W}{mK}\right)} + \left(\frac{1}{7821.6}\ \frac{m^2K}{W}\right)$$

$$+ 0.0007\frac{m^2K}{W} = 1.73 \times 10^{-3}\ \frac{m^2K}{W}$$

The calculated U-value based on the tube inside the surface area of the designed heat exchanger:

$$U_i = 578.84\ \frac{W}{m^2\ K}$$

The calculated overall heat transfer coefficient based on the designed specification is higher than the assumed value; consequently, the design is booming, and the heat exchanger is over-specified. To be more accurate, increase the baffle spacing slightly to decrease the shell heat transfer coefficient, assuming a new baffle spacing, $b = 0.254$ m (note that the maximum baffle spacing should not exceed shell side ID, $D_s = 0.7366$ m).

$$A_{cf} = \frac{D_s}{P_t} \times \text{clearance} \times b$$

Substitution,

$$A_{cf} = \left(\frac{0.7366}{0.0254}\right)(0.00635)(0.254) = 0.0468\ m^2$$

The shell side mass velocity, G_o

$$G_o = \frac{m_o}{A_{cf}} = \frac{(12.6\ kg/s)}{0.0468\ m^2} = 269.23\ \frac{kg}{m^2 s}$$

The shell side Reynolds number (N_{Reo})

$$N_{Reo} = \frac{D_e G_o}{\mu_o} = \frac{(0.024\ m)\left(269.23\dfrac{kg}{m^2.s}\right)}{\left(3.5 \times 10^{-3}\dfrac{Kg}{m.s}\right)} = 1846.15$$

The shell side Prandtl number (N_{Pro})

$$N_{Pro} = \frac{C_{po}\mu_o}{k_o} = \frac{\left(2.72 \times 10^3 \frac{J}{kgK}\right)\left(3.5 \times 10^{-3} \frac{kg}{m.s}\right)}{\left(0.2767 \frac{W}{m.K}\right)} = 34.41$$

Substituting values of Reynolds number and Prandtl number in the shell heat transfer equations and neglecting the effect of change in viscosity,

$$Nu_o = \frac{h_o D_e}{k_o} = 0.36(1846.15)^{0.55}(34.4)^{1/3}(1)^{0.14} = 73.3$$

The heat transfer coefficient, h_o

$$h_o = Nu_o \frac{k_o}{D_e} = 73.3 \left(\frac{0.2767 \frac{W}{m.K}}{0.024\ m}\right) = 844.8 \frac{W}{m^2\ K}$$

The overall heat transfer coefficient,

$$\frac{1}{U_i} = \frac{D_i}{D_o h_o} + \frac{D_i \Delta x}{D_{LM} k_w} + \frac{1}{h_i} + R_{fi}$$

Substituting needed value to the above equation,

$$\frac{1}{U_i} = \frac{0.016\ m}{0.019\ m\left(844.8\frac{W}{m^2 K}\right)} + \frac{(0.016\ m)(0.00165\ m)}{(0.017\ m)\left(51.89\frac{W}{mK}\right)} + \left(\frac{1}{7821.6\frac{W}{m^2.K}}\right)$$

$$+ 0.0007 \frac{m^2.K}{W} = 0.00184 \frac{m^2.K}{W}$$

The overall all heat transfer coefficient calculated based on heat exchanger design arrangement is

$$U_i = 544 \frac{W}{m^2 m},\quad U_o = U_i \frac{D_i}{D_o} = \left(\frac{0.01575\ m}{0.01905\ m}\right)\left(544\frac{W}{m^2 K}\right) = 450 \frac{W}{m^2 K}$$

The calculated value based on the exchanger design specification is close to the assumed corrected value, and so the designed shell and tube heat exchanger is successful.

PRESSURE DROP

The pressure drop for the flow of liquid without phase change through a circular tube is given by the following equations [7]:

$$-\Delta P_o = P_{in} - P_{out} = K_s \frac{2N_R f' G_o^2}{g_c \rho_c \phi}$$

The value of N_R

$$N_{reversals} = \frac{L}{b} = \frac{5.4864}{0.254} = 21.6$$

The value of K_s,
$$K_s = 1.1 \times N_{reversals} = 1.1 \times 21.6 = 23.76$$
Number of tubes at the centerline (N_{cl})

$$N_{cl} \cong \frac{D_s}{P_i} = \frac{0.7366}{0.0254} = 29$$

The number of tube rows across which the shell fluid flows, N_R, equals the total number of tubes at the center plane minus the number of tube rows that pass through the baffles' cut portions. For 25% cut baffles, consider N_R as 50% of the number of the tubes at the center plane.

$$N_R = 0.5 \times 29 = 14.5 \sim 15$$

The modification friction factor (f')

$$f' = b_2 \left(\frac{D_o G_o}{\mu_o} \right)^{-0.15}$$

where b_2 for square pitch is,

$$b_2 = 0.044 + \frac{0.08 x_L}{(x_T - 1)^{0.43 + 1.13/x_L}}$$

x_T is the transverse pitch ratio to flow to tube OD, and x_L is the ratio of pitch parallel to tube OD, for square pitch $x_L = x_T$

$$x_L = x_T = \frac{P_t}{D_o} = \frac{0.0254}{0.01905} = 1.33$$

Hence,

$$b_2 = 0.044 + \frac{0.08 \times 1.333}{(1.333 - 1)^{0.43 + 1.13/1.333}} = 0.48$$

Substitute b_2

$$f' = b_2 \left(\frac{D_o G_o}{\mu_o} \right)^{-0.15} = 0.48 \left(\frac{(0.01905 \text{ m})(269.23 \text{ kg/m}^2\text{s})}{3.5 \times 10^{-3} \frac{kg}{m.s}} \right)^{-0.15} = 0.161$$

Substituting into the shell pressure drop equation:

$$-\Delta P_s = K_s \frac{2N_R f' G_o^2}{g_c \rho_o \phi} = 23.76 \frac{2(15)(0.161)\left(269.23 \frac{kg}{m^2 s} \right)^2}{\left(1\frac{kg.m}{N.s^2} \right)\left(1098.867 \frac{kg}{m^3} \right)(1)} = 7570 \frac{N}{m^2} = 7.57 \text{ kPa}$$

Estimating the pressure drop on the tube side is much easier than the calculated pressure drop on the shell side. Calculate the pressure drop in the tube side using the following equations [7]:

$$-\Delta P_i = 1.2 \frac{N_P f_D G_i^2 L}{2g_c \rho_i D_i \phi}$$

where

$$f_D = \left(1.82\log_{10}N_{Rei} - 1.64 \right)^{-2}$$

Substitute values of Reynolds number

$$f_D = \left(1.82\log_{10}35,800 - 1.64 \right)^{-2} = 0.0226$$

Tube side mass velocity

$$G_i = \rho_i u_i = 1000 \frac{kg}{m^3} \times 1.524 \frac{m}{s} = 1524 \frac{kg}{m^2 \, s}$$

Substituting in pressure drop equation,

$$-\Delta P_t = 1.2\frac{N_p f_D G_i^2 L}{2g_c \rho_i D_i \phi} = 1.2\frac{4(0.0226)\left(1523.32\frac{kg}{m^2.s}\right)^2(5.4864\ m)}{2\left(1\frac{kg.m}{N.s^2}\right)\left(1000\frac{kg}{m^3}\right)(0.01575\ m)\times 1}$$

$$= 43843\frac{N}{m^2} = 43.84\ kPa$$

Table 4.8 lists a summary of the specifications of the designed shell and tube heat exchanger for the case described in Example 4.2.

The overall heat transfer coefficient (U_o) obtained by manual calculation is different from the value obtained with PRO/II. Here the U-value is based on the outside surface area, which gives the same results found on tubes outside and IDs.

$$\frac{1}{U_o} = \frac{1}{h_o} + \frac{A_o \Delta x}{A_{LM} k_w} + \left(\frac{A_o}{A_i}\right)\frac{1}{h_i} + \left(\frac{A_o}{A_i}\right)R_{fi}$$

TABLE 4.8
Design Specifications for the Case Presented in Example 4.3

	Shell Side	Tube Side
Components	Ethylene glycol (100%)	Water (100%)
Mass flow rate (kg/h)	45359	117934
Temperature (°C)	121/59	32.2/48.9
Pressure (bar)	1.0	1.0
Pass	1	4
Shell ID (in.)	0.7366	
Tubes: OD/ID/P, (in.)		0.019/0.0157/0.0254
Tube configuration		Square
Length (m)		1.67
Total number of tubes		440
Number of baffles		20
Baffle spacing (m)		0.00516
Fouling factor (W/m.K)	0.0	0.0123
Pressure drop (kPa)	7.6	44
LMTD (°C)	42	
F	0.85	
U_o (W/m²°C)	449	
Duty (kW)	2285.95	

Substitute gave parameters,

$$\frac{1}{U_o} = \frac{1}{\left(844.7\frac{W}{m^2\,K}\right)} + \frac{\left(144.5\ m^2\right)\left(0.00165\ m\right)}{\left(131.33\ m^2\right)\left(51.887\frac{W}{mK}\right)} + \left(\frac{144.5\ m^2}{120\ m^2}\right)\frac{1}{\left(7802\frac{W}{m^2K}\right)}$$

$$+ \left(\frac{144.5\ m^2}{120\ m^2}\right) \times 7.04 \times 10^{-4}\frac{m^2K}{W} = 1.22 \times 10^{-3}\frac{m^2K}{W}$$

Taking the inverse

$$U_o = 820.3\frac{W}{m^2\,K}$$

Using Donohue Equation 4.18, the outside heat transfer coefficient (h_o)

$$\frac{h_o D_o}{k_o} = 0.2\left(\frac{D_o G_e}{\mu_o}\right)^{0.6}\left(\frac{C_{po}\mu_o}{k_o}\right)^{0.33}\left(\frac{\mu_o}{\mu_w}\right)^{0.14}$$

The cross-sectional areas for flow, S_c

$$D_o = 0.01905\ m \quad D_s = 0.7366\ m$$

$$P_t = 0.0254\ m \quad b = 0.254\ m$$

The area of crossflow, S_c

$$S_c = b\ D_s\left(1 - \frac{D_o}{P_t}\right) = \left(0.0254\ m\right)\left(0.737\ m\right)\left(1 - \frac{0.01905\ m}{0.0254\ m}\right) = 0.0468\ m^2$$

The number of tubes in the baffle window is approximately equal to the fractional area of the window (f_b) multiplied by the total number of tubes, for a 25% baffle, $f_b = 0.1955$ [3].

$$N_p = 0.1955 \times 440 = 86$$

The area for flow in baffle window,

$$S_b = f_b\frac{\pi D_s^2}{4} - N_b\frac{\pi D_o^2}{4} = 0.1955\frac{\pi\left(0.7366\ m\right)^2}{4} - 86\frac{\pi\left(0.01905\right)^2}{4} = 0.0588\ m^2$$

The mass velocities are

$$G_c = \frac{\dot{m}_o}{S_c} = \frac{12.6 \text{ kg/s}}{0.0468 \text{ m}^2} = 269.25 \frac{\text{kg}}{\text{m}^2 s}$$

$$G_b = \frac{\dot{m}_o}{S_b} = \frac{12.6 \text{ kg/s}}{0.0588 \text{ m}^2} = 214.28 \frac{\text{kg}}{\text{m}^2 s}$$

where

$$G_e = \sqrt{G_b G_c} = \sqrt{\left(269.25 \frac{\text{kg}}{\text{m}^2 s}\right)\left(214.28 \frac{\text{kg}}{\text{m}^2 s}\right)} = 240.2 \frac{\text{kg}}{\text{m}^2 s}$$

The Donohue equation is

$$\frac{h_o D_o}{k_o} = 0.2 \left(\frac{D_o G_e}{\mu_o}\right)^{0.6} \left(\frac{C_{p_o} \mu_o}{k_o}\right)^{0.33} \left(\frac{\mu_o}{\mu_w}\right)^{0.14}$$

Substitution,

$$\frac{h_o (0.019 \text{ m})}{\left(0.277 \frac{W}{m.K}\right)} = 0.2 \left(\frac{(0.019 \text{ m})\left(240.2 \frac{Kg}{m^2 s}\right)}{\left(0.0035 \frac{Kg}{m.s}\right)}\right)^{0.6} \left(\frac{\left(2720 \frac{J}{Kg.K}\right)\left(0.0035 \frac{Kg}{m.s}\right)}{\left(0.2767 \frac{W}{m.K}\right)}\right)^{0.33} (1)^{0.14}$$

Rearranging,

$$\frac{h_o (0.01905 \text{ m})}{\left(0.2767 \frac{J}{m.s.K}\right)} = 0.2 (1307.4)^{0.6} (34.41)^{0.33} (1)^{1.14} = 47.64$$

The shell heat transfer coefficient, h_o

$$h_o = 691.94 \frac{W}{m^2.K}$$

Hence

$$\frac{1}{U_o} = \frac{1}{h_o} + \frac{A_o \Delta x}{A_{LM} k_w} + \left(\frac{A_o}{A_i}\right)\frac{1}{h_i} + \left(\frac{A_o}{A_i}\right) R_{fi}$$

Substituting given parameters

$$\frac{1}{U_o} = \left(\frac{1}{691.94\frac{W}{m^2.K}}\right) + \frac{\left(144.5\ m^2\right)\left(0.001651\ m\right)}{\left(131.55\ m^2\right)\left(51.887\frac{W}{m.K}\right)} + \left(\frac{144.5\ m^2}{120\ m^2}\right)\left(\frac{1}{7801.9\frac{W}{m^2K}}\right)$$

$$+ \left(\frac{144.5\ m^2}{120\ m^2}\right)\left(0.0007\frac{m^2.K}{W}\right)$$

Simplifying further,

$$\frac{1}{U_o} = 1.445 \times 10^{-3} + 3.5 \times 10^{-5} + 1.543 \times 10^{-4} + 7.04 \times 10^{-4} = 2.48 \times 10^{-3}\frac{m^2.K}{W}$$

The overall heat transfer coefficient, U_o

$$U_o = 403\frac{W}{m^2K}$$

Corrected area,

$$A_i = \frac{Q_{req}}{U_i F\ \Delta T_{lm}} = \frac{\left(2285.4\ kW\right)}{\left(403W/m^2K\right)\left(0.85\right)\left(42\ K\right)} = 159\ m^2$$

HYSYS/UNISIM SIMULATION

Start UniSim, add the component (water, ethylene glycol), and then choose the suitable fluid package (NRTL). While on Design/Parameters page, select the Steady State Rating from the pull-down menu under Heat Exchanger Model. For this mode of calculations, calculate the pressure drop in the tube side and shell side by UniSim/Hysys. Figure 4.22 presents the constructed process flowsheet using Hysys/UniSim. Calculate the overall heat transfer coefficient estimated using UniSim from the heat duty equation (i.e., $U_o = Q/(A_o°F\Delta T_{LM})$ and not from outside and inside the film heat transfer coefficient, the case in PRO/II and Aspen Plus. The discrepancy of the exit side shell temperature calculated by UniSim (T for stream S4 = 138) is due to the variation in mass heat capacity (C_p). The manually estimated heat capacity is $C_p = 4.22$ kJ/kg.K.

PRO/II SIMULATION

The data provided to the process flowsheet is as follows:

Calculation type: Tube outlet temperature is 322.04 K, number of tubes per shell is 440, and shell ID is 0.7366 m.

Tubes: Outside tube diameter is 0.01905 m (0.75 inches), ID is 0.015748 m (0.62 inches), tube pitch is 0.0254 m (1 inch) square pitch, and tube length is 5.4864 m.

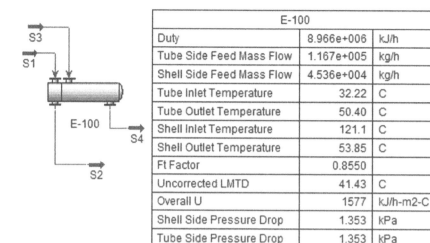

E-100		
Duty	8.966e+006	kJ/h
Tube Side Feed Mass Flow	1.167e+005	kg/h
Shell Side Feed Mass Flow	4.536e+004	kg/h
Tube Inlet Temperature	32.22	C
Tube Outlet Temperature	50.40	C
Shell Inlet Temperature	121.1	C
Shell Outlet Temperature	53.85	C
Ft Factor	0.8550	
Uncorrected LMTD	41.43	C
Overall U	1577	kJ/h-m2-C
Shell Side Pressure Drop	1.353	kPa
Tube Side Pressure Drop	1.353	kPa

FIGURE 4.22 UniSim generated process flow diagram and stream summary for the heat exchanger design case designated in Example 4.3. The selected property package is NRTL-SRK.

Baffles: Fractional cut is 0.25, baffle spacing is 0.254 m (10 inches).
Configuration: Countercurrent, horizontal, number of tube passes/shell is 4.
Nozzle: Shell nozzle inlet and outlet diameter is 0.2032 m, tube side nozzle inlet, and outlet diameter is 0.1524 m.

The value obtained by the Donohue equation is closer to the U (fouling) obtained with Pro II and is shown in Figure 4.23. The discrepancy is mainly due to the slight dissimilarity in the physical properties used in PRO/II and manual calculations.

ASPEN SIMULATION

Start Aspen Plus, while in the property mode, add the component (water, ethylene glycol), and then select the appropriate fluid package (NRTL).

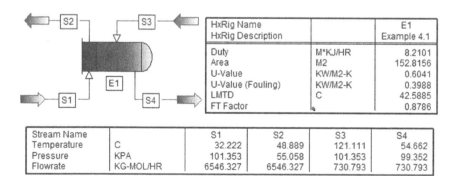

HxRig Name		E1
HxRig Description		Example 4.1
Duty	M*KJ/HR	8.2101
Area	M2	152.8156
U-Value	KW/M2-K	0.6041
U-Value (Fouling)	KW/M2-K	0.3988
LMTD	C	42.5885
FT Factor		0.8786

Stream Name		S1	S2	S3	S4
Temperature	C	32.222	48.889	121.111	54.662
Pressure	KPA	101.353	55.058	101.353	99.352
Flowrate	KG-MOL/HR	6546.327	6546.327	730.793	730.793

FIGURE 4.23 PRO/II generated the process flow diagram and stream summary for the exchanger design case described in Example 4.3.

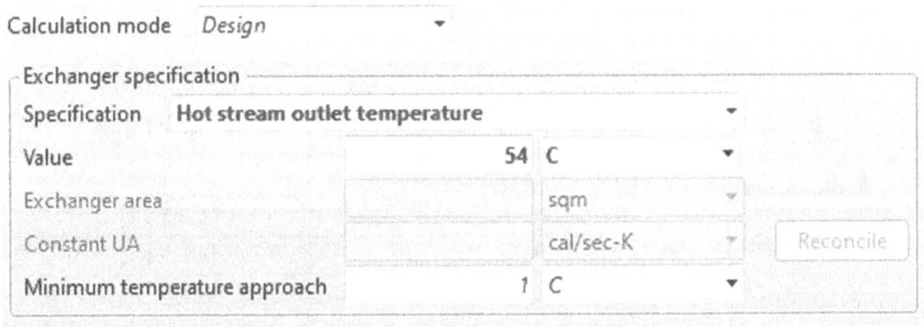

Calculation mode	Design	▾	

Exchanger specification

Specification	Hot stream outlet temperature		▾	
Value		54	C	▾
Exchanger area			sqm	▾
Constant UA			cal/sec-K	▾
Minimum temperature approach		1	C	▾

Reconcile

FIGURE 4.24 Aspen Plus setup page of the heat exchanger for the case described in Example 4.3.

The Aspen Plus HeatX unit is the fundamental heat exchanger algorithm used in the rigorous design, which calculates energy balance, the pressure drop, and the exchanger area. Select the icon by pressing the Heat Exchangers tab in the model library. The block requires hot and cold streams. The feed streams are fully specified by double clicking on each stream and filling in temperature, pressure, flow rate, and compositions. While on the Setup page, select the Calculation type Detailed. The hot fluid is on the shell side. Figure 4.24 shows the cold stream outlet temperature.

Aspen Plus will calculate the LMTD correction factor for the exchanger. The default setting is the correction based on the geometry. Now click on the Pressure drop tab at the top of the page, and a new input page will appear; Calculate from geometry is the preferred option. Enter both the hot and cold side streams pressure drop. The input page for the overall heat exchanger appears next. There are several options to calculate the U-value. For this example, from Film coefficients, calculate U from individual heat transfer coefficient (h_o, h_i). This option requires input and that page appears by clicking on the Film Coefficients tab at the top of the screen. The method to calculate the individual heat transfer coefficients is defined on the film coefficients page. Here Calculate from geometry is selected. Notice, however, that both sides (hot and cold) of the exchanger need to be specified. After the heat transfer coefficient calculation is determined, the next step is to set out the heat exchanger's geometry (Figure 4.25). On the geometry page:

Shell: The number of tube passes is 4, inside shell diameter is 0.7366 m.
Tubes: The tubes are 440 and 5.4864 m long, the pipes inside and outside diameter are 0.02/0.0157 m or nominal diameter (OD = 0.02 m, BWG = 18), the pitch is 0.0254 m.
Baffles: The number of baffles is 20, baffle cut 0.25, and baffle spacing 0.254 m.
Nozzle: Shell side nozzle inlet and outlet diameter is 0.2032 m, and tube side nozzle inlet and outlet diameter is 0.1524 m, supplied from the manual calculations.

When a blue mark appears near the tab, the input sheet is complete. The simulation is now ready to run. Click on Next and run the simulation. The summary

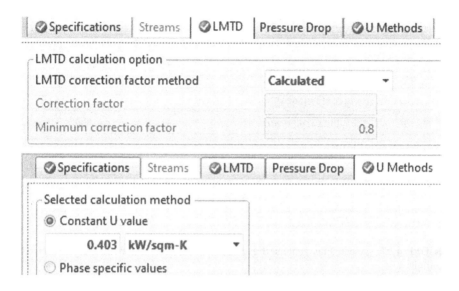

FIGURE 4.25   Aspen Plus required LMTD and U Methods values in the heat exchanger for the case described in Example 4.3.

result page of the heat is exchanger as shown in Figure 4.26. Figure 4.27 lists the Exchanger Details page. The actual exchanger area is calculated based on the number of tubes, length, and outside diameter.

$$A_{actual} = (\pi D_o L) N_t = (\pi \times 0.01905 \text{ m} \times 5.4864 \text{ m}) \times 440 = 144.47 \text{ m}^2$$

The required exchanger area calculated with Aspen is

$$A_{req} = \frac{Q_{req}}{U_o F \Delta T_{LM}} = \frac{2164.4 \times 10^3}{358.156 \times 42.63} = 141.76 \text{ m}^2$$

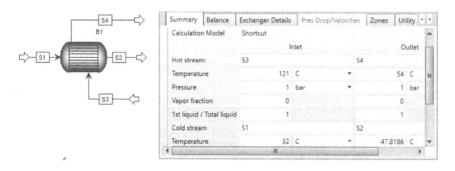

FIGURE 4.26   Aspen Plus generated the process flow diagram and stream summary for the exchanger-designed case described in Example 4.3.

Heat duty [cal/sec]	512194
Calculated heat duty [cal/sec]	512194
Required exchanger area [sqm]	141.37
Actual exchanger area [sqm]	141.37
Average U (Dirty) [cal/sec-sqcm-K]	0.00962549
Average U (Clean)	
UA [cal/sec-K]	13607.6
LMTD (Corrected) [C]	37.6403
LMTD correction factor	0.883912

FIGURE 4.27 Aspen Plus generated exchanger details design for the case presented in Example 4.3.

SuperPro Designer Simulation

With the SuperPro designer, a simple energy balance is performed, as shown in Figure 4.28.

Aveva Process Simulation

1. Start Aveva Process Simulation. Drag the heat exchanger onto the simulation area (canvas).
2. Connect inlet stream to the source, specify temperature, pressure, flow rate, and composition. Connect the outlet of the exchanger to sinks.

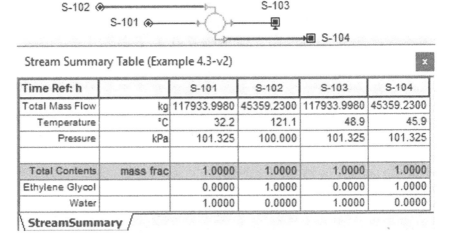

Stream Summary Table (Example 4.3-v2)

Time Ref: h		S-101	S-102	S-103	S-104
Total Mass Flow	kg	117933.9980	45359.2300	117933.9980	45359.2300
Temperature	°C	32.2	121.1	48.9	45.9
Pressure	kPa	101.325	100.000	101.325	101.325
Total Contents	mass frac	1.0000	1.0000	1.0000	1.0000
Ethylene Glycol		0.0000	1.0000	0.0000	1.0000
Water		1.0000	0.0000	1.0000	0.0000

StreamSummary

FIGURE 4.28 SuperPro generated the process flow diagram and stream summary for the heat exchanger design case described in Example 4.3.

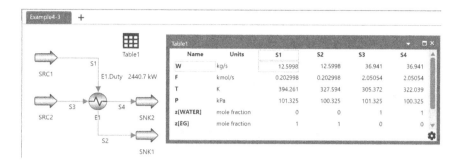

FIGURE 4.29 Process flowsheet and stream table properties simulated with Aveva Process Simulation for the case described in Example 4.3.

3. Change the inlet and outlet temperatures and adjust the overall heat transfer coefficient if needed. Aveva Process Simulation has calculated the area for the exchanger, as well as the overall utility requirement (Figure 4.29).

CONCLUSION

The PRO/II simulation results were the closest to the manual calculations; Aspen Plus results come in the second rank, followed by Hysys/UniSim and Aveva Process Simulation.

Example 4.4: Demineralized Water-Raw Water Exchanger

Design a shell and tube heat exchanger needed to heat water from 23.85°C to 26.67°C using 18.9 kg/s of demineralized water entering the exchanger shell side at 35°C and exiting at 29.44°C. Assume that the tube side fouling resistance is 1.76×10^{-4} (m²s K/J).

SOLUTION

MANUAL CALCULATIONS

Table 4.9 shows the shell side and tube side fluid's physical properties at average temperatures, and Figure 4.30 is a schematic diagram of 1–2 shell and tube heat exchanger. The required heat duty, Q_{req}

TABLE 4.9
Physical Properties at Average Temperatures

Parameters	Shell Side Demineralized water (90°F)	Tube Side Raw water (78°F)
P (kg/m³)	99.955	99.955
C_p (kJ/kg K)	4.187	4.229
μ(cP)	0.81	0.92
K (W/m K)	0.623	0.628

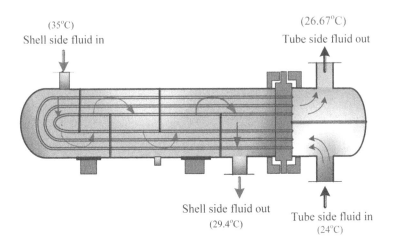

Shell side fluid out
(29.4°C)   Tube side fluid in
                        (24°C)

FIGURE 4.30 A schematic diagram of 1–2 shell and tube heat exchanger for the case described in Example 4.4.

$$Q_{req} = m_o Cp_o \Delta T_o = \left(18.9\frac{kg}{s}\right)\left(4.1868\frac{kJ}{kg.K}\right)(35-29.44)°C = 439.97\frac{kJ}{s} \sim 440 \ kW$$

The tube mass flow rate, m_t

$$\dot{m}_t = \frac{Q_{req}}{C_{p_i}\Delta T_i} = \frac{(439.97 kJ/s)}{\left(4.229\frac{kJ}{kg.K}\right)(26.67-23.85)K} = 34.68\frac{kg}{s}$$

The log mean temperature difference, ΔT_{LM}

$$\Delta T_{LM} = \frac{(308.15-300)-(302.6-297.04)}{\ln(308.15-300/302.6-297.04)} = 6.773 \ K$$

Consider an exchanger configuration as one shell pass and two tube passes; 1–2 pass, the correction factor, F_{1-2}

$$F_{1-2} = \frac{\left[\sqrt{R^2+1}/(R-1)\right]\ln\left[(1-S)/(1-SR)\right]}{\ln\left[A+\sqrt{R^2+1}/A-\sqrt{R^2+1}\right]}$$

where,

$$A = \frac{2}{S}-1-R$$

Calculate the values of R, S, and A as follows:

$$R = \frac{308.15 - 302.6}{300 - 297} = 2.0$$

$$S = \frac{300 - 297}{308.15 - 297} = 0.25$$

$$A = \frac{2}{0.25} - 1 - 2 = 5.0$$

Substituting values of R, S, and A into the following equation to calculate F_{1-2},

$$F_{1-2} = \frac{\left[\sqrt{2^2 + 1}/2 - 1\right] \ln\left[1 - 0.25/1 - 0.25 \times 4\right]}{\ln\left[5.0 + \sqrt{2^2 + 1}/5.0 - \sqrt{2^2 + 1}\right]} = 0.94$$

A typical heat transfer coefficient for the system demineralized water and raw water is within the range 1,703.5–2,839.13 W/m² K. Assume the overall heat transfer coefficient is 2,271.3 W/m² K

$$A_i = \frac{Q_{req}}{U_i F \Delta T_{LM}} = \frac{\left(439.65 \times 10^3 \, \text{J/s}\right)}{\left(2271.3 \frac{W}{m^2 K}\right)(0.94)(6.773K)} = 30.4 \ m^2$$

The tubes' cross-sectional area/pass, A_{ci}

$$A_{ci} = \frac{F_i}{u_i} = \frac{\dot{m}_i / \rho_i}{u_i} = \frac{\left(34.65 \frac{kg}{s}\right)\left(1000 \frac{kg}{m^3}\right)}{\left(1.524 \frac{kg}{m.s}\right)} = 0.0227 \ m^2/pass$$

Choosing a tube's outside diameter, OD = 0.01905 m (0.75 in), 16 BWG tubing, ID = 0.01575 m (0.62 in) arranged on a 1-inch square pitch and tube length 3.048 m (10 ft). The data required before estimating the total number of tubes and number of tube passes is the inside cross-sectional area/tube, A_{tc}

$$A_{tc} = \frac{\pi D_i^2}{4} = \frac{\pi \times (0.01575 \ m)^2}{4} = 1.95 \times 10^{-4} \ m^2/tube$$

The total number of tubes per pass, N_t

$$N_t = \frac{A_{ci}}{A_{tc}} = \frac{0.0227 \ m^2/pass}{1.95 \times 10^{-4} \ m^2/tube} \cong 117 \ tube/pass$$

A single-tube internal surface, A_t

$$A_t = \pi D_i L = \pi \times 0.01575 \ m \times 3.048 \ m = 0.1508 \ m^2$$

Number of tube passes, N_p

$$N_p = \frac{A_i}{A_t \times N_t} = \frac{30.4 \ m^2}{0.1508 \left(\dfrac{m^2}{tube}\right)\left(117\dfrac{tube}{pass}\right)} = 1.723 \ passes$$

The number of tube passes rounded up to 2; hence, $N_p = 2$, accordingly, the corrected total heat transfer surface area (A_i),

$$A_i = N_p N_t (\pi D_i L) = (2)(117)(\pi(0.01575 \ m)(3.049 \ m)) = 35.3 \ m^2$$

The overall heat transfer coefficient should be corrected based on the new overall heat transfer area because of the fixed required heat duty. The new overall heat transfer coefficient, U_i

$$U_i = \frac{Q_{req}}{A_i F \Delta T_{lm}} = \frac{439.65 \ (KW)}{(35.3 \ m^2)(0.94)(6.773 \ K)} = 1786.69 \frac{W}{m^2.K}$$

The new value of the overall heat transfer coefficient is still within the design range (1,703.5–2,839.13 W/m² K). The designed exchanger's overall heat transfer coefficient should be calculated based on the tube and shell heat transfer coefficients. Hence the tube heat transfer coefficient, h_i [7]

$$Nu_i = \frac{h_i D_i}{k_i} = 0.027 \ N_{Rei}^{0.8} N_{Pri}^{1/3} \left(\frac{\mu}{\mu_w}\right)^{0.14}$$

where Reynolds number,

$$N_{Rei} = \frac{D_i \rho_i u_i}{\mu_i} = \frac{(0.01575 \ m)\left(1000\dfrac{kg}{m^3}\right)\left(1.5229\dfrac{m}{s}\right)}{\left(0.92 \times 10^{-3}\dfrac{kg}{m.s}\right)} = 26{,}072$$

Prandtl number,

$$N_{Pri} = \frac{C_{p_i}\mu_i}{k_i} = \frac{\left(4.1868\dfrac{J}{kg.K}\right)\left(0.00092\dfrac{kg}{m.s}\right)}{\left(0.6278\dfrac{W}{m.K}\right)} = 6.13$$

Substituting values of Reynolds number and Prandtl number in tube heat transfer coefficient, h_i, [7]

$$Nu_i = \frac{h_i D_i}{k_i} = 0.027 \, N_{Rei}^{0.8} N_{Pri}^{1/3} \left(\frac{\mu}{\mu_w}\right)^{0.14} = 0.027 (26072)^{0.8} (6.13)^{1/3} \times 1 = 168.57$$

The tube heat transfer coefficient, h_i

$$h_i = 168.57 \left(\frac{k_i}{D_i}\right) = (168.57) \left(\frac{0.6278 \frac{W}{m.K}}{0.01575 \, m}\right) = 6719 \frac{W}{m^2 K}$$

The shell heat transfer coefficient (h_o) is calculated using the heat transfer equation from Kern [3]

$$Nu = \frac{h_o D_e}{k_o} = 0.36 \, N_{Reo}^{0.55} N_{Pro}^{1/3} \left(\frac{\mu_o}{\mu_w}\right)^{0.14}$$

where $N_{Reo} = D_e \, G_o/\mu_o$ and $N_{Pro} = C_{po} \, \mu_o/k_o$. The effective hydraulic diameter (D_e):

$$D_e = \frac{4\left(P_t^2 - \left(\pi D_o^2/4\right)\right)}{\pi D_o} = \frac{4 \times \left(0.0254^2 - \left(\pi (0.01905)^2/4\right)\right)}{\pi \times 0.01905} = 0.024 \, m$$

To calculate the mass velocity normal to tubes at the exchanger's centerline (G_o), one should first calculate the cross-sectional area between baffles and shell axis (A_{cf}). There are 256 tube, 0.01905 m OD, and 0.0254 m square pitch for two tube passes; the shell diameter found from the tube sheet layouts (Table 4.3) or Kern [3]. The closest value for 254 tubes is 270, the shell ID is 0.53975 m, assuming the baffle spacing is 0.254 m.

$$A_{cf} = \frac{D_s}{P_t} \times \text{clearance} \times b$$

Substitute required values to calculate, A_{cf}

$$A_{cf} = \left(\frac{0.54 \, m}{0.0254 \, m}\right)(0.00635 \, m)(0.254 \, m) = 0.0343 \, m^2$$

The shell side mass velocity, G_o

$$G_o = \frac{m_o}{A_{cf}} = \frac{18.9 \, (kg/s)}{0.0343 \, m^2} = 551.02 \frac{kg}{m^2 \, s}$$

The shell side Reynolds number, N_{Reo}

$$N_{Reo} = \frac{D_e G_o}{\mu_o} = \frac{(0.024 \text{ m})\left(551.02\dfrac{kg}{m^2 s}\right)}{\left(0.81\times10^{-3}\dfrac{kg}{m.s}\right)} = 16,343$$

The shell side Prandtl number, N_{Pro}

$$N_{Pro} = \frac{C_{po}\mu_o}{k_o} = \frac{\left(4.1868\times10^3\dfrac{J}{Kg.K}\right)\left(0.00081\dfrac{kg}{m^2.s}\right)}{\left(0.6226 \text{ W/m}^2 K\right)} = 5.44$$

Substituting values of Reynolds number and Prandtl number in the heat transfer equations and neglecting the effect of change in viscosity,

$$Nu_o = \frac{h_o D_e}{k_o} = 0.36(16343)^{0.55}(5.44)^{1/3}(1)^{0.14} = 131.5$$

The heat transfer coefficient, h_o

$$h_o = Nu_o\frac{k_o}{D_e} = 131.5\left(\frac{0.6226\dfrac{W}{m.K}}{0.024 \text{ m}}\right) = 3411.33\frac{W}{m^2.K}$$

The overall heat transfer coefficient, U_i

$$\frac{1}{U_i} = \frac{D_i}{D_o}\left(\frac{1}{h_o}\right) + \frac{D_i\Delta x}{D_{LM}k_w} + \frac{1}{h_i} + R_{fi}$$

Substituting looked-for values in the above equation,

$$\frac{1}{U_i} = \frac{0.0158 \text{ m}}{0.019 \text{ m}}\left(\frac{1}{3411\dfrac{W}{m^2 K}}\right) + \frac{(0.0158 \text{ m})(0.0017 \text{ m})}{(0.017 \text{ m})\left(51.887\dfrac{W}{m.K}\right)} + \left(\frac{1}{6719\dfrac{W}{m^2 K}}\right)$$

$$+ 0.000176\frac{m^2 K}{W} = 5.963\times10^{-4}\frac{m^2 K}{W}$$

Hence the overall heat transfer coefficient based on the tube inside the area (U_i):

$$U_i = 1677\frac{W}{m^2 K}$$

Based on the outside area, the overall heat transfer coefficient (U_o)

$$\frac{1}{U_o} = \frac{1}{h_o} + \frac{A_o \Delta x}{A_{LM} k_w} + \left(\frac{A_o}{A_i}\right)\frac{1}{h_i} + \left(\frac{A_o}{A_i}\right)R_{fi}$$

where $A_i = 38.55$ m^2 and $A_o = 46.637$ m^2, the log mean area, A_{LM}

$$A_{LM} = \frac{A_o - A_i}{\ln(A_o/A_i)} = \frac{(46.637 - 38.55)}{\ln(46.637/38.55)} = 42.465 \text{ m}^2$$

Hence,

$$\frac{1}{U_o} = \frac{1}{3411.33} + \frac{(46.64)(0.00166)}{(42.465)(57.887)} + \left(\frac{46.637}{38.550}\right)\left(\frac{1}{6719}\right) + \left(\frac{46.637}{38.55}\right)(0.000176)$$

$$= 7.176 \times 10^{-4} \frac{\text{m}^2\text{K}}{\text{W}}$$

Accordingly, the overall heat transfer coefficient based on the outside area (U_o)

$$U_o = 1393.5 \text{ W/m}^2\text{K}$$

The calculated overall heat transfer coefficient (U_o) based on the designed heat exchanger specifications is less than the assumed value; consequently, the designed heat exchanger is not successful and underspecified. Accordingly, increasing the shell heat transfer coefficient (h_o) increases the overall heat transfer coefficient by decreasing the baffle spacing; b, (0.1778 m).

$$A_{cf} = \frac{D_s}{P_t} \times \text{clearance} \times b$$

Substitute required values

$$A_{cf} = \frac{0.59 \text{ m}}{0.0254 \text{ m}} \times (0.00635 \text{ m})(0.1778 \text{ m}) = 0.0262 \text{ m}^2$$

The shell side mass velocity, G_o

$$G_o = \frac{m_o}{A_{cf}} = \frac{(18.9 \text{ kg/s})}{0.0262 \text{ m}^2} = 720.7 \frac{\text{kg}}{\text{m}^2\text{s}}$$

The shell side Reynolds number, N_{Reo}

$$N_{Reo} = \frac{D_e G_o}{\mu_o} = \frac{(0.024 \text{ m})(720.7 \text{ kg/m}^2\text{s})}{\left(0.81 \times 10^{-3} \frac{Kg}{m.s}\right)} = 21,423$$

Substituting values of Reynolds number and Prandtl number in the heat transfer equations and neglecting the effect of change in viscosity,

$$\mathrm{Nu}_o = \frac{h_o D_e}{k_o} = 0.36(21423)^{0.55}(5.44)^{1/3}(1)^{0.14} = 152$$

The heat transfer coefficient, h_o

$$h_o = \mathrm{Nu}\frac{k_o}{D_e} = 152\frac{\left(0.6226\dfrac{W}{m.K}\right)}{0.024\ m} = 3943.13\frac{W}{m^2.K}$$

The new value of U_i

$$\frac{1}{U_i} = \frac{0.01575\ m}{(0.019\ m)\left(3943.13\dfrac{W}{m^2.K}\right)} + \frac{(0.01575\ m)(0.00165\ m)}{(0.0173\ m)\left(51.887\dfrac{W}{m.K}\right)} + \frac{1}{\left(6719\dfrac{W}{m^2 K}\right)}$$

$$+\left(0.000176\frac{m^2 K}{W}\right) = 5.635\times10^{-4}\frac{m^2.K}{W}$$

Hence, the new overall heat transfer coefficient, U_i is 1,775 W/m².K
The corrected exchanger heat transfer area,

$$A_i = \frac{Q_{req}}{U_i F\,\Delta T_{LM}} = \frac{\left(439.65\times10^3\ J/s\right)}{\left(1775\dfrac{W}{m^2 K}\right)(0.94)(6.773 K)} = 38.9\ m^2$$

PRESSURE DROP

The pressure drop for the flow of liquid without phase change through a circular tube is given by the following equations [7]:

$$-\Delta P_o = P_{in} - P_{out} = K_s\frac{2N_R f'G_o^2}{g_c\rho_c\phi}$$

$$N_{reversals} = \frac{L}{b} = \frac{3.048\ m}{0.1778\ m} = 17.14$$

The value of K_s

$$K_s = 1.1\times N_{reversals} = 1.1\times17.14 = 18.86$$

Number of tubes at the centerline, N_{cl}

$$N_{cl} \cong \frac{D_s}{P_t} = \frac{0.53975\ m}{0.0254\ m} = 21.25$$

The number of tube rows across which the shell fluid flows (N_R) equals the total number of tubes at the center plane minus the number of tube rows that pass through the baffles' cut portions. For 25% cut baffles, N_R is approximately 50% of the tubes at the center plane.

$$N_R = 0.5 \times 21.25 = 10.625 \sim 11$$

The modification friction factor, f'

$$f' = b_2 \left(\frac{D_o G_o}{\mu_o} \right)^{-0.15}$$

where b_2 for square pitch is

$$b_2 = 0.044 + \frac{0.08 X_L}{\left(X_T - 1\right)^{0.43+1.13/X_L}}$$

x_T is the transverse pitch ratio to flow to the tube outside diameter, and x_L is the ratio of pitch parallel to the tube outside diameter. For square pitch $x_L = x_T$

$$X_L = X_T = \frac{P_t}{D_o} = \frac{0.0254\ m}{0.01905\ m} = 1.33$$

Hence,

$$b_2 = 0.044 + \frac{0.08 \times 1.333}{\left(1.333 - 1\right)^{0.43+1.13/1.333}} = 0.48$$

The f' is calculate as follows,

$$f' = b_2 \left(\frac{D_o G_o}{\mu_o} \right)^{-0.15} = 0.48 \left(\frac{0.01905\ m \times 720.7\ kg/m^2 s}{0.81 \times 10^{-3} \frac{kg}{m.s}} \right)^{-0.15} = 0.11$$

Substituting in the shell pressure drop equation, ΔP_s,

$$-\Delta P_s = K_s \frac{2N_R f' G_o^2}{g_c \rho_c \phi} = 18.86 \frac{2(11)(0.11)\left(720.7\frac{kg}{m^2 s}\right)^2}{\left(1\frac{kg.m}{N.s^2}\right)\left(1000\frac{kg}{m^3}\right)(1)} = 23706.4\frac{N}{m^2} = 23.7 \text{ kPa}$$

Estimating the pressure drop on the tube side is much easier than the calculated pressure drop on the shell side. The pressure drop in the tube side (ΔP_t)[7]:

$$-\Delta P_t = 1.2\frac{N_p f_D G_i^2 L}{2g_c \rho_i D_i \phi}$$

where

$$f_D = \left(1.82 \log_{10} N_{Rei} - 1.64\right)^{-2}$$

Substitution,

$$f_D = \left(1.82 \log_{10} 26{,}072 - 1.64\right)^2 = 0.0244$$

Tube side mass velocity

$$G_i = \rho_i u_i = 1000\frac{kg}{m^3} \times 1.524\frac{m}{s} = 1524\frac{kg}{m^2 s}$$

Substituting in pressure drop equation,

$$-\Delta P_t = 1.2\frac{N_p f_D G_i^2 L}{2g_c \rho_i D_i \phi}$$

Substitution,

$$-\Delta P_t = 1.2\frac{2(0.0244)\left(1524\frac{kg}{m^2 s}\right)^2(3.048 \ m)}{2\left(1\frac{kg \ m}{N.s^2}\right)\left(1000\frac{kg}{m^3}\right)(0.01575 \ m)(1)} = 13148\frac{N}{m^2} = 13.15 \text{ kPa}$$

Table 4.10 summarized the specifications of the designed heat exchanger.

UNISIM/HYSYS SIMULATION

The heat exchanger model used is the steady-state rating. In the fluid package, the NRTL method is the most suitable. The heat exchanger is one shell, two passes, and the manual calculation provides some of the required data by UniSim/Hysys. Figure 4.31 predicted the case described in Example 4.4.

TABLE 4.10

Specifications of the Designed Shell and Tube Heat Exchanger Described in Example 4.4

	Shell Side	Tube Side
Components	Water (100%)	Water (100%)
Mass flow rate (kg/s)	18.89	34.68
Temperature (°C)	35/29.44	23.85/26.67°C
Pressure (bar)	1.0	1.0
Pass	1	2
Shell ID (cm)	53.975	
Tubes: OD/ID/P_t (cm)		1.90/1.58/2.54
Tube configuration		Square
Length (m)		3.048
Total number of tubes		256
Number of baffles		15
Baffle spacing (cm.)	17.78	
Fouling factor (m² K/W)	0.0	0.00018
Pressure drop (kPa)	28.27	13.17
LMTD (°C)	6.85	
F factor	0.94	
U_o (W/m² K)	1320	
Duty (kW)	440	

E-100		
Duty	1.588e+006	kJ/h
Tube Side Feed Mass Flow	1.361e+005	kg/h
Shell Side Feed Mass Flow	6.804e+004	kg/h
Tube Inlet Temperature	23.89	C
Tube Outlet Temperature	26.67	C
Shell Inlet Temperature	35.00	C
Shell Outlet Temperature	29.49	C
Control UA	2.453e+005	kJ/C-h
Ft Factor	0.9427	
Uncorrected LMTD	6.869	C
Overall U	5253	kJ/h-m2-C
Tube Side Pressure Drop	25.80	kPa
Shell Side Pressure Drop	79.96	kPa

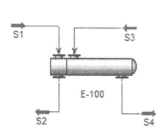

FIGURE 4.31 UniSim generated the process flow diagram and stream summary for the shell and tube heat exchanger for the case described in Example 4.4. The used property package is NRTL-SRK.

PRO/II SIMULATION

Using Pro II and selecting a rigorous heat exchanger *(Rigorous HX)*, the fluid package is NRTL. Shell side and tube side inlet streams need to be fully specified. *Fixed Duty* calculation type is used ($Q = 439,606.6$ W). The process passes per shell is 2. The overall U-value estimated is 1703.5 W/m²K as an initial guess on the Film Coefficients Data page. The scale factors are the default values. Figure 4.32 shows the process flow diagram and stream summary. The tubes OD, ID, Pitch, and tube length enter on the tube page. In the current example, values are 0.01905 m, 0.015748 m, 0.0254 m square-90°, and 3.048 m, respectively. The number of tube side fouling resistance is 0.000176 m²K/W and zero for the shell side. In the baffle geometry data, 0.25 for the fraction cut and 0.254 m. for center spacing. The nozzle data are those of PRO/II default values. After filling in all these data, the simulation is ready to run, and the output data should look like that in Figure 4.32.

ASPEN SIMULATION

Start Aspen Plus, click New, and create a blank simulation. Enter the ID of the component: type water, and then click enter. Select the HeatX heat exchanger under Exchangers in the Model Pallete.

Connect two inlet streams and two exit streams with the exchanger's red arrows while in the connection mode.

Steam-TA, the property estimation model, is the suitable fluid package for measuring the required thermodynamic fluid properties.

Double click on stream S1 in the primary flowsheet and specify the stream to 23.89°C, the pressure to 1 bar, total flow basis to mass, and flow rate at 18.9 kg/s, for composition, gives a value of one for water. Next in the navigation pane. Double click on the B1 folder. In the exchanger specification box in the specification drop-down menu, select hot stream outlet temperature and specify the value to 29.44°C. Now, switch to the Pressure Drop tab. Specify the outlet pressure of the hot side to be −0.3 bar. Then, change to the cold side, and similarly, specify that left pressure to be −0.3 bar. Next, switch to the U-Methods tab, and the constant U value should be approximately 0.0203 in the B1 folder; toward the bottom, select summary, Run the simulation. Now, in the B1 folder, select the Thermal Results. Switch to the exchanger Details tab, and the calculated heat today should be approximately 104,845 calories per second.

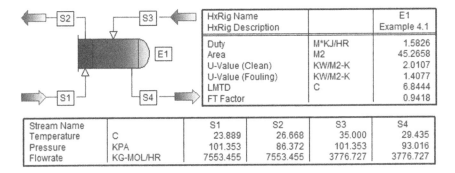

HxRig Name		E1
HxRig Description		Example 4.1
Duty	M*KJ/HR	1.5826
Area	M2	45.2658
U-Value (Clean)	KW/M2-K	2.0107
U-Value (Fouling)	KW/M2-K	1.4077
LMTD	C	6.8444
FT Factor		0.9418

Stream Name		S1	S2	S3	S4
Temperature	C	23.889	26.668	35.000	29.435
Pressure	KPA	101.353	86.372	101.353	93.016
Flowrate	KG-MOL/HR	7553.455	7553.455	3776.727	3776.727

FIGURE 4.32 PRO/II generated the process flow diagram and stream summary for the designed shell and tube heat exchanger for the case described in Example 4.4.

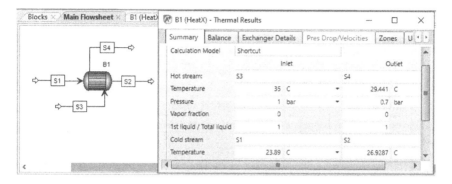

FIGURE 4.33 Aspen Plus generated the process flow diagram and stream summary for the heat exchanger design case described in Example 4.4.

The manual calculation provides the tube mass flow rate. While in the *Setup* page and Specifications menu, select Design and hot stream outlet temperature to 29.44°C for the type of calculations. For Exchanger specification, select Exchanger duty. Select the U-Methods for the film coefficient; measured from exchanger geometry. Figure 4.33 depicts the process flowsheet and stream table properties. Figure 4.34 predicts the results of the simulated detailed exchanger specifications.

SuperPro Simulation

Figure 4.35 shows the simulated result of the shell and tube heat exchanger performed by the SuperPro designer.

Aveva Process Simulation

1. Start Aveva Process Simulation. Simply drag the heat exchanger onto the simulation.
2. Connect inlet stream to the source, specify temperature, pressure, flow rate, and composition. Connect the outlet of the exchanger to sink.

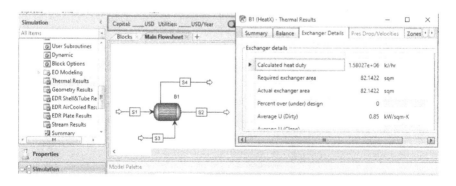

FIGURE 4.34 Exchanger detailed design results generated by Aspen Plus for the case described in Example 4.4.

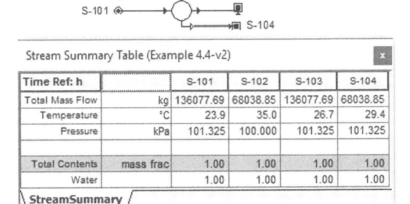

FIGURE 4.35 Process flow diagram of a shell and tube heat exchanger and stream table properties generated by SuperPro expressed in Example 4.4.

3. Change the inlet and outlet temperatures and adjust the overall heat transfer coefficient if needed in only a few steps. Aveva Process Simulation has calculated the area for the exchanger, as well as the overall utility requirement.

Figure 4.36 shows the generated process flow diagram and stream summary of the case described in Example 4.4.

CONCLUSION

The simulation results were close to the manual calculation.

Example 4.5: Hot Water Heats a Cold Water

Hot water stream flowing at 12.6 kg/s, a temperature of 394.261 K, and a pressure of 103.4 kPa heat a cold water stream (324.81 K and the pressure is 206.84 kPa) in a shell and tube heat exchanger. The outlet temperature of the cold and hot streams are

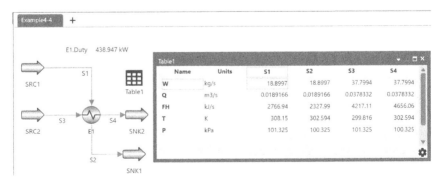

FIGURE 4.36 Aveva Process Simulation generated the process flow diagram and stream summary for the shell and tube heat exchanger designed for the case defined in Example 4.4.

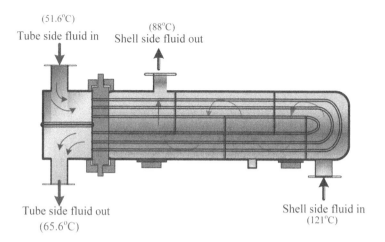

(51.6°C)
Tube side fluid in

(88°C)
Shell side fluid out

Tube side fluid out
(65.6°C)

Shell side fluid in
(121°C)

FIGURE 4.37 The schematic diagram described the 1–2 type shell and tube heat exchanger for Example 4.5.

338.7 K and 360.93 K, respectively. A hot-water stream enters the heat exchanger's shell side and the cold-water stream into the exchanger's tube side. Design a coun-ter-current shell and tube heat exchange that achieves the required heat duty. The assumed fouling factor for the tube side and shell side is 0.00035 m²K/W. Compare manual calculations with available software packages, UniSim/Hysys, PRO/II, Aspen Plus, SuperPro, and Aveva Process Simulation.

SOLUTION

MANUAL CALCULATIONS

Figure 4.37 shows a schematic diagram of 1–2 shell and tube heat exchanger for the hot-water–the cold-water system. Table 4.11 listed the physical properties data for the shell side and tube side fluids at average temperatures. The required heat duty, Q

$$Q_{req} = m_o C_{p_o} \Delta T_o = \left(12.6\frac{kg}{s}\right)\left(4.1868\frac{kJ}{kg.K}\right)(394.261-360.928)K = 1758.44 \ kW$$

TABLE 4.11
Physical Properties at Average Temperatures

Parameters	Shell Side Hot water (104.4°C)	Tube Side Coldwater (59°C)
ρ (kg/m³)	96	98
C_p(kJ/kg.K)	4.187	4.228
μ(cP)	0.27	0.47
k (W/m.K)	0.675	0.657

The mass flow rate of water mass in the tube side, $\dot{m}_i$

$$\dot{m}_i = \frac{Q_{req}}{C_{pi}\Delta T_i} = \frac{(1758.44 \text{ kJ/s})}{\left(4.23\dfrac{\text{kJ}}{\text{kg K}}\right)(338.706 - 324.817)\text{K}} = 30 \frac{\text{kg}}{\text{s}}$$

The log means temperature difference, ΔT_{LM}

$$\Delta T_{LM} = \frac{(250-150)-(190-125)}{\ln(250-150/190-125)} = \frac{35}{\ln(100/65)} = 81.25°F$$

The correction factor, F for 1–2 shell/tube passes or multiple of 2 tube passes, is calculated from the following equation:

$$F_{1-2} = \frac{\left[\sqrt{R^2+1}/R-1\right]\ln[1-S/1-SR]}{\ln\left[A+\sqrt{R^2+1}/A-\sqrt{R^2+1}\right]}$$

where

$$A = \frac{2}{S}-1-R$$

The calculated R-value

$$R = \frac{394.26-360.928}{338.706-324.817} = 2.4$$

The calculated S-value

$$S = \frac{338.706-324.817}{394.26-324.817} = 0.2$$

The calculated A-value

$$A = \frac{2}{0.2}-1-2.4 = 6.6$$

Substituting values of R, S, and A to calculate the temperature correction factor, F_{1-2}

$$F_{1-2} = \frac{\left[\sqrt{2.4^2+1}/2.4-1\right]\ln[1-0.2/1-0.2\times2.4]}{\ln\left[6.6+\sqrt{2.4^2+1}/6.6-\sqrt{2.4^2+1}\right]} = 0.96$$

A typical overall heat transfer coefficient for the water–water heat exchanger is within the range of 140–260 Btu/ft^2 h°F (795–1,475 W/m^2°C). Assume overall heat transfer coefficient is the average of the ranged values 200 Btu/ft^2 h°F 1,136 W/m^2°C

$$A_i = \frac{Q_{req}}{U_i F \Delta T_{LM}} = \frac{(1758.44 \text{ kW})}{\left(1.136 \frac{\text{kW}}{\text{m}^2\text{K}}\right)(0.96)(45.137 \text{ K})} = 35.73 \text{ m}^2$$

The tubes cross-sectional area/pass, A_{ci}

$$A_{ci} = \frac{F_i}{u_i} = \frac{m_i/\rho_i}{u_i} = \frac{30(\text{kg/s})/(983.5 \text{ kg/m}^3)}{\left(1.524 \frac{\text{kg}}{\text{m.s}}\right)} = 0.02 \text{ m}^2/\text{pass}$$

Utilizing 0.01905 inch tube OD, 16 BWG tubing (ID = 0.01575 m) arranged on a 0.0254 m. square pitch and 3.6576 m tube length. The cross-sectional area/tube, A_{tc}

$$A_{tc} = \frac{\pi D_i^2}{4} = \frac{\pi \times (0.01575 \text{ m})^2}{4} = 1.95 \times 10^{-4} \ m^2/\text{tube}$$

The total number of tubes per pass, N_t

$$N_t = \frac{A_{ci}}{A_{tc}} = \frac{0.0202 \text{ m}^2}{1.95 \times 10^{-4} \text{ m}^2/\text{tube}} = 104 \text{ tube/pass}$$

A single tube internal surface area, A_t

$$A_t = \pi D_i L = \pi \times 0.01575 \times 3.6576 = 0.181 \text{ m}^2$$

The number of tube passes, N_p

$$N_p = \frac{A_i}{A_t \times N_t} = \frac{35.73 \text{ m}^2}{(0.181 \text{ m}^2/\text{tube})\left(104 \frac{\text{tube}}{\text{pass}}\right)} = 1.9 \cong 2 \text{ passes}$$

The number of tube passes rounded up to 2 (N_p = 2). Accordingly, the corrected total tubes inside surface heat transfer area, A_i:

$$A_i = N_p N_t (\pi D_i L) = (2)(104)(\pi)(0.01575 \ m)(3.6576 \ m) = 37.64 \text{ m}^2$$

The tube outside surface area, A_o

$$A_o = N_p N_t (\pi D_o L) = 2(104)(\pi)(0.01905 \ m)(3.6576 \ m) = 45.53 \ m^2$$

The corrected overall heat transfer coefficient, U_i

$$U_i = \frac{Q}{A_i F \Delta T_{lm}} = \frac{(1758.44 \times 10^3 \ W)}{(37.64 \ m^2)(0.96)(45.137 \ K)} = 1078.14 \frac{W}{m^2 K}$$

The corrected overall heat transfer coefficient is within the design range (795–1,476 W/m²K). The assumed value should match the U value estimated from the heat exchanger design specifications that depend on the film heat transfer coefficient of tube side and shell side, fouling factor and metal resistance.

$$\frac{1}{U_i} = \frac{A_i}{A_o h_o} = \frac{A_i \Delta x}{A_{LM} k_w} = \frac{1}{h_i} + R_{fi} + R_{fo} \frac{A_i}{A_o}$$

The tube heat transfer coefficient, h_i

$$Nu_i = \frac{h_i D_i}{k_i} = 0.027 N_{Rei}^{0.8} N_{Pri}^{1/3} \left(\frac{\mu}{\mu_w} \right)^{0.14}$$

The Reynolds number of the tube side, N_{Rei}

$$N_{Rei} = \frac{D_i \rho_i u_i}{\mu_i} = \frac{(0.01575 \ m)\left(983.3 \frac{kg}{m^3} \right)\left(1.524 \frac{m}{s} \right)}{\left(0.00047 \frac{kg}{m.s} \right)} = 50,218$$

The Prandtl number of the tube side, N_{Pri}

$$N_{Pri} = \frac{C_{pi} \mu_i}{k_i} = \frac{\left(4228.67 \frac{J}{kg \ K} \right)\left(0.00047 \frac{kg}{m.s} \right)}{\left(0.657 \frac{W}{m \ K} \right)} = 3.02$$

The tube heat transfer coefficient, hi [7],

$$Nu_i = \frac{h_i D_i}{k_i} = 0.027 N_{Rei}^{0.8} N_{Pri}^{1/3} \left(\frac{\mu}{\mu_w} \right)^{0.14} = 0.027(50218)^{0.8} (3.02)^{\frac{1}{3}} (1)^{0.14} = 225$$

Hence, the tube heat transfer coefficient h_i,

$$h_i = \frac{225 \times k_i}{D_i} = 225 \times \frac{\left(0.657 \frac{W}{m.K}\right)}{(0.01575\ m)} = 9385.7 \frac{W}{m^2.K}$$

The shell heat transfer coefficient, h_o, is calculated using the heat transfer equation from Kern [3]:

$$Nu_o = \frac{h_o D_e}{k_o} = 0.36 N_{Reo}^{0.55} N_{Pro}^{1/3} \left(\frac{\mu}{\mu_w}\right)^{0.14}$$

where the Reynolds number, N_{Reo}

$$N_{Reo} = \frac{D_e G_o}{\mu_o}$$

The Prandtl number, N_{Pro}

$$N_{Pro} = \frac{C_{P_o} \mu_o}{k_o}$$

The effective hydraulic diameter, D_e

$$D_e = \frac{4\left(P_t^2 - (\pi D_o^2/4)\right)}{\pi D_o} = \frac{4 \times \left(0.0254^2 - (\pi(0.01905)^2/4)\right)}{\pi \times 0.01905} = 0.024\ m$$

The shell side mass velocity normal to tubes at the centerline of the exchanger (G_o) requires estimating the cross-sectional area between baffles and shell axis (A_{cf}). For 2 tube passes, 104 tubes per pass (Total = 208 tubes), 0.01905 m. OD, 0.0254 m. square pitch. The tube sheet layouts can find the diameter of the shell side [3]. The closest available value for the number of tubes that got Ds value for 208 tubes is 220, belongs to shell ID is 0.48895 m, and assume baffle spacing 0.254 m.

$$A_{cf} = \frac{D_s}{P_t} \times clearance \times b$$

Substitution,

$$A_{cf} = \left(\frac{0.48895\ m}{0.0254\ m}\right)(0.00635\ m)(0.254\ m) = 0.031\ m^2$$

The shell side mass velocity, G_o

$$G_o = \frac{m_o}{A_{cf}} = \frac{12.6 \text{ kg/s}}{0.031 \text{ m}^2} = 405.82 \frac{\text{kg}}{\text{m}^2\text{s}}$$

The shell side Reynolds number, N_{Reo}

$$N_{Reo} = \frac{D_e G_o}{\mu_o} = \frac{(0.02403 \text{ m})\left(405.82 \dfrac{\text{kg}}{\text{m}^2\text{s}}\right)}{\left(0.00027 \dfrac{\text{kg}}{\text{m.s}}\right)} = 36,125$$

The shell side Prandtl number, N_{Pro}

$$N_{Pro} = \frac{C_{po}\mu_o}{k_o} = \frac{\left(0.00419 \dfrac{\text{J}}{\text{kg.K}}\right)\left(0.00027 \dfrac{\text{kg}}{\text{m.s}}\right)}{\left(0.6745 \dfrac{\text{W}}{\text{m.K}}\right)} = 1.675$$

Substituting values of Reynolds number and Prandtl number in the shell heat transfer equations and neglecting the effect of change in viscosity,

$$Nu_o = \frac{h_o D_e}{k} = 0.36(36,125)^{0.55}(1.675)^{1/3}(1)^{0.14} = 137$$

The shell heat transfer coefficient, h_o

$$h_o = Nu\frac{k_o}{D_e} = 137 \frac{\left(0.6745 \dfrac{\text{W}}{\text{m.K}}\right)}{0.02403 \text{ m}} = 3845.46 \frac{\text{W}}{\text{m}^2\text{.K}}$$

Substituting estimated values of shell side and tube heat transfer coefficients in the following equation:

$$\frac{1}{U_i} = \frac{A_i}{A_o h_o} + \frac{A_i \Delta x}{A_{LM} k_w} + \frac{1}{h_i} + R_{fi} + R_{fo}\frac{A_i}{A_o}$$

The tube inner heat transfer area, A_i

$$A_i = \pi D_i L = \pi(0.01575 \text{ m})(3.6576 \text{ m}) = 0.181 \text{ m}^2$$

The tube outside surface area, A_o

$$A_o = \pi D_o L = \pi (0.01905 \ m)(3.6576 \ m) = 0.2189 \ m^2$$

The tube's wall thermal conductivity, $k_w = 51.887$ W/m.K. The tube wall thickness, Δx

$$\Delta x = \frac{0.01905 - 0.0157}{2} = 1.65 \times 10^{-3} \ m$$

The log mean area, A_{LM}

$$A_{LM} = \frac{A_o - A_i}{\ln (A_o / A_i)} = \frac{0.2189 \ m^2 - 0.187 \ m^2}{\ln (0.2189 / 0.187)} = 0.2 \ m^2$$

Substitution,

$$\frac{1}{U_i} = \left(\frac{0.181}{0.219} \right) \left(\frac{1}{3845.46} \right) + \left(\frac{0.181}{0.219} \right) \frac{\left(1.65 \times 10^{-3} \right)}{(51.887)} + \left(\frac{1}{9386} \right)$$

$$+ 0.00035 + 0.00035 \left(\frac{0.181}{0.219} \right) = 9.583 \times 10^{-4} \ \frac{m^2 K}{W}$$

The overall heat transfer coefficient based on the inside area, U_i

$$U_i = 1043.5 \frac{W}{m^2 K}$$

The overall designed heat transfer coefficient is less than the assumed value; consequently, the designed heat exchanger is not successful. The heat transfer coefficient of the shell side increases by decreases baffle spacing. The value should not be less than a fifth of the shell ID and not exceeding the shell side diameter. Assuming baffle spacing equal to 0.127 m:

$$A_{cf} = \frac{D_s}{P_t} \times \text{clearance} \times b$$

Substitute required parameters,

$$A_{cf} = \left(\frac{0.48895 \ m}{0.0254 \ m} \right) (0.00635 \ m)(0.187 \ m) = 0.0155 \ m^2$$

The shell side mass velocity, G_o

$$G_o = \frac{m_o}{A_{cf}} = \frac{(12.6 \text{ kg/s})}{0.0155 \text{ m}^2} = 811.64 \frac{kg}{m^2 s}$$

The shell side Reynolds number, N_{Reo}

$$N_{Reo} = \frac{D_e G_o}{\mu_o} = \frac{(0.024 \text{ m})(811.64 \text{ kg/m}^2 s)}{\left(0.00027 \frac{kg}{m.s}\right)} = 72,202$$

Shell side Prandtl number, N_{Pro}

$$N_{Pro} = \frac{C_{po} \mu_o}{k_o} = \frac{\left(4186.8 \frac{J}{kg.K}\right)\left(0.00027 \frac{kg}{m.s}\right)}{\left(0.6745 \frac{W}{m.K}\right)} = 1.675$$

Substituting values of shell side Reynolds number and Prandtl number in the heat transfer equations and neglecting the effect of change in viscosity,

$$Nu = \frac{h_o D_e}{k} = 0.36(72,202)^{0.55}(1.675)^{1/3}(1)^{0.14} = 201$$

The heat transfer coefficient, h_o

$$h_o = Nu \frac{k_o}{D_e} = 201 \frac{\left(0.675 \frac{W}{m.K}\right)}{(0.024 \text{ m})} = 5648.94 \frac{W}{m^2 .K}$$

The overall heat transfer coefficient, U_i

$$\frac{1}{U_i} = \frac{D_i}{D_o h_o} + \frac{D_i \Delta x}{D_{LM} k_w} + \frac{1}{h_i} + R_{fi} + R_{fo}\left(\frac{D_i}{D_o}\right)$$

Substituting estimated values $(h_o, h_i, k_w, R_{fi}, R_{fo})$ as follows:

$$\frac{1}{U_i} = \frac{0.01575}{(0.01905)(5648.94)} + \frac{(0.01575)(0.00165)}{(0.01735)(51.887)} + \left(\frac{1}{9386}\right)$$

$$+ 0.00035 + 0.00035\left(\frac{0.0158}{0.0191}\right) = 8.9 \times 10^{-4} \frac{m^2 K}{W}$$

The calculated overall heat transfer coefficient based on the internal tube area (U_i):

$$U_i = 1124 \frac{W}{m^2.K}$$

The value is very close to the corrected overall heat transfer coefficient (1078.87 W/m² K); thus, the designed exchanger is successful. Based on the outside tube area (U_o):

$$\frac{1}{U_o} = \frac{1}{h_o} + \frac{A_o \Delta x}{A_{LM} k_w} + \left(\frac{A_o}{A_i}\right)\frac{1}{h_i} + \left(\frac{A_o}{A_i}\right) R_{fi} + R_{fo}$$

Substituting values:

$$\frac{1}{U_o} = \frac{1}{5648.94} + \frac{(0.2189)(0.00165)}{(0.2)(51.887)} + \left(\frac{0.2189}{0.181}\right)\frac{1}{9385.7} + \left(\frac{0.2189}{0.181}\right)(0.00035)$$

$$+ 0.00035 = 1.076 \times 10^{-3} \frac{m^2 K}{W}$$

The overall heat transfer coefficient based on outside tube surface area, U_o

$$U_o = 929.4 \frac{W}{m^2 K}$$

Pressure Drop

The pressure drop for the flow of liquid without phase change through a circular tube is given by the following equations [7]:

$$-\Delta P_o = P_{in} - P_{out} = K_s \frac{2 N_R f' G_o^2}{g_c \rho_c \phi}$$

$$N_{reversals} = \frac{L}{b} = \frac{3.6576 \ m}{0.127 \ m} = 28.8$$

The value of K_s

$$K_s = 1.1 \times N_{reversals} = (1.1)(28.8) = 31.68$$

Number of tubes at the centerline (N_{cl})

$$N_{cl} \cong \frac{D_s}{P_t} = \frac{0.48895 \ m}{0.0254 \ m} = 19.25$$

The number of tube rows across which the shell fluid flows (N_R) equals the total number of tubes at the center plane minus the number of tube rows that pass through the baffles' cut portions. For 25% cut baffles, N_R is 50% of the tubes at the center plane.

$$N_R = 0.5 \times 19.25 = 9.625 \sim 10$$

The modification friction factor (f')

$$f' = b_2 \left(\frac{D_o G_o}{\mu_o} \right)^{-0.15}$$

where b_2 for square pitch is

$$b_2 = 0.044 + \frac{0.08 x_L}{(x_T - 1)^{0.43 + 1.13/x_L}}$$

XT is the transverse pitch ratio to flow through the tube's outside diameter, and x_L is the ratio of pitch parallel to the tube's outside diameter, for square pitch; $x_T = x_L$.

$$x_L = x_T = \frac{P_t}{D_o} = \frac{0.0254}{0.01905} = 1.33$$

Hence,

$$b_2 = 0.044 + \frac{0.08 \times 1.333}{(1.333 - 1)^{0.43 + 1.13/1.333}} = 0.48$$

The f' value,

$$f' = b_2 \left(\frac{D_o G_o}{\mu_o} \right)^{-0.15} = 0.48 \left(\frac{(0.01905 \ m)\left(811.64 \frac{kg}{m^2 s} \right)}{\left(0.00027 \frac{kg}{m.s} \right)} \right)^{-0.15} = 0.093$$

The pressure drop in the shell side, ΔP_s

$$-\Delta P_s = K_s \frac{2 N_R f' G_o^2}{g_c \rho_c \phi} = 31.68 \frac{2(10)(0.093)\left(811.64 \frac{kg}{m^2 s} \right)^2}{\left(1 \frac{kg.m}{N.s^2} \right)\left(961.108 \frac{kg}{m^3} \right) \times 1} = 40,388 \frac{N}{m^2} = 40.4 \ kPa$$

The pressure drop in the tube side, ΔP_t [7]

$$-\Delta P_t = 1.2 \frac{N_P f_D G_i^2 L}{2g_c \rho_i D_i \phi}$$

where

$$f_D = \left(1.82\log_{10} N_{Rei} - 1.64\right)^{-2}$$

Substitution,

$$f_D = \left(1.82\log_{10} 50,218 - 1.64\right)^{-2} = 0.021$$

Tube side mass velocity, G_i

$$G_i = \rho_i u_i = \left(983.534 \frac{kg}{m^3}\right)\left(1.524 \frac{m}{s}\right) = 1498.9 \frac{kg}{m^2 s}$$

The pressure drop in the tube side of the heat exchanger, ΔP_i

$$-\Delta P_t = 1.2 \frac{N_P f_D G_i^2 L}{2g_c \rho_i D_i \phi}$$

Substituting required values in the pressure drop equation,

$$-\Delta P_t = 1.2 \frac{2(0.021)\left(1498.9 \frac{kg}{m^2 s}\right)^2 (3.048\ m)}{2\left(1 \frac{kg\,m}{N.s^2}\right)\left(983.534 \frac{kg}{m^3}\right)(0.015748\ m)(1)} = 11,141 \frac{N}{m^2} = 11.14\ kPa$$

Table 4.12 shows a summary of the heat-exchange design specification.

Hysys/UniSim Simulation
UniSim obtains the process flowsheet as shown in Figure 4.38, results are close to the manual calculations with some deviation in the shell side exit temperature and duty, and this is due to the discrepancy of heat capacity values between manual estimates and UniSim value. The used property package is NRTL-SRK.

PRO/II Simulation
Figure 4.39 shows the process flowsheet and the stream table properties obtained using PRO/II software. Based on the outside area, U_o is 0.80 kW/m²°C.

TABLE 4.12

Design Specifications of the Case Described in Example 4.5

	Shell Side	Tube Side
Components	Hot water	Coldwater
Mass flow rate (kg/s)	12.6	30.0
Temperature (°C)	121/88	51.67/65.56
Pressure (bar)	1.0	1.0
Pass	1	2
Shell ID inch (cm)	$19\frac{1}{4}$(48.895)	
Tubes: OD/ID/P_t (cm)		1.905/1.575/2.54
Tube configuration		Square pitch
Length, ft (m)		12(3.6576)
Total number of tubes		208
Number of baffles		28
Baffle spacing, inch(cm)	5(12.7)	
Fouling factor (m²°C/W)	0.00035	0.00035
Pressure drop (kPa)	40.4	11.14
LMTD (°C)	45.14	
F factor	0.96	
U_o (W/m²°C)	895	
Duty (kW)	1758.43	

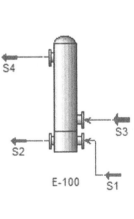

E-100		
Duty	6.368e+006	kJ/h
Tube Side Feed Mass Flow	1.089e+005	kg/h
Shell Side Feed Mass Flow	4.536e+004	kg/h
Tube Inlet Temperature	51.67	C
Tube Outlet Temperature	65.56	C
Shell Inlet Temperature	121.1	C
Shell Outlet Temperature	87.73	C
Control UA	1.470e+005	kJ/C-h
Ft Factor	0.9604	
Uncorrected LMTD	45.10	C
Overall U	3229	kJ/h-m2-C
Tube Side Pressure Drop	19.84	kPa
Shell Side Pressure Drop	46.99	kPa

S4

S3

S2

E-100   S1

FIGURE 4.38 UniSim simulation results generated for the heat exchanger case described in Example 4.5. The used property package is NRTL-SRK.

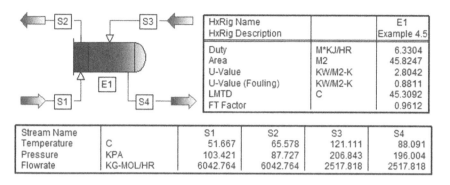

Stream Name		S1	S2	S3	S4
Temperature	C	51.667	65.578	121.111	88.091
Pressure	KPA	103.421	87.727	206.843	196.004
Flowrate	KG-MOL/HR	6042.764	6042.764	2517.818	2517.818

FIGURE 4.39 PRO/II simulation results generated for the heat exchanger case described in Example 4.5.

ASPEN PLUS SIMULATION

The shell and tube heat exchange designed using Aspen Plus is like that in Example 4.4. Figure 4.40 depicts the process flowsheet and stream table properties. Figure 4.41 predicts the specifications of the designed heat exchanger for the case presented in Example 4.5.

SUPERPRO SIMULATION

Figure 4.42 shows the stream summary results performed by SuperPro Designer for the design case presented in Example 4.5.

AVEVA PROCESS SIMULATION

1. Start Aveva Process Simulation. Simply drag the heat exchanger onto the simulation.
2. Connect inlet stream to the source, specify temperature, pressure, flow rate, and composition. Connect the outlet of the exchanger to sink.
3. Change the inlet and outlet temperatures and adjust the overall heat transfer coefficient if needed in only a few steps. Aveva Process Simulation has calculated the area for the exchanger, as well as the overall utility requirement.

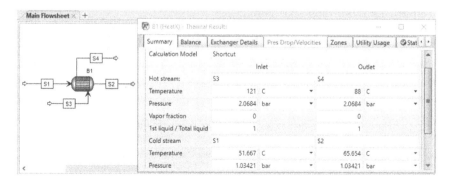

FIGURE 4.40 Aspen Plus generated process flow diagram and streams summary for the heat exchanger case described in Example 4.5.

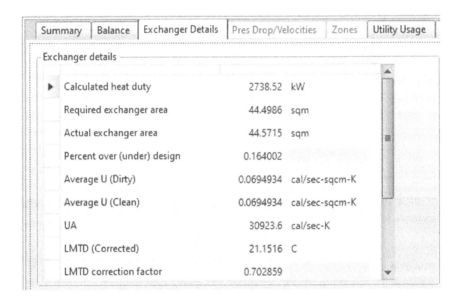

FIGURE 4.41 Aspen Plus generated the heat exchanger design data generated by for the case described in Example 4.5.

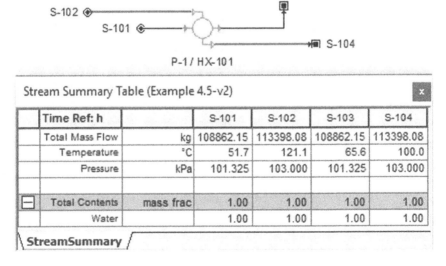

FIGURE 4.42 SuperPro generated the simulation results for the case described in Example 4.5.

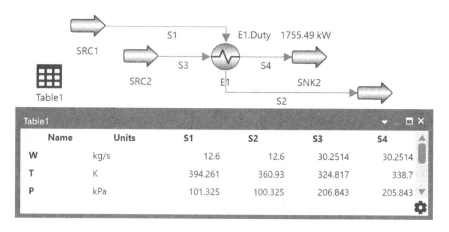

FIGURE 4.43 Aveva Process Simulation predicted results generated for the case described in Example 4.5.

CONCLUSION

The Aveva process simulation results were the closest to the manual calculations.

PROBLEMS

4.1 HOT-WATER–COLD-WATER HEAT EXCHANGER

Design a shell and tube heat exchanger for hot water at 12.6 kg/s and 344.261 K is cooled with 25.2 kg/s of cold water at 305.372 K, which is heated to 322.04 K in a countercurrent shell and tube heat exchanger. Assume that the exchanger has 6.096 m steel tubes (thermal conductivity of steel is 45 W/(m.K), 0.01905 m. OD, and 0.015748 m. ID.) The tubes are on 0.0254 m square pitch (The appropriate fluid package is NRTL-Ideal).

4.2 HEATING OF NATURAL GAS WITH HOT WATER

Hot water at 388.706 K is used to heat 14.5 kg/s of natural gas (60% methane, 25% ethane, and 15% propane) at 3,447.38 KPa from 300 K 308.15 K. The heating water is available at 394.261 K and 620.528 kPa with a flow rate of 3.8 kg/s. Hot water is flowing on the shell side. Assuming that the fouling factor for water is $3.52 \times 10^{-4} \frac{m^2.K}{W}$. Design a shell and tube heat exchanger for this purpose. (The proper fluid package is NRTL-SRK).

4.3 COOLING DIETHANOLAMINE SOLUTION IN SWEETENING PLANT

Design a shell and tube heat exchanger to cool 6.3 kg/s of DEA solution (0.2 mass fractions DEA/0.8 water) from 335.372 K to 318.15 K by using water at 298.15 K heated to 310.928 K. Assume that the tube inside fouling resistance is 3.52×10^{-4} m²K/W, and shell side fouling resistance is 3.52×10^{-4} m²K/W. Compare design results with Example 4.1 (The suitable fluid package is NRTL-SRK).

4.4 COOLING ETHYLENE GLYCOL IN THE DEHYDRATION PROCESS

Design a shell and tube heat exchanger for 12.6 kg/s of ethylene glycol (EG) at 394.26K cooled to 327.594 K using cooling water heated from 305.37 K to a temperature of 322.039 K. The shell side fouling resistance is 7.04×10^{-4} m²K/W, and tube side fouling resistance is also 7.04×10^{-4} m²K/W. Compare results with Example 4.2. (The suitable fluid package is NRTL-SRK).

4.5 RICH GLYCOL: LEAN GLYCOL HEAT EXCHANGER

Rich glycol (TEG) from the absorber at 4.14 kg/s and 291.483 K is exchanging heat with 4.03 kg/s lean glycol from the air cooler at 333.15 K and leaving 301.483 K. Design a shell and tube heat exchanger for this purpose. (The suitable fluid package is NRTL-SRK).

4.6 SHELL SIDE HEAT TRANSFER COEFFICIENT (MCCABE P. 441)

A shell and tube heat exchanger is used to heat 12.6 kg/s of benzene is heated from 277.594 K to 299.817 K in the shell side heat exchanger using 12.6 kg/s hot water 355.372 K., The exchanger contains 828 tubes, 0.01905 m. OD, 3.6576 m long on 0.0254 m. square pitch. The baffles are 25% cut, and baffle spacing is 0.3048 m. the shell side is 0.889 m. Calculate the shell heat transfer coefficient using the Donohue Equation 4.18 and compare it with Equation 4.14 (the proper fluid package is NRTL).

4.7 HEAT EXCHANGER FOR ETHYLBENZENE AND STYRENE

Design a shell-and-tube heat exchanger to preheat a stream of 8.34 kg/s containing equal amounts of ethylbenzene and styrene from 283.15 K to 370.372 K. Heat supply is medium saturated steam at 388.706 K at a flow rate of 10 kg/s and atmospheric pressure. Additional data:

Density = 856 kg m³
Viscosity = 0.4765 cP
Specific heat = 0.428 kcal/kg°C
Thermal conductivity = 0.133 kcal/h m°C
The suitable fluid package is NRTL.

4.8 HEATING OF METHANOL LIQUID MIXTURE

A liquid mixture of 134.2 kmol/h (0.97 methanol, 0.02 ethanol, 0.01 water) enters a tube side of a shell and tube heat exchange at 25°C, and 1,520 kPa is heated to 155°C using 1,000 kgmol/h high-pressure steam at 15 atm. The exit high-pressure steam temperature is 160°C. Using UniSim/Hysys software package (PRSV), determine the inlet steam temperature. Neglect heat losses and pressure drop in the heat exchanger.

4.9   COOLING OF DIMETHYL ETHER (DME)

The DME gas mixture 100 kgmol/h (50 mol% DME, 35% H_2O, 10% methanol, 5% ethanol) exits a reactor at 250°C and 1,470 kPa is utilized to heat a methanol gas stream of 134 kmol/h (0.97 methanol, 0.02 ethanol, 0.01 water) from 25°C and 1520 kPa to 200°C. Using UniSim/Hysys software package (PRSV), determine the DME exit stream temperature. Neglect heat losses and pressure drop in the heat exchanger.

4.10   COOLING OF BENZENE/TOLUENE LIQUID MIXTURE

The saturated liquid mixture (12 kgmol/h) of benzene and toluene (50% benzene) at 1 atm cooled to a temperature of 30°C using cooling water in a shell and tube heat exchanger. The cooling water inlet temperature is 15°C and 1 atm pressure, and the temperature increases in the heat exchanger to 25°C. Neglect heat losses and pressure drop in the heat exchanger. Using UniSim/Hysys, determine the cooling water inlet flow rate. Use the hot stream into the tube side and the cooling water into the shell side.

REFERENCES

1. Sadik, K. and L. Hongtan, 2020. Heat Exchangers: Selection, Rating and Thermal Design, 4th edn, CRC Press, New York, NY.
2. Perry, R. H. and D. W. Green, 2018. Perry's Chemical Engineers' Handbook, 9th edn, McGraw-Hill, New York, NY.
3. Kern, D. Q., 1950. Process Heat Transfer, 7th edn, McGraw-Hill, New York, NY.
4. Holman, J. P., 2014. Heat Transfer, 5th edn, McGraw-Hill, New York, NY.
5. Peters, M. S., K. D. Timmerhaus and R. E. West, 2003. Plant Design and Economics for Chemical Engineers, 7th edn, McGraw-Hill, New York, NY.
6. Geankoplis, C. J., D. H. Lepek and A. Hersel, 2018. Transport Processes and Unit Operations, 5th edn, Prentice-Hall, NJ.
7. Seider, W. D., J. D. Seader, D. R. Lewin and S. Widagdo, 2016. Product and Process Design Principles, Synthesis, Analysis, and Evaluation, 4th edn, John Wiley, New York, NY.
8. Coulson, J. M., J. F. Richardson, J. R. Backhurst and J. H. Harker, 1999. Chemical Engineering Fluid Flow, Heat Transfer and Mass Transfer, Vol. 1, 6th edn, Butterworth and Heinemann, Oxford.
9. McCabe, W. L., J. C. Smith and P. Harriott, 2005. Unit Operations of Chemical Engineering, 7th edn, McGraw-Hill, New York, NY.
10. Hewitt, G. F., 2013. Handbook of Heat Exchanger Design, 2nd edn, Beggel House, New York, NY.

5 Reactor Design

At the end of this chapter, students should be able to:

1. Compute the volume of continuous stirred-tank reactor (CSTR) and plug flow reactor (PFR) to achieve a specific conversion.
2. Calculate the volume of catalytic reactors to attain an actual conversion.
3. Design isothermal and nonisothermal reactors involve multiple reactions.
4. Analyze reactors' performance in which single and multiple reactions are occurring.

5.1 INTRODUCTION

Chemical reactors are vessels designed to contain chemical reactions. Chemical engineers design reactors to ensure that the reaction proceeds with the highest efficiency toward the desired output product, producing the highest yield while requiring the least amount of money to purchase and operate. Ordinary operating expenses include energy input, energy removal, and raw material costs. Energy changes can come in the form of heating or cooling. The three major types of reactors covered in this chapter are CSTR, PFR, and packed bed reactor (PBR). Conversion reactors are used to solve the material and energy balance of single or multiple reactions. In a CSTR, one or more fluid reagents flow into a tank reactor equipped with an impeller while the reactor effluent exits the reactor. The impeller stirs the reagents to ensure proper mixing.

Simply dividing the tank's volume by the average volumetric flow rate through the tank gives the residence time or the average amount of time a discrete quantity of reagent spent inside the tank; the chemical kinetics help calculate the reaction's expected percent completion. The inlet mass flow rate must equal the mass outlet flow rate (steady-state); otherwise, the tank will overflow or go empty (transient state). In a PFR, one or more fluid reagents are pumped through a pipe or tube, and the chemical reaction proceeds as the reagents travel through the PFR. In this type of reactor, the changing reaction rate creates a gradient concerning the distance traversed; at the inlet of the PFR, the reaction rate is very high, but as the concentrations of the reagents decrease and the concentration of the product(s) increases, the reaction rate slows down. PBRs are tubular reactors filled with solid catalyst particles, most often used to catalyze gas reactions. The chemical reaction takes place on the surface of the catalyst. The advantage of using a PBR is the higher conversion per weight of catalyst than other reactors. The reaction rate is based on the solid catalyst's amount rather than the reactor's volume [1, 2].

5.2 PLUG FLOW REACTOR (PFR)

The PFR generally consists of a bank of tubes. The reactants continuously consumed while flowing through the reactor; consequently, there will be an axial variation in concentration. By contrast, considering the plug flow motion, there is no change in the

DOI: 10.1201/9781003167365-5

FIGURE 5.1   Schematic diagram of a plug flow reactor (PFR).

mass or energy in the radial direction, and axial mixing is negligible. The schematic diagram of a gas-phase reaction taking place in the PFR is shown in Figure 5.1. Several assumptions were made to the PFR system to simplify the problem, such as the plug flow motion. The steady-state assumption, the constant physical property (reasonable for some liquids but a 20% error for polymerizations; valid for gases only if there is no pressure drop, no net change in the number of moles, nor any significant temperature change), and constant tube diameter. The PFR model can simulate multiple reactions and reactions involving changing temperatures, pressures, and densities of the flow. Although these complications are not considered in this chapter, they are often relevant to industrial processes [3].

The reaction taking place is:

$$A \rightarrow B + C$$

The molar flow rate of component A,

$$\frac{dF_A}{dV} = r_A, \text{ where } r_A = -kC_A \tag{5.1}$$

The molar flow rate of component B,

$$\frac{dF_B}{dV} = r_B \tag{5.2}$$

The molar flow rate of component C,

$$\frac{dF_C}{dV} = r_C, \text{ where } r_A = -r_B = -r_C \tag{5.3}$$

As a function of change in conversion of liquid phase reaction,

$$F_A = F_{A0}(1 - X) \tag{5.4}$$

Differentiating both sides of Equation 5.4,

$$dF_A = -F_{A0} \, dX \tag{5.5}$$

Substituting into Equation 5.1,

$$\frac{dX}{dV} = \frac{-r_A}{F_{A0}}$$

(5.6)

Rate of reaction, for the first order concerning component A

$$r_A = -kC_A$$

(5.7)

Concentration as a function of conversion, x

$$C_A = \frac{F_A}{\upsilon_A}$$

(5.8)

Gas volumetric flow rate

$$\upsilon = \upsilon_0 (1 + \varepsilon X) \frac{T}{T_0} \frac{P_0}{P}$$

(5.9)

The value of $\varepsilon = y_{Ao}\delta$
    Change in stoichiometric coefficient

$$\delta = \sum \upsilon_i, \ \delta = \sum \upsilon_i = 1 + 1 - 1 = 1$$

Hence, the concentration of component A,

$$C_A = \frac{F_A}{\upsilon} = \frac{F_{A0}(1-X)}{\upsilon_0 (1+X)(T/T_0)(P_0/P)}$$

(5.10)

Substituting Equation 5.10 into Equation 5.7 at constant pressure

$$-r_A = \frac{kC_{A0}(1-X)}{(1+X)} \frac{T_0}{T}$$

(5.11)

The change in reaction temperature as a function of the reactor volume:

$$\frac{dT}{dV} = \frac{(-r_A)(-\Delta H_{rxn})}{F_{A0}\left(\sum \theta_i C_{pi} + X\Delta C_p \right)}$$

(5.12)

The heat of reaction (ΔH_{rxn}) at temperatures other than the reference temperature (25°C):

$$\Delta H_{rxn} = \Delta H_{rxn}^{\circ} + \Delta a \left(T - T_r \right) + \frac{\Delta b}{2} \left(T^2 - T_r^2 \right) + \frac{\Delta c}{3} \left(T^3 - T_r^3 \right) \qquad (5.13)$$

The second form of the energy balance,

$$\frac{dT}{dV} = \frac{\left(-r_A \right) \left(-\Delta H_{rxn} \right)}{F_A C_{pA} + F_B C_{pB} + F_C C_{pC}} \qquad (5.14)$$

5.3   PACKED BED REACTORS (PBRs)

As previously mentioned, in PBRs, the reaction rate is based on the solid catalyst's amount rather than the reactor's volume. The steady-state PBR design equation:

$$\frac{dF_A}{dW} = r_A' \qquad (5.15)$$

Figure 5.2 shows the schematic diagram of a PBR.

The Ergun equation predicts the pressure drop along the length of a packed bed given the fluid velocity, packing size, viscosity, and density of the fluid [2].

$$\frac{dP}{dz} = \frac{G}{\rho D_p} \left(\frac{1-\phi}{\phi^3} \right) \left[\frac{150 \left(1 - \phi \right)}{D_p} + 1.75G \right] \qquad (5.16)$$

where G is the superficial mass velocity, kg/(m^2 s), ρ is the fluid density, μ is the fluid viscosity, ϕ is the void fraction, and D_p is the particle diameter.

5.4   CONTINUOUS STIRRED-TANK REACTOR (CSTR)

CSTR (Figure 5.3) is usually used to handle liquid-phase reactions. The behavior of a CSTR modeled an ideal perfectly mixed reactor. The CSTR model is often used to simplify engineering calculations approached in practice as industrial-size

FIGURE 5.2   Schematic diagram of a packed bed reactor (PBR) utilized for the catalytic reactions.

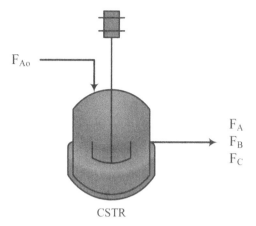

CSTR

FIGURE 5.3 A schematic diagram of the continuous stirred-tank reactor (CSTR) frequently used in liquid-phase reactions.

reactors. Assuming perfect or ideal mixing, the steady-state material balance for the CSTR reactor

$$F_{A0}X + r_A V = 0 \tag{5.17}$$

The liquid phase concentration of component A in the exit stream as a function of conversion,

$$C_A = C_{A0}(1-X) \tag{5.18}$$

The second-order reaction rate concerning component A,

$$r_A = -kC_A^2 \tag{5.19}$$

The CSTR reactor volume a function of fractional conversion,

$$V_{CSTR} = \frac{F_{A0}X}{kC_{A0}^2(1-X)^2} \tag{5.20}$$

Example 5.1: CSTR Reactor Volume

Find the reactor volume that achieves 94% conversion of ethanol for the following liquid-phase reaction in an isothermal CSTR.

$$A + B \rightarrow C + D$$

The reactants are ethanol (A) and diethylamine (B), and the products are water (C) and triethylamine (D). The inlet molar flow rate is 50 kmol/h ethanol, 50 kmol/h diethylamine, and 100 kmol/h water. The feed stream temperature is at 50°C, and the operating pressure is 3.5 bar. The reaction is in second-order concerning ethanol.

$$r_A = -kC_A^2, k = Ae^{-E/RT}$$

The activation energy (E) is 1.0×10^4 kJ/mol, the temperature (T) is 50°C; the pre-exponential factor (A) is 4,775 L/mol h; the operating pressure (P) is 3.5 atm; the molar density of ethanol is 16.6 mol/L; diethylamine is 9.178 mol/L, and water is 54.86 mol/L, the gas constant (R) is 8.314 J/mol K. Compare the volume obtained from the manual calculation with the available software packages, such as UniSim/Hysys, PRO/II, Aspen Plus, SuperPro Designer, and Aveva Process Simulation.

SOLUTION

Manual Calculations

First, calculate the average molar density of the feed stream, ρ

$$\frac{1}{\rho} = \sum \frac{x_i}{\rho_i} = \frac{0.25}{16.6} + \frac{0.25}{9.178} + \frac{0.5}{54.86} = \frac{0.0514}{1}$$

Taking the inverse value $1/(1/\rho)$, the average density,

$$\rho = \frac{1}{0.0514} = 19.45 \text{ mol/L}$$

The steady-state material balance on the CSTR reactor is:

$$F_{A0}X + (-r_A)(V) = 0$$

The concentration of component A as a function of conversion, $C_A = C_{A0}(1-X)$
The rate of consumption of the reactant A, $r_A = -kC_A^2$
The design equation for the CSTR reactor volume as a function of conversion

$$V_{CSTR} = \frac{F_{A0}X}{kC_{A0}^2(1-X)^2}$$

The initial feed concentration, C_{A0}

$$C_{A0} = \frac{50 \text{ } kmol/h}{9.901 \text{ } m^3/h} = \frac{50}{9.901} \frac{kmol}{m^3} = 5.05 \frac{mol}{L}$$

TABLE 5.1
Polymath Program for the Manual Solution,
the Case Described in Example 5.1

```
# Example 5.1 CSTR
Fain = 50,000 # mol/h
x = 0.94
CA0 = 5.05 # mol/L
A = 4,775 # L/mol h
Ea = 10,000 # J/mol
R = 8.314
T = 50 + 273.15 # K
CA = CA0 × (1 – x)
k = A × exp(-Ea/(R × T))
ra = k × CA ^ 2
f(V) = V – (Fain × x/ra)
V(min) = 100 # liter
V(max) = 500 # liter
```

The volume of the reactor (V_{CSTR}) required to achieve 94% conversion of the limiting reactant (A),

$$V_{CSTR} = \frac{F_{A0}X}{kC_{A0}^2(1-X)^2} = \frac{50,000 \; mol/h(0.94)}{115.28 \; L/mol \; h(5.05 \; mol/L)^2(1-0.94)^2} = 4440.8 \; L$$

Table 5.1 listed the code of the polymath program for the set of equations in the manual calculations. The obtained fractional conversion is approximately 0.94 for a CSTR volume of 4,433 liters.

UniSim/Hysys Simulation

The CSTR reactor in the UniSim object palette can be used for this purpose as follows:

1. Start UniSim and create a new case.
2. Add all components involved in the reaction system, click on Add Pure, and select the components: ethanol, diethylamine, triethylamine, and water. Close the selected components window.
3. Select a suitable fluid package by clicking on the Fluid Pkgs tab; in this case, the non-random two-liquid (NRTL) is the most appropriate fluid package. Close the fluid package window.
4. Under Flowsheet in the toolbar menu, the Reactions Package is to be selected. Choose the Kinetic reaction rate.
5. Add the components ethanol, diethylamine, triethylamine, and water to the reaction. Type the stoichiometric coefficients (–1) for ethanol and diethylamine (consumed) and (1) for triethylamine and water (produced).

FIGURE 5.4 UniSim required kinetic reaction data, stoichiometry, basis, and parameters for the case described in Example 5.1.

The forward order automatically defaults to the stoichiometric number 1; for this case, it is different than how we defined our reaction data. Assume no reverse reactions. Change the reaction order to 2 concerning ethanol and 0 concerning diethylamine for the forward reaction order since the reaction is irreversible; under Rev Order, type zero for all components.

6. On the Basis page. Ethanol is the base component (mol/L) and the reaction rate (mol/L h). The reaction is taking place in the liquid phase.
7. On the Parameters page. Enter the values of A = 4775 L/mol h, E = 10,000 kJ/kmol (Figure 5.4).
8. Click on Add Set to bring the set under Current Reaction Sets. Close the window and click on Enter Simulation Environment.
9. Select the graphic that is a CSTR from the object palette and add it to the process flow diagram (PFD) workspace. Connect the energy stream, feed, and product streams.
10. In UniSim, the isothermally operating reactor necessitates adding an energy stream to the process. Consequently, heat must be added or removed to keep the temperature constant.
11. Click on the Rating tab and set the volume to 4,433 liters (the same value obtained with polymath calculations for comparison).
12. Click on the feed stream and set the pressure to 354.5 kPa and T = 50°C. Feed molar flow rate is 200 kmol/h. Click on Composition and then type 0.25 for ethanol, 0.25 for diethylamine, and 0.5 for water.
13. Double click on the reactor again and click on the Reaction tab. Add the global rxn set to the available reactions if the default name was not renamed. The reaction set status fields should change to green when fully specified.
14. Click on Reactions and then on Results; the percent conversion is 93.85%. The reaction is in the liquid phase, and so most of the product will exit in stream 3. Click on the Workbook icon in the toolbar, and then Right-click anywhere on the simulation area and select Add Workbook Table; click on Select on the popup menu to form the table shown in Figure 5.5. Modify the content of the table from the Workbook/Setup. Since the reaction is in the liquid phase, no vapor is leaving with stream 2. The UniSim calculated fractional conversion (X) is 0.94, calculated as follows:

$$X = \frac{50 - 3.0}{50} = 0.94$$

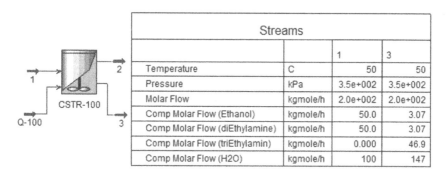

Streams			
		1	3
Temperature	C	50	50
Pressure	kPa	3.5e+002	3.5e+002
Molar Flow	kgmole/h	2.0e+002	2.0e+002
Comp Molar Flow (Ethanol)	kgmole/h	50.0	3.07
Comp Molar Flow (diEthylamine)	kgmole/h	50.0	3.07
Comp Molar Flow (triEthylamin)	kgmole/h	0.000	46.9
Comp Molar Flow (H2O)	kgmole/h	100	147

FIGURE 5.5 UniSim generated the process flow diagram and stream summary for the isothermal reaction in CSTR for the case described in Example 5.1.

PRO/II SIMULATION

The CSTR in PRO/II can only work for liquids and vapor phases; one must declare all components for just the vapor and liquid phase; otherwise, a warning message will pop up when running the system. After selecting all species involved in the system. Enter the reaction as follows:

1. Click on the Input in the toolbar, then click on Reaction Data.
2. Select Power Law from the pull-down menu for the Calculation Method.
3. Click on Enter Data and add the reactant and product stoichiometry.
4. Double click on the block diagram and then click on Unit Reaction Definition.
5. Click on kinetic data and enter the pre-exponential factor's values (A = 4,775) and Activation Energy (E = 10,000 kJ/kmol), and the base component (Ethanol).
6. Click on Reaction order and set the value to 2 for ethanol only.

Figure 5.6 shows the process flowsheet and stream summary. The fraction conversion (X) calculated as follows:

$$X = \frac{50 - 3.0}{50} = 0.94$$

The result is precisely the same as that obtained from UniSim software package.

ASPEN SIMULATION

1. Start Aspen Plus and create a new case by choosing a blank simulation. Choose the Components option in the data browser window (or click on Next) to add chemical components. Insert the chemicals involved in the reaction: Ethanol, Diethylamine, Triethylamine, and Water. Use Find to search for long compounds and then click on the Add button.
2. Click on the method and select the base property method (NRTL). Since these components are liquids, NRTL thermodynamic package is the most convenient fluid package.

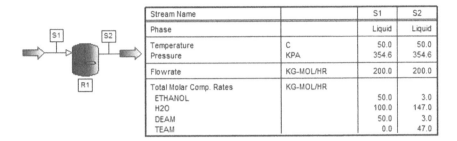

Stream Name		S1	S2
Phase		Liquid	Liquid
Temperature	C	50.0	50.0
Pressure	KPA	354.6	354.6
Flowrate	KG-MOL/HR	200.0	200.0
Total Molar Comp. Rates	KG-MOL/HR		
ETHANOL		50.0	3.0
H2O		100.0	147.0
DEAM		50.0	3.0
TEAM		0.0	47.0

FIGURE 5.6 PRO/II generated process flow diagram and stream summary for the isothermal CSTR reaction for the case described in Example 5.1.

3. Install CSTR reactor under Reactors in the Model Palate, and connect inlet and exit streams.

4. Double click on the feed stream, and specify the feed stream conditions (323.15 K, 3.5 bar) and compositions (0.25 ethanol, 0.25 Diethylamine, and 0.5 water).

5. Input the reactor specifications; double click on the reactor block. The reactor Data Browser opens. Specify an isothermal reactor (50°C) and the reactor volume to 4,433 liters; the value obtained from manual calculations.

6. Add the reactions to complete the specifications of the CSTR.

7. Choose the Reactions block in the browser window and then click on Reactions. Click New on the window that appears. A new dialog box opens; enter a reaction ID and specify the reaction as Power Law. Then click on OK.

8. The kinetic data are essential to make Aspen converge. Mainly specifying accurate units for pre-exponential factor (k) is very important. The value must be in the SI units. Kinetic data values required by Aspen Plus:

$$k = 4775 \frac{L}{mol.h} \left(\frac{1000 \ mol}{1.0 \ kmol} \right) \left(\frac{h}{3600 \ s} \right) \left(\frac{m^3}{1000 \ L} \right) = 1.326 \frac{m^3}{kmol.s}$$

Figure 5.7 shows the result of the process flowsheet and the stream summary. The fractional conversion achieved using Aspen Plus (X),

$$X = \frac{50 - 2.885}{50} = 0.942$$

SUPERPRO DESIGNER SIMULATION

Register the components involved in the reaction: ethyl alcohol, diethylamine, water, and triethylamine. Select the CSTR reactor from the Unit Procedure menu, select Continuous chemical reaction, Kinetic, and then in CSTR. Connect feed and product streams. Double click on the feed stream and specify feed flow rate and feed stream conditions (50°C, 3.5 atm). Double click on the reactor, and in the reaction page (click on Reaction tab), define the reaction

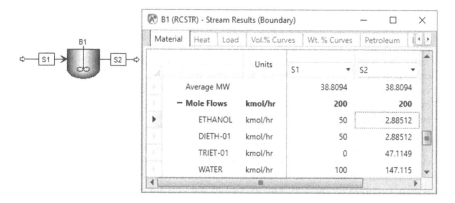

FIGURE 5.7 Aspen Plus generated process flowsheet and stream table for the CSTR reactor, the reaction item described in Example 5.1.

stoichiometry (by double clicking on Reactions). The kinetic parameters have to be determined; to do that, double click on the symbol R (view/Edit kinetic Rate) and fill in the kinetic parameters. On the operating conditions page, the exit temperature is 50°C (isothermal operation, reference component is ethyl alcohol, K1 = K2 = 1, reaction order for ethyl alcohol is 2, others are 0, frequency factor, A = 1.326, Activation energy, E = 10,000 J/mol). The reactor's volume can be set by right clicking on the Reactor icon and then selecting Equipment Data. If the ratio of work volume to vessel volume is 90%, then the vessel volume is divided by 0.9 to achieve the actual working volume of 4,433 liters (i.e., volume = 4,433/0.9 = 4,926 liters). Solving for mass and energy balances should lead to the result shown in Figure 5.8. The fractional conversion (X) achieved using SuperPro is

$$X = \frac{50 - 3.10}{50} = 0.938$$

Stream Summary Table (Example 5.1-v2)

Time Ref: h			S-101	S-102
Total Mass Flow		kmol	200.00	200.00
Temperature		°C	50.0	50.000
Pressure		bar	3.546	3.546
Total Contents		mole frac	1.00	1.000
Diethylamine			0.25	0.016
Ethyl Alcohol			0.25	0.016
Triethylamine			0.00	0.235
Water			0.50	0.734

S-101

P-1 / R-101   S-102

Kinetic Reaction

StreamSummary

FIGURE 5.8 SuperPro generated the isothermal CSTR process flowsheet and stream table properties for the case described in Example 5.1.

Aveva Process Simulation

The CSTR is a kinetic reactor model available only in the model library; the CSTR reactor is one of the five reactor models currently in the Process model library. The other reactor models are conversion, equilibrium, Gibbs, and the PFRs. So, use the kinetic rate equation to calculate how much other feed reactants are converted into product components. The reactor uses the replaceable submodel functionality within Aveva Process Simulation to allow users to switch to the reactor's set of reactions at any time.

Begin by dragging a CSTR onto the simulation environment (Canvas). Connect the Source feed stream to the reactor inlet, and then connect the reactor effluent stream to a new sink. Aveva Process Simulation can standardize your workflow with Display simulation variables directly in the workspace. We will be able to display variables to the workflow so we can see how the effluent composition as we make changes to the reactor's properties look like. Notice that from the start, everything in the simulation is square and solved. That is one of the significant parts of Aveva Process Simulation, and it can help speed up the simulation building process. Now we can begin to configure the reactor:

1. Specify the feed stream molar rate (200 kmol/h), molar composition (0.5 ethanol, 0.25 diethylamine, and 0.25 water), temperature (50°C), and pressure (3.5 bar).
2. Double click on the reactor to see the properties window. Start by changing the parameters in the calculation options tab, isothermal operation
3. Drag and drop default kinetics (rename to DME_Kinetics) into the example work area and add the equations listed in Figure 5.9.
4. Set the conversion to 0.94 and keep reactor volume unknown. Specify either diameter or length of the tubes and keep the default number of tubes.
5. Under Reactor Data, set the volume to 4.433 m^3, D to 2 m. The Operations menu, set the $T_2 = 313$ K. Select the kinetic model from Example 5.1 subfolder (DME_kinetics).

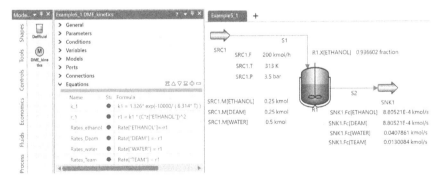

FIGURE 5.9 Aveva Process Simulation generates the CSTR process flow diagram and streams summary for the case described in Example 5.1.

CONCLUSIONS

Manual calculations obtained the conversion results of 0.94, and the conversion results obtained with the software packages – UniSim/Hysys (0.939), PRO/II (0.939), Aspen Plus (0.942), SuperPro Designer (0.938), and Aveva Process Simulation (0.937) – are almost the same.

Example 5.2: Production of Acetone

Pure acetone fed into an adiabatic PFR at a mass flow rate of 7.85 kg/h. The feed stream's inlet temperature and pressure are 760°C and 162 kPa (1.6 atm), respectively. The reaction took place in a vapor phase. Acetone (CH_3COCH_3) reacts to ketene (CH_2CO) and methane (CH_4). The following reaction takes place:

$$CH_3COCH_3\left(A\right) \rightarrow CH_2CO\left(B\right) + CH_4\left(C\right)$$

The reaction is in first-order concerning acetone, and the specific reaction rate constant is expressed by

$$k\left(s^{-1}\right) = 8.2 \times 10^{14} exp\left[\frac{-2.845 \times 10^5 \left(kJ/kmol\right)}{RT}\right]$$

Manually calculate the conversion using a reactor volume of 5 m³ and compare it with the five available software packages: UniSim, PRO/II, Aspen Plus, SuperPro, and Aveva Process Simulation.

SOLUTION

MANUAL CALCULATIONS

Considering the steady-state material balance for the PFR, the molar flow rate of ethanol,

$$\frac{dF_A}{dV} = r_A$$

The molar flow rate of ketene,

$$\frac{dF_B}{dV} = r_B$$

The molar flow rate of methane,

$$\frac{dF_C}{dV} = r_C$$

The change in conversion with a reactor volume

$$\frac{dX}{dV} = \frac{-r_A}{F_{A0}}$$

The rate of reaction (first-order concerning acetone)

$$r_A = -kC_A \text{ and } r_A = -r_B = -r_c$$

Gas volumetric flow rate, v

$$v = v_0(1 + \varepsilon X)\frac{T}{T_0}\frac{P_0}{P}, \varepsilon = y_{A0} * \delta$$

Changes in stoichiometric coefficients

$$\delta = \sum v_i = \sum v_i = 1 + 1 - 1$$

Since we have pure acetone; $y_{A0} = 1$
    Concentration as a function of conversion (X):

$$C_A = \frac{F_A}{v} = \frac{F_{A0}(1-X)}{v_0(1+X)(T/T_0)(P_0/P)}$$

Substituting the equations in the reaction rate and rearranging,

$$-r_A = \frac{kC_{A0}(1-X)}{(1+X)}\frac{T_0}{T}$$

The change in temperature as a function of reactor volume (adiabatic operation)

$$\frac{dT}{dV} = \frac{(-r_A)(-\Delta H_{rxn})}{F_{A0}\left(\sum \theta_i C_{pi} + X\Delta C_p\right)}$$

The heat of reaction at temperatures other than reference temperature (25°C),

$$\Delta H_{rxn} = \Delta H_{rxn}^0 + \Delta a(T - T_r) + \frac{\Delta b}{2}(T^2 - T_r^2) + \frac{\Delta c}{3}(T^3 - T_r^3)$$

The values of $\Delta a = 6.8$
    The values of $\frac{\Delta b}{2} = -5.75 \times 10^{-3}$

The values of $\dfrac{\Delta c}{3} = -1.27 \times 10^{-6}$

The standard heat of reaction, $\Delta H_{rxn}^0 = 80{,}770 \; (kJ/kmol)$

The change in the specific heat (ΔC_p):

$$\Delta C_p = 6.8 - 11.5 \times 10^{-3} T - 3.81 \times 10^{-6} T^2 \left(\dfrac{J}{mol.K} \right)$$

The heat capacity of ethanol (Cp_A):

$$Cp_A = 26.63 + 0.183 \; T - 45.86 \times 10^{-6} \; T^2 \left(\dfrac{J}{mol.K} \right)$$

The set of the equations are solved using the polymath program shown in Table 5.2. The second method for the change in reaction temperature with reactor volume (adiabatic operation)

$$\dfrac{dT}{dV} = \dfrac{(-r_A)(-\Delta H_{rxn})}{F_A C_{pA} + F_B C_{pB} + F_C \; C_{pC}}$$

The components' specific heat capacities (J/mol K):

$$Cp_A = 26.63 + 0.183 \; T - 45.86 \times 10^{-6} T^2$$

TABLE 5.2
Polymath Code for Example 5.2 (Method 1)

Polymath Program

```
# Example 5.2 (Method 1)
d(X)/d(V) = −ra / FA0
X(0) = 0
T(0) = 1035
ra = −k × CA0 × (1 − X) × (T0 / T) / (1 + X)
DHrxn = DHrxn0 + Da × (T − Tr) + (Db / 2) × (T ^ 2 − Tr ^ 2) + (Dc / 3) × (T ^ 3 − Tr ^ 3)
FA = FA0 × (1 − X)
FB = FA0 × X
FC = FA0 × X
DHrxn0 = 80770 # J/mol
Cpi = 26.63 + 0.183 × T − 45.86E-6 × T ^ 2
Dcp = 6.8 − 11.5E-3 × T − 3.81E-6 × T ^ 2
k = 8.2E14 × exp(-34222 / T)
CA0 = 18.85 # mol/m^3
Da = 6.8
Db = −(5.75E-3) × 2
Dc = −(1.27E-6) × 3
FA0 = 0.03754
```

The heat capacities of component B

$$Cp_B = 20.04 + 0.0945\ T - 30.95 \times 10^{-6} T^2$$

The heat capacity of component C

$$Cp_c = 13.39 + 0.077\ T - 18.71 \times 10^{-6} T^2$$

The initial concentration, C_{A0}

$$C_{A0} = \frac{P_{A0}}{RT} = \frac{162\ kPa}{8.314\ kPa.m^3/kmol.K\,(760 + 273.15\ K)(1\ kmol/1000\ mol)}$$

$$= 18.87 \frac{mol}{m^3}$$

The ethanol initial molar flow rate, F_{A0}

$$F_{A0} = \left(\frac{7.850\ kg/h}{58\ kg/kmol}\right) \frac{1000\ mol}{1\ kmol} \frac{1\ h}{3600\ s} = 0.0376 \frac{mol}{s}$$

The set of equations used in the method are solved utilizing the Polymath program. The Polymath program for methods 1 and 2 is shown in Tables 5.2 and 5.3, respectively. Results show that both ways give the exact single-pass conversion. A reactor volume of 5 m³ is required to achieve a percent conversion of 52%. Running the Polymath program for both methods (Tables 5.2 and 5.3) resulted in a single pass conversion, X = 0.52.

UNISIM/HYSYS SIMULATION

Perform UniSim process flowsheet (PFD) using the PFR from the object palette, and identify feed stream conditions, temperature 760°C, pressure 162 kPa, and mass flow rate 7.850 kg/h. The suitable fluid package is the Benedict–Webb–Rubin–Starling (BWRS) property package. Under Flowsheet, select Reaction package and choose the kinetic type of reactions. The base component is acetone; the chemical reaction occurs in the vapor phase (Figure 5.10). The results are generated when the Ready message appears. Figure 5.11 shows the PFD of the PFR and the stream summary. The fractional conversion (X) is 0.5 calculated as follows:

$$X = \frac{0.135 - 0.068}{0.135} = 0.50$$

PRO/II SIMULATION

The PFR and the CSTR in PRO/II can work for liquid and vapor phases only; consequently, all components involved in the reaction are vapor and liquid

TABLE 5.3
Polymath Program for Example 5.2 (Method 2)

Polymath Program

$d(X)/d(V) = -ra / FA0$

$X(0) = 0$

$d(T)/d(V) = -ra \times (-DHrxn) / (FA \times CpA + FB \times CpB + FC \times CpC)$

$T(0) = 1035$

$ra = -k \times CA0 \times (1 - X) \times (T0 / T) / (1 + X)$

$DHrxn = DHrxn0 + Da \times (T - Tr) + (Db / 2) \times (T ^ 2 - Tr ^ 2) + (Dc / 3) \times (T ^ 3 - Tr ^ 3)$

$FA = FA0 \times (1 - X)$

$FB = FA0 \times X$

$FC = FA0 \times X$

$DHrxn0 = 80770 \text{ \# J/mol}$

$CpA = 26.63 + 0.183 \times T - 45.86E\text{-}6 \times T ^ 2$

$CpB = 20.04 + 0.0945 \times T - 30.95E\text{-}6 \times T ^ 2$

$CpC = 13.39 + 0.077 \times T - 18.71E\text{-}6 \times T ^ 2$

$k = 8.2E14 \times exp(-34222 / T)$

$CA0 = 18.85 \text{ \# mol/m}^3$

$Da = 6.8$

$Db = -(5.75E\text{-}3) \times 2$

$Dc = -(1.27E\text{-}6) \times 3$

$FA0 = 0.03754$

phases (exclude solid). After selecting the species involved in the process, click · on Components Phases, and change the component's phase to liquid and vapor. For the Thermodynamic Data, select BWRS, which is the most suitable thermodynamic fluid package for such compounds. Build the PFR process flowsheet and specify inlet stream conditions (760°C, 162 kPa). The inlet mass flow rate of acetone is 7.85 kg/h. Under the Input menu, select Reaction Data, use the Power Law, and enter stoichiometric coefficients. Double click on the reactor in the PFD area

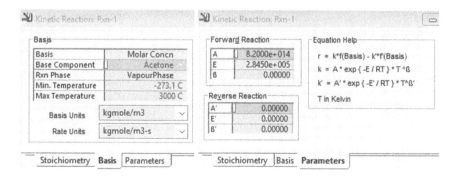

FIGURE 5.10 UniSim required data for the kinetic reaction; Basis and parameters in the reaction menu for the case presented in Example 5.2.

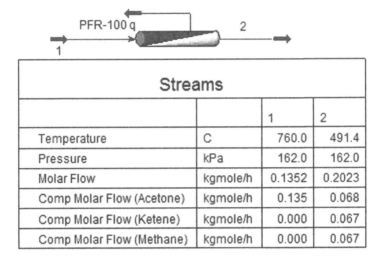

FIGURE 5.11   UniSim generated process flowsheet and streams summary for the reaction case described in Example 5.2.

and select the reaction set name. Click on Unit Reaction Definitions and specify the kinetic data as shown in Figure 5.12. Press Reactor Data, indicate 6.366 m for the reactor length and 1 m for the diameter (total volume of 5 m³). Figure 5.13 explores the process flowsheet and stream table properties. The UniSim calculated fractional conversion (X) is 0.45, calculated as follows:

$$X = \frac{0.135 - 0.074}{0.135} = 0.45$$

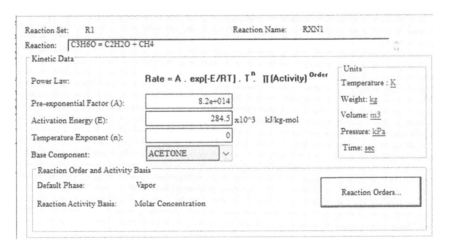

FIGURE 5.12   The PRO/II required kinetic parameters from the reaction presented in Example 5.2.

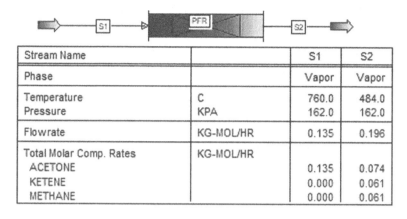

Stream Name		S1	S2
Phase		Vapor	Vapor
Temperature	C	760.0	484.0
Pressure	KPA	162.0	162.0
Flowrate	KG-MOL/HR	0.135	0.196
Total Molar Comp. Rates	KG-MOL/HR		
ACETONE		0.135	0.074
KETENE		0.000	0.061
METHANE		0.000	0.061

FIGURE 5.13 PRO/II generated process flowsheet and stream table for the adiabatic PFR reactor described in Example 5.2.

ASPEN PLUS SIMULATION

The following procedures help to build the PFR with Aspen Plus:

1. Start Aspen Plus, click on New, and create a blank simulation. Use the SI units for all reactors with kinetics.
2. Start adding chemical components. Enter the active compounds: acetone, ketene, and methane. Use the Find option to find the component's ID and then add it.
3. Next, under Properties, Specifications, select your base Property method. Since these components are liquids, use the BWRS thermodynamic package (Peng–Robinson did not work for this example).
4. Build the process flowsheet for the PFR. Select the PFR reactor from the Model palette to the simulation area, and add inlet and exit streams. Specify the feed stream. Input the reactor specifications; double click on the reactor block. The reactor Data Browser opens. Specify an adiabatic reactor and the reactor volume to 5 m³ (1 m diameter and 6.366 m length), the value obtained from hand calculations. Specify the reactions to complete the specifications of the PFR.
5. Choose the Reactions block in the browser window. Then click on Reactions, click on New on the window that appears. In the new dialog box opened window, enter a reaction ID and specify the reaction as Power Law. Then click on Ok. The kinetic data are essential to make Aspen Plus converge. Mainly establishing accurate units for the pre-exponential factor, the parameters A and k are significant. The value must be in the SI unit (Figure 5.14). Figure 5.15 shows the reactor PFD and stream summary. The fractional conversion (X) using Aspen Plus is 0.45, calculated as follows:

$$X = \frac{0.135 - 0.074}{0.135} = 0.45$$

1) ACETONE --> KETENE(MIXED) + METHANE(MIXED)   ▼

Reacting phase   **Vapor**   ▼   Rate basis   *Reac (vol)*   ▼

Power Law kinetic expression
If To is specified   Kinetic factor $=k(T/To)^n\, e^{-(E/R)[1/T-1/To]}$
If To is not specified   Kinetic factor $=kT^n\, e^{-E/RT}$

Edit Reactions

k   8.2e+14

n   0

E   284500 kJ/kmol   ▼

Solids

To   C   ▼

FIGURE 5.14   Aspen Plus required data for the Power Low kinetic reactions used for Example 5.2.

SuperPro Designer Simulation

The process is adiabatic; the Enthalpy for Reaction Heat is 1,389 kJ/kg, for Reference Component select acetone. The reaction rate constants are A = 8.2 × 1,014, E = 284,500 kJ/kmol. The user defines the size of the reactor on the Equipment Data page using the Rating Mode as 5,000 L. Number of units can be set as 100 to overcome errors. Results should appear like those in Figure 5.16. The single-pass conversion (X) is 0.59, calculated as follows:

$$X = \frac{0.135 - 0.055}{0.135} = 0.59$$

Aveva Process Simulation

Begin by dragging a PFR onto the simulation area (Canvas). Connect the feed stream to the reactor inlet, and then connect the reactor effluent stream to a new sink. Aveva Process Simulation can consolidate workflow by displaying simulation variables directly in the workspace. We will add variable displays to our workspace,

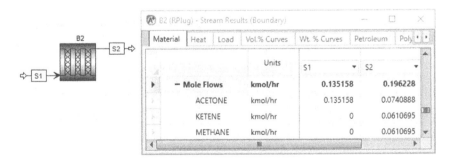

	Units	S1	S2
▶ **— Mole Flows**	kmol/hr	0.135158	0.196228
ACETONE	kmol/hr	0.135158	0.0740888
KETENE	kmol/hr	0	0.0610695
METHANE	kmol/hr	0	0.0610695

FIGURE 5.15   Aspen Plus generated a process flow diagram and stream summary for the case described in Example 5.2.

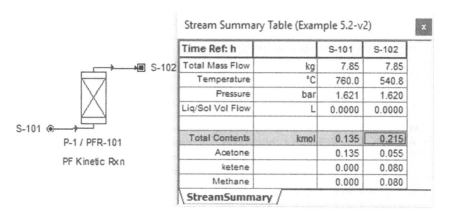

FIGURE 5.16 SuperPro generated a process flow diagram and stream table for the case described in Example 5.2.

so we can see how the effluent composition as we make changes to the reactor's properties. Notice that from the start, everything in the simulation is square and solved. That is one of the significant parts of Aveva Process Simulation, and it can help speed up the simulation building process. Now we can begin to configure the reactor.

Double click on the reactor to see the properties window. Start by changing the parameters in the calculation options tab. There are a few ways through which we can define our product stream better, the default settings for a PFR asked us to specify the number of tube lengths and diameter. Since we do not know how many tubes we will need, we need to find another variable to identify to reach our simulation. Assuming that we see each tube's length and diameter, specify the number of pipes. Under Calculation options, set operation mode to adiabatic and the volume (V) to 5 m³, Diameter (D) to 0.5 m, number of tubes (Nt) to 10. The pulldown menu near Reaction Model Type selects the kinetic model (kin_AA) from Example 5.2 subfolder. Figure 5.17

FIGURE 5.17 Aveva Process Simulation generated process flow diagram and stream summary for the case described in example 5.2.

shows the set of reaction equations, process flowsheet, and streams summary. The fractional conversion (X) generated by Aveva Process Simulation is:

$$X = \frac{0.1352 - 0.064}{0.1352} = 0.53$$

CONCLUSIONS

On drawing a comparison between the manual calculations and simulation results, it becomes clear that Aveva Process Simulation (0.53), UniSim (0.5), PRO/II (0.45), and Aspen Plus (0.45) give the closest conversion to manually calculated results. By contrast, there is a discrepancy between the results simulated with the SuperPro Designer (0.59).

Example 5.3: Packed Bed Reactors (PBRs)

The following reaction took place in styrene production from ethylbenzene's dehydrogenation in a PBR.

$$C_6H_5 - C_2H_5 \rightarrow C_6H_5 - CH = CH_2 + H_2$$

The feed consists of 217.5 mol/s ethylbenzenes, 2610 mol/s of inert steam. The reaction is isothermal at a constant temperature of 607°C and a pressure of 1.378 bar. The reaction rate is in first-order concerning ethylbenzene. The reactor's volume is 160 m³ (3 m in length) and a void fraction of 0.445. The catalyst particle diameter is 0.0047 m, and the particle density is 2,146.3 kg/m³. The first-order reaction defines the rate law,

$$r = -kP_{EB}$$

The specific reaction rate constant:

$$k\left(\frac{\text{mol}}{g_{\text{cat}}.s.kP_a}\right) = 7.491 \times 10^{-2} \exp\left[\frac{21874 \text{ cal/mol}}{\left(1.987 \frac{\text{cal}}{\text{mol.K}}\right)T}\right]$$

Determine the achieved fraction conversion with the current reactor volume. Compare the manually calculated results with simulated results from UniSim, PRO/II, Aspen Plus, SuperPro, and Aveva Process Simulation.

SOLUTION

MANUAL CALCULATIONS

The steady-state material balances performed as follows: the Ethylbenzene material balance,

$$\frac{dF_{EB}}{dw} = r'_{EB}$$

Styrene material balance,

$$\frac{dF_s}{dw} = r_s = -r'_{EB}$$

Hydrogen material balance,

$$\frac{dF_H}{dw} = r_H = -r'_{EB}$$

Calculate the pressure drop using the Ergun equation,

$$\frac{dP}{dz} = \frac{G}{\rho D_P}\left(\frac{1-\phi}{\phi^3}\right)\left[\frac{150(1-\phi)}{D_P} + 1.75G\right]$$

where G is the superficial mass velocity, kg/(m²s), ρ is the fluid density, μ is the fluid viscosity, ϕ is the void fraction, and D_P is the particle diameter.

$$G = \frac{F_{EB0}MW_{EB} + F_w MW_w}{\pi D^2/4} = \frac{(217.5)(106.17) + (2610)(18)}{\pi(9.73\ m)^2/4} = 0.941\ kg/m^2 s$$

ρ is the fluid density, the mass flow rate

$$m = \rho V = m_0 = \rho_0 V_0 \Rightarrow v = v_0\left(\frac{P_0}{P}\right)\left(\frac{T}{T_0}\right)\left(\frac{F_T}{F_{T0}}\right)$$

Fluid density (ρ) as a function of temperature and pressure,

$$\rho = \rho_0\left(\frac{v_0}{v}\right) = \rho_0\left(\frac{P}{P_0}\right)\left(\frac{T_0}{T}\right)\left(\frac{F_{T0}}{F_T}\right)$$

The initial fluid density (ρ_0),

$$\rho_0 = \frac{P\bar{M}_W}{RT}$$

The average molecular weight, $\bar{M}_W$

$$\bar{M}_W = 106.17\left(\frac{217.7}{217.5 + 2610}\right) + 18\left(\frac{2610}{217.5 + 2610}\right)\frac{g}{mol} = 24.79\ g/mol$$

Substitute the average molecular weight in the initial density calculation

$$\rho_0 = \frac{P\bar{M}_W}{RT} = \frac{1.378 \times 10^5 Pa \left(24.79 \frac{g}{mol}\right)}{\frac{8.314\ Pam^3}{mol.K}(880\ K)} = 466.89\ g/m^3$$

The initial calculated density (ρ_o):

$$\rho_o = 466.89 \frac{g}{m^3}\left(\frac{1\ kg}{1000\ g}\right) = 0.467\ kg/m^3$$

The value of the dimensionless constant (β_0):

$$\beta_0 = \frac{G(1-\phi)}{\rho_0 D_P \phi^3}\left[\frac{150(1-\phi)}{D_P}\mu + 1.75G\right]$$

Substituting values:

$$\beta_0 = \frac{0.94 \frac{kg}{m^2 s}(1-0.445)}{\left(0.467 \frac{kg}{m^3}\right)(0.0047\ m)(0.445)^3}\left[\frac{150(1-0.445)}{0.0047\ m}3\times10^{-5} + 1.75(0.94)\right]$$

Hence

$$\beta_0 = 5881\frac{kg}{m^2 s^2}$$

Calculating the value of (β):

$$\beta = \frac{\beta_0}{A_c(1-\phi)\rho_c}$$

Substituting known values:

$$\beta = \frac{5881.29\ kg/m^2 s^2}{\left(\pi(9.43)^2/4\right)m^2(1-0.445)\times 2146.3\ kg/m^3} = 0.0664\frac{Pa}{kg}$$

The pressure drop as a function of the weight of the catalyst (dP/dw):

$$\frac{dP}{dw} = -\beta\left(\frac{P_0}{P}\right)\left(\frac{T}{T_0}\right)\left(\frac{F_T}{F_{T_0}}\right)$$

The total molar flow rate (F_T):

$$F_T = F_{WO} + F_{EB} + F_S + F_H$$

The initial partial pressure of ethylbenzene ($P_{EB,0}$):

$$P_{EB,0} = \frac{F_{EB}}{F_T} \times P = \left(\frac{217.5}{2827.5}\right)1.378 \times 10^5 \, P_a \left(\frac{1 \, kPa}{1000 Pa}\right) = 10.6 \, kPa$$

The initial pressure, P_0

$$P_0 = 1.378 \times 10^5 \, Pa = 137.8 \, kPa$$

At the steady-state and adiabatic operation, no shaft works. The change in temperature with variable catalyst weight,

$$\frac{dT}{dw} = \frac{(-r_A)(-\Delta H_{rxn})}{F_A C_{pA} + F_B C_{pB} + F_C C_{pC}}$$

Expected catalyst weight (w) obtained from UniSim simulation (V = 163 m³)

$$w = V(1-\phi)\rho_c$$

Substituting known values

$$w = 163 \, m_R^3 \, \frac{(1-0.445)m_{Cat}^3}{m_R^3} \, \frac{2{,}146 \, k \, g_{cat}}{m_{Cat}^3} = 194{,}138 \, kg$$

The set of equations derived manually is solved using the polymath software. The isothermal case's single-pass conversions are 0.91 (Table 5.4) and 0.61 for the adiabatic situation (Table 5.5).

UNISIM SIMULATION

Isothermal Operation

The unit of reaction rates in Hysys is in "mol/vol/s." The change of measurement units from gram catalyst given in the reaction rate of Example 5.3 to the UniSim units (mol/vol/s) is as follows:

$$A\left(\frac{mol}{L.s.kPa}\right) = \left(\frac{mol}{g_{cat}.s.kPa}\right)(\rho_{cat})\left(\frac{1-\varepsilon}{\varepsilon}\right)$$

TABLE 5.4
Packed Bed Reactor Polymath Program (Isothermal Operation), the Calculated Conversion Is 0.91

List of Equations

Isothermal operation
d(FEB)/d(W) = −rEB
FEB(0) = 217.5
d(FS)/d(W) = rEB
FS(0) = 0
d(FH)/d(W) = rEB
FH(0) = 0
d(P)/d(W) = -B × P0 / P × FT/FT0
P(0) = 1.378E + 05
# Explicit equations as entered by the user
FW = 2,610
T = 880
rho_cat = 2,146
phi = 0.445 # void fraction
k = 7.491e-2 × exp(-21874/1.987/T)
FT = FEB + FS + FH + FW
P0 = 1.378e5
pEB = FEB/FT × P/1,000
rEB = k × pEB

FEB0 = 217.5 # mol/s
MWEB = 106.17
FW0 = 2,610 # mol/s
MWw = 18
FT0 = FEB0 + FW0
D = 9.73 # m
pi = 3.14
Ac = (pi × D ^ 2/4)
G = (FEB0 × MWEB + FW0 × MWw)/(Ac × 1,000) # kg/s
rh0 = 0.467 # kg/m³ average density
Dp = 4.7E-3 # m
mu = 3E-5 # kg/m s
B01 = (G × (1 − phi)/(rh0 × Dp × phi ^ 3))
B0 = B01 × ((150 × (1 − phi) × mu/Dp) + 1.75 × G)
B = B0/(Ac × rho_cat × (1 − phi))/1,000
X = (217.5 − FEB)/217.5
W(0) = 0
W(f) = 1.948e8

---

where ρ_{cat} is the catalyst density, ε is the void fraction

$$A = 7.491 \times 10^{-2} \left(\frac{mol}{g_{cat}.s.kPa} \right) \left(2146 \frac{g_{cat}}{L} \right) \left(\frac{1 - 0.445}{0.445} \right) \frac{m^3 cat}{m^3 gas} = 200.49 \frac{mol}{L.s.kPa}$$

The rate of reaction in mol/(L.s.kPa) is:

$$r_{EB} \left(\frac{mol}{L_{gas} s.kPa} \right) = -200.49 \left(\frac{mol}{L.s.kPa} \right) exp \left[-\frac{21874 \frac{cal}{mol}}{RT} \right]$$

The gas constant, R.,

$$R = 1.987 \frac{cal}{mol. K}$$

UNISIM/HYSYS SIMULATION
Isothermal Operation

Select a new case in UniSim, add the components involved in the process, and then select PRSV as a fluid package. Enter the simulation environment, select

TABLE 5.5
Packed Bed Reactor Polymath Program (Adiabatic Operation), the Calculated Conversion Is 0.61

List of Equations

Adiabatic operation

$d(FEB)/d(W) = -rEB$

$FEB(0) = 217.5$

$d(FS)/d(W) = rEB$

$FS(0) = 0$

$d(FH)/d(W) = rEB$

$FH(0) = 0$

$d(P)/d(W) = -B \times P0/P \times FT/FT0$

$P(0) = 1.378E + 05$

$d(T)/d(W) = rEB \times (-DHrxn)/(FEB \times CpEB +$
$FS \times CpS + FH \times CpH + FW0 \times CpW)$

$T(0) = 880$

$DHrxn = DHrxn0 + CpH + CpS - CpEB$

$DHrxn0 = 1.2E5$ # J/mol

$CpEB = -43.006 + 0.7067 \times T - 4.81E-4 \times$
$T \wedge 2 + 1.3E-7 \times T \wedge 3$

$CpS = -36.91 + 0.665 \times T - 4.85E-4 \times$
$T \wedge 2 + 1.408E-7 \times T \wedge 3$

$CpH = 28.84 + 0.00765E-2 \times (T - 273) +$
$0.3288E-5 \times (T - 273) \wedge 2$ # C

$Tr = 298$ # K

$CpW = 33.46 + (0.688e-2) \times (T - 273) +$
$0.7604e-5 \times (T - 273) \wedge 2$ # K

$Da = 28.84 - 36.91 + 43.006$

$Db = (0.00765E-2 + 0.665 - 0.7067)$

$Dc = 0.3288E-5 - 4.85E-4 + 4.81E-4$

$FW = 2,610$

$rho\_cat = 2,146$

$phi = 0.445$ # void fraction

$k = 7.491e-2 \times exp(-21874/1.987/T)$

$FT = FEB + FS + FH + FW$

$P0 = 1.378e5$

$pEB = FEB/FT \times P/1,000$

$rEB = k \times pEB$

$FEB0 = 217.5$ # mol/s

$MWEB = 106.17$

$FW0 = 2,610$ # mol/s

$MWw = 18$

$FT0 = FEB0 + FW0$

$D = 9.73$ # m

$pi = 3.14$

$Ac = (pi \times D \wedge 2 / 4)$

$G = (FEB0 \times MWEB + FW0 \times MWw)/(Ac \times 1,000)$ # kg/s

$rh0 = 0.467$ # kg/m³ average density

$Dp = 4.7E-3$ # m

$mu = 3E-5$ # kg/m s

$B01 = (G \times (1 - phi)/(rh0 \times Dp \times phi \wedge 3))$

$B0 = B01 \times ((150 \times (1 - phi) \times mu/Dp) + 1.75 \times G)$

$B = B0/(Ac \times rho\_cat \times (1 - phi))/1,000$

$X = (217.5 - FEB)/217.5$

$W(0) = 0$

$W(f) = 1.948e8$

material stream, and specify feed conditions, temperature, pressure, and flow rates. The composition page specifies the mole fraction of feed components or specifies feed flow rates by clicking on Basis (Basis: partial pressure; Base component: E-Benzene; Rxn Phase: vapor phase). Specify the stoichiometric reaction coefficients after selecting the kinetic reaction type. In the reaction basis, specify the reaction basis unit, rate units, and reaction phase. Enter reaction parameters, pre-exponential factor ($A = 200.48$), and activation energy ($E_a = 91521$). The number of segments affects the results accuracy, and as the number of segments increases, the results become more accurate. The default value in UniSim is 20. Set the volume to 163 m³ and the particle diameter to 0.0047 m. The inlet temperature and exit temperature are the same for the isothermal operation, but the software calculates the heat duty. The set operator in UniSim sets the exit temperature the same as the inlet temperature (Figure 5.18). The isothermal operation fractional conversion of ethylbenzene (X):

$$X = \frac{783 - 51}{783} = 0.935 (94\%)$$

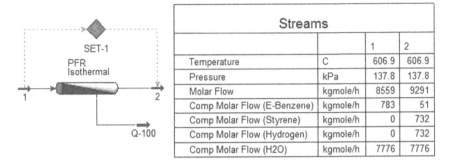

FIGURE 5.18 The isothermal ($T_{in} = t_{out}$) operation results generated by UniSim for the case described in Example 5.3.

Adiabatic Operation

In UniSim, the isothermal operation (Figure 5.19) changes into the adiabatic model. For the adiabatic process, delete the exit stream's temperature, click on the energy stream, and set the duty to zero. The fraction conversion of the adiabatic mode is less than that in isothermal because the reaction is endothermic and needs heat for the conversion to proceed further; hence the UniSim calculated conversion (X):

$$X = \frac{783 - 310.55}{783} = 0.60$$

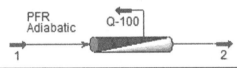

Streams			
		1	2
Temperature	C	606.9	497.2
Pressure	kPa	137.8	137.8
Molar Flow	kgmole/h	8559	9031
Comp Molar Flow (E-Benzene)	kgmole/h	783	311
Comp Molar Flow (Styrene)	kgmole/h	0	472
Comp Molar Flow (Hydrogen)	kgmole/h	0	472
Comp Molar Flow (H2O)	kgmole/h	7776	7776

FIGURE 5.19 UniSim generated adiabatic operation results ($duty, Q = 0$) for the case described in Example 5.3.

PRO/II SIMULATION
Isothermal Operation Mode

PFRs and CSTRs in PRO/II can only work for liquids and vapor phases; consequently, all components involved in the reaction are vapor and liquid phases (exclude solid). After selecting components, click on Components Phases, and change the component phases to liquid and vapor only. For the Thermodynamic Data, select Peng–Robinson, the most commonly used thermodynamic fluid package for hydrocarbons. Construct the PFR process flowsheet and specify inlet stream conditions (880 K, 137.8 kPa). Under the Input menu, select Reaction Data, use the Power Law, and enter stoichiometric coefficients. Double click on the reactor in the PFD area and select the reaction set name. Click on Unit Reaction Definitions and specify the kinetic data and the reaction's order (Figure 5.20). The reactor volume is 160 m^3, and the diameter is 3 m. Click on Reaction Data and specify the reactor length and diameter (22.65 m, 3.0 m). The temperature is at 880 K (isothermal operation). Neglecting the pressure drop, the volume of the reactor is 160 m^3. The pre-exponential factor is:

$$k = 0.20048 \frac{kmol}{m^3 s.\ Pa}$$

The final process flowsheet and stream table properties are shown in Figure 5.21. The fractional conversion for the isothermal operation:

$$x = \frac{217.5 - 18.7}{217.5} = 0.91.4\,(91\%)$$

PRO/II SIMULATION
Adiabatic Operation Mode

For adiabatic operation mode in PRO/II, the Fixed Duty radio button will be selected and set to zero (adiabatic operation mode). Figure 5.22 shows the process flowsheet and stream table property of the adiabatic operation. Note that the

Kinetic Data		
Power Law:	Rate = A . exp[-E/RT] . T^n. $\prod$ [Activity] ^{Order}	**Units** Temperature : K
Pre-exponential Factor (A):	200.48	Weight: kg
Activation Energy (E):	21.874 x10^3 cal/g-mol	Volume: m3
Temperature Exponent (n):	0	Pressure: kPa
Base Component:	EBENZENE	Time: sec
Reaction Order and Activity Basis		
Default Phase:	Vapor	
Reaction Activity Basis:	Partial Pressure	Reaction Orders...

FIGURE 5.20   PRO/II kinetic data required for the reaction described in Example 5.3.

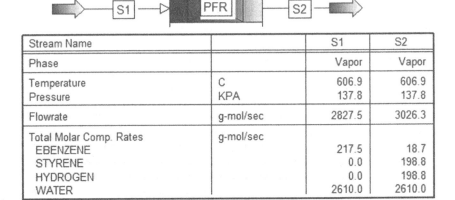

Stream Name		S1	S2
Phase		Vapor	Vapor
Temperature	C	606.9	606.9
Pressure	KPA	137.8	137.8
Flowrate	g-mol/sec	2827.5	3026.3
Total Molar Comp. Rates	g-mol/sec		
EBENZENE		217.5	18.7
STYRENE		0.0	198.8
HYDROGEN		0.0	198.8
WATER		2610.0	2610.0

FIGURE 5.21 PRO/II generated process flow diagram and stream summary for the isothermal operation case presented in Example 5.3.

product temperature decreases because the reaction is endothermic. The fractional conversion (X) of the adiabatic operation mode obtained by PRO/II is:

$$X = \frac{217.5 - 83.5}{217.5} = 0.60\,(60\%)$$

ASPEN PLUS SIMULATION

Create a new case in Aspen Plus, enter the reaction components: ethylbenzene, styrene, hydrogen, and steam (H_2O) as the inert vapor. The reactor length and diameter (22.65 m, 3.0 m). The Peng–Robinson is a suitable fluid package. The reaction is in

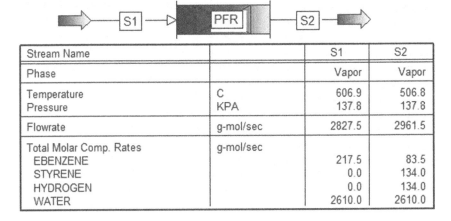

Stream Name		S1	S2
Phase		Vapor	Vapor
Temperature	C	606.9	506.8
Pressure	KPA	137.8	137.8
Flowrate	g-mol/sec	2827.5	2961.5
Total Molar Comp. Rates	g-mol/sec		
EBENZENE		217.5	83.5
STYRENE		0.0	134.0
HYDROGEN		0.0	134.0
WATER		2610.0	2610.0

FIGURE 5.22 PRO/II generates the adiabatic operation mode for the case reaction case presented in Example 5.3.

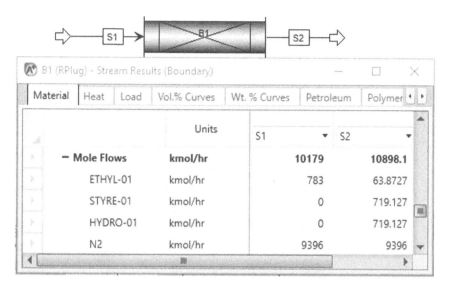

FIGURE 5.23 Aspen Plus generates the process flow diagram and stream summary for the isothermal operation case described in Example 5.3.

the vapor phase. The basis is vapor pressure. The kinetic data are: reaction phase: vapor, $k = 0.20048$, $E = 21874$ cal/mol, basis: partial pressure. Figure 5.23 shows the process flowsheet and stream table for isothermal operation. The fractional conversion for the isothermal operation method (X):

$$X = \frac{783 - 63.87}{783} = 0.918 \; (91.8\%)$$

Adiabatic Operation Mode

Figure 5.24 shows the adiabatic operation (Reactor type: Adiabatic reactor), the process flowsheet, and the stream table. Since the reaction is endothermic, the fractional conversion of adiabatic operation is less than that of isothermal operation. The fractional conversion for the adiabatic operation method (X):

$$X = \frac{783 - 328.797}{783} = 0.58 (58\%)$$

SUPERPRO DESIGNER SIMULATION

Isothermal

The heat of the reaction is in the units of kJ/kg. The residence time is 1.09 h (volume of reactor 160 m³). Use the design mode of operation in the simulation (Figure 5.25). The fractional conversion (X) of the isothermal operational method,

$$X = \frac{783 - 50.41}{783} = 0.94$$

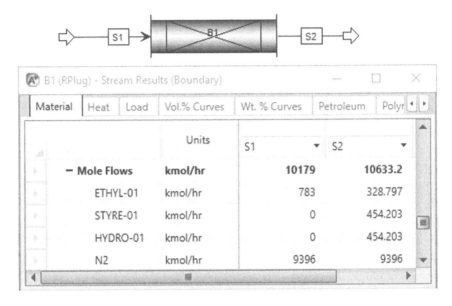

FIGURE 5.24 Aspen Plus generates the process flow diagram and stream summary for the adiabatic operation case described in Example 5.3.

Adiabatic Operation

Everything was kept constant as isothermal operation, except for adiabatic operation mode, the heat duty was changed to 0, and do not specify the exit temperature. By contrast, provide the heat of the reaction. The results are supposed to be as shown in Figure 5.26. The fractional conversion (X) of the adiabatic operation mode,

$$X = \frac{783 - 300.6}{783} = 0.61(61.6\%)$$

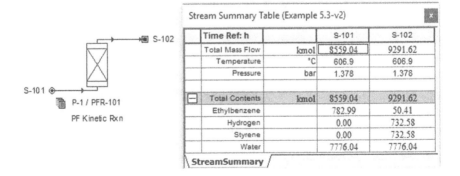

FIGURE 5.25 PFD and stream summary generated by SuperPro for the isothermal operation case presented in Example 5.3.

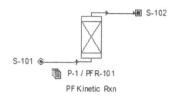

Stream Summary Table (Example 5.3-2)				x
Time Ref: h			S-101	S-102
Total Mass Flow	kmol		8559.04	9041.43
Temperature	°C		606.9	492.4
Pressure	bar		1.378	1.378
Total Contents	kmol		8559.04	9041.43
Ethylbenzene			782.99	300.60
Hydrogen			0.00	482.39
Styrene			0.00	482.39
Water			7776.04	7776.04
\ **StreamSummary** /				

FIGURE 5.26 SuperPro generated the process flow diagram and streams summary for the adiabatic operation reaction in Example 5.3.

AVEVA PROCESS SIMULATION

For this example, we will create an acetic anhydride synthesis reaction in adiabatic PFR.

1. Create a custom fluid for the system, the proper components, and the thermal methods selected.
2. Let us start by placing the Source and PFR on the canvas. Once the model is connected, the simulation is ready (squared, and solved). The Source is set up with a default fluid, and the PFR with the default reaction submodel, that is both designed to work with one another. The default reaction submodel set reaction rate of all components to zero.
3. We will eventually replace this default reaction submodel with our reaction submodel. To begin creating our reaction submodel, we need to make sure that we are in the model writing role.
4. The process of creating our reaction submodel is considered modal writing and uses the same shape to develop as any other model in the program.
5. One can copy an existing reaction submodel as the starting point for our new chemical reaction. We will use the defKinetics, this is usually the default reaction set for the PFR, and the PFR requires kinetic information to work correctly. Right click and choose Edit to start customizing the submodel.
6. Provide a user name and description in the general section of the submodel. This reaction sets an acetic anhydride formed as an easy reference for the explanation. Type whatever one thinks might be helpful to include, such as the equation's reaction rate and description.
7. Next, we need to enter our parameters. Parameters are variables that tend not to change with changing process conditions and other models.
8. What fluid the reaction side needs to use for the reaction, and a prefactor and activation energy for both the forward and reverse reactions.
9. For the fluid definition, we will use our custom fluid that was already defined. To enter the other parameters, we need to click this plus sign to add new parameters, then provide their information.
10. We will need to provide the parameters name, what type of variable orders, and a default starting condition or value; depending on the type

of variable, we may need to define the unit of measure or set the parameter to be an array by making a selection in the dim column.

11. Optionally, we can enter a description, while we can use the information in this word document for quick entry. You will notice that as we enter our parameter information.

12. Once the data entry is satisfied, the read status circle will turn green. The next area that we need to edit is the variable section; it is the area where we define the variables used in the equations.

13. Depending on the type of variable, we may need to define the unit of measure or set the variable to be an array by selecting the dim column. There are also options for making the variables and favorite, requiring the user to check the input data, make the variable and variant variable, and make the variable specified by default in each operation mode. We can also enter a description for each variable; we will again use this word document for quick data entry.

14. The last section that we need to edit is the equation section, where we will find all of our equations for the reaction we are trying to implement for this reaction set. Optionally, we can enter a description of the equation.

15. It is unnecessary to rearrange the equation in any fashion to force one variable we are trying to solve for it to be alone on one side of the equation. In this manner, Process Simulation is exceptionally flexible. Also, the equations' writing falls elementary equation writing standards, similar to what you would do in a cell when writing the formula for ease of data entry.

16. Notice that our first equations are the stoichiometry equations. The third equation is the kinetic rate equation. The new reaction submodel is ready, and the system method in the custom fluid is Peng–Robinson. We can drag and drop the Defluid on the Source to use our new reaction submodel; we can select it from the pulldown at the top of the reactors data entry window.

Isothermal Operation

In the isothermal operation (Figure 5.27), change the process to isothermal and specify the exit temperature. Set the volume to 162 m^3, the number of tubes (Nt) to 500, tube diameter (D) to 0.5 m, and exit temperature (T_2) to 880 K.

Adiabatic Operation

In adiabatic mode (Figure 5.28), everything in the isothermal method remains the same except the operation changed to adiabatic and uncheck the cell that belongs to exit temperature T_2.

Conclusions

Comparison of the simulation results and manual calculations are close to each other. SuperPro Designer required external data, such as heat of reaction.

Example 5.4: Ethane-cracking Process

Ethane reacts using thermal cracking of light hydrocarbons to produce ethylene. Ethylene is among the essential base petrochemicals that form the building block of the petrochemical industry. Thus, any improvement in the process of ethylene production may enhance the industrial economic output. An industrial plant uses

FIGURE 5.27   PFD and stream summary generated by Aveva Process Simulation for the adiabatic operation case in Example 5.3.

4,788 kgmol/h of ethane and 1,915 kmol/h steam at 3 atm pressure and 800°C to produce ethylene. The following reactions, along with the reaction rate constants, are shown in Table 5.6. Estimate the fraction conversion of the isothermal dehydrogenation of ethane to ethylene in a 20 m³. Compare the calculated results with simulation prediction using the following software packages: UniSim, PRO/II, Aspen Plus, SuperPro, and Aveva Process Simulation.

SOLUTION

MANUAL CALCULATIONS

The change in the molar flow rate of component j as a function of reactor length,

$$\frac{dF_j}{dV} = \sum \alpha_{ij} r_i$$

The molar flow (F_j) is the rate of component j, and α_{ij} is the stoichiometric coefficient of component j in reaction i. The reaction rate j,

$$r_i = k_i \prod_j C_i$$

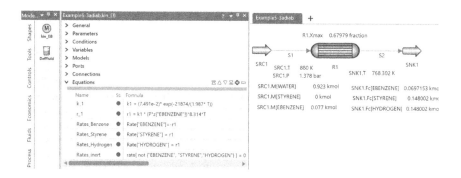

FIGURE 5.28   Aveva Process Simulation generated process flow diagram and stream summary for the adiabatic operation presented in example 5.3.

TABLE 5.6
Reaction Scheme of Ethane Steam Cracking Process [3–5]

Reaction	Reaction Rate (mol/L s)	Rate Constant
$C_2H_6 \xrightarrow{K_1} C_2H_4 + H_2$	$r_1 = k_1 C_{C_2H_6}$	$k_1 = 4.65 \times 10^{13} e^{-65210/RT}$
$2C_2H_6 \xrightarrow{K_2} C_3H_8 + CH_4$	$r_2 = k_2 C_{C_2H_6}^2$	$k_2 = 3.85 \times 10^{11} e^{-65250/RT}$
$C_3H_6 \xrightarrow{K_3} C_2H_2 + CH_4$	$r_3 = k_3 C_{C_3H_6}$	$k_3 = 9.81 \times 10^{8} e^{-36920/RT}$
$C_2H_2 + C_2H_4 \xrightarrow{K_4} C_4H_6$	$r_4 = k_4 C_{C_2H_2} C_{C_2H_4}$	$k_4 = 1.03 \times 10^{12} e^{-41260/RT}$
$C_2H_4 + C_2H_6 \xrightarrow{K_5} C_3H_6 + CH_4$	$r_5 = k_5 C_{C_2H_4} C_{C_2H_6}$	$k_5 = 7.08 \times 10^{13} e^{-60430/RT}$

Source: Data from references [3–5].

Reaction rate constants (k_i):

$$k_i = A_i exp \frac{-E_i}{RT}$$

Component molar concentrations of component j,

$$C_j = \left(\frac{F_j}{F_T} \right) \left(\frac{P}{RT} \right)$$

F_T is the total molar flow rate, P is the reactor pressure, T is the absolute temperature, and R is the gas constant. The overall molar flow rate (F_T),

$$F_T = F_{steam} + \sum_i F_i$$

The above set of equations are easily solved using the Polymath program shown in Table 5.7.

UNISIM/HYSYS SIMULATION

Start a new case in UniSim, add the active component, perform UniSim process flowsheet, and select Reaction Packages under Flowsheet. Enter the five reactions involved, all as vapor phase. All the reactions are assumed to be irreversible elementary reactions. Peng–Robinson is a suitable fluid package. The reactor's volume is 20 m³ (the value used for the manual calculations). On the kinetic reaction and the Basis page, set the Basis Units to mol/liter and Rate Units as mol/liter/second. The data's values for the pre-exponential factor, A (4.65×10^{13}), and units into account for the activation energy E (2.7284×10^5 J/mol) are on the parameters page. Figure 5.29 shows the results predict by UniSim. The fractional conversion of ethane, $X = 0.54$.

TABLE 5.7
Polymath Program Code Written for the Multiple Reactions in PFR Described in Example 5.4, the Calculated Conversion Is 0.6

Polymath Code

```
# Example 5.4 (Isothermal operation)
k1 = 4.65 × 10 ^ 13 × exp(-65,210/(R × T))
k2 = 3.85 × 10 ^ 11 × exp(-65,250/(R × T))
k3 = 9.81 × 10 ^ 8 ×* exp(-36,920/(R × T))
k4 = 1.03 × 10 ^ 12 × exp(-41,260/(R × T))
k5 = 7.08 × 10 ^ 13 × exp(-60,430/(R × T))
P = 3 # atm
R = 1.987
T = 1,073.15 # K
d(FC2H6)/d(V) = -(k1 × C2H6) - (2 × k2 × C2H6 ^ 2)
    - (k5 × C2H6 × C2H4)
FC2H6(0) = 1,330 # mol/s
FC2H60 = 1,330
d(FC4H6)/d(V) = k4 × C2H2 × C2H4
FC4H6(0) = 0
d(FC3H8)/d(V) = k2 × C2H6 ^ 2
FC3H8(0) = 0
d(FC3H6)/d(V) = -k3 × C3H6 + k5 × C2H4 × C2H6
FC3H6(0) = 0
d(FC2H2)/d(V) = k3 × C3H6 - k4 × C2H2 × C2H4
FC2H2(0) = 0
d(FC2H4)/d(V) = k1 × C2H6 - k4 × C2H4 × C2H2 - k5
    × C2H4 × C2H6

FC2H4(0) = 0
d(FH2)/d(V) = k1 × C2H6
FH2(0) = 0
d(FCH4)/d(V) = k2 × C2H6 ^ 2 + k3 × C3H6 +
    k5 × C2H6 × C2H4
FCH4(0) = 93
Fsteam = 0.4 × FC2H60
FT = FC2H6 + FC4H6 + FC3H8 + FC3H6 +
    FC2H2 + FC2H4 + FCH4 + FH2 + Fsteam
CT = P/(0.08314 × T)
C2H6 = FC2H6 × CT/FT
C3H6 = FC3H6× CT/FT
C3H8 = FC3H8 × CT/FT
C4H6 = FC4H6 × CT/FT
C2H2 = FC2H2 × CT/FT
C2H4 = FC2H4 × CT/FT
CH4 = FCH4 × CT/FT
H2 = FH2 × CT/FT
x = (FC2H60 - FC2H6)/FC2H60
V(0) = 0
V(f) = 25,000 # in liters
```

PRO/II SIMULATION

Build in the process flowsheet for the PFR as in previous examples. Select all the components involved in the process (nine species). Peng–Robinson is a suitable thermodynamic fluid package. The volume of the PFR is 20 m³. Form the Reaction Data under Input builds in the five involved reactions (Figure 5.30). Double click on the reactor and then click on Unit Reaction Definitions and fill in the reaction rate constants shown in Figure 5.31 for the first reaction. Click on Reaction Order and set the order as 1.0 for ethane and 0.0 for others. Change the unit of time into second. After all the given data are specified, the system is ready to run. Running the system provides the results that appear like those in Figure 5.32. The fractional conversion (X) is 0.55 calculated as follows:

$$X = \frac{4788 - 2164}{4788} = 0.55\,(55\%)$$

ASPEN PLUS SIMULATION

Like the previous procedure, construct the process flowsheet for a PFR in Aspen Plus. In the data browser, specify the feed stream properties. Specify inlet reactions

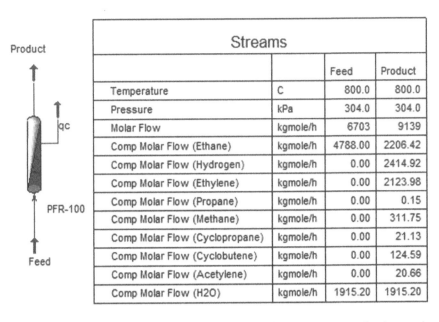

FIGURE 5.29 UniSim generated process flowsheet and stream summary for the reaction process described in Example 5.4.

stoichiometry and parameters as shown in Figure 5.33 for the first reaction. For the thermodynamic data, select the Peng–Robinson. The reactor is assumed iso-thermal. The reactor's volume is 20 m³ (6 m long, 2.06 m diameter). The process flowsheet and stream property table are both shown in Figure 5.34. The single-pass conversion is 0.52.

SuperPro Designer Simulation

Similar procedure as in previous examples, select the PFR in SuperPro. Figure 5.35 shows the kinetic parameters of the first reaction. Figure 5.36 depicts the simulation results. The fractional conversion of ethane (X):

$$X = \frac{4788 - 2184.724}{4788} = 0.544$$

RX1	C2H6 = H2 + C2H4	H...	E...	K...
RX2	2.00 C2H6 = C3H8 + CH4	H...	E...	K...
RX3	C3H6 = CH4 + C2H2	H...	E...	K...
RX4	C2H4 + C2H2 = C4H6	H...	E...	K...
RX5	C2H6 + C2H4 = CH4 + C3H6	H...	E...	K...
	Reactants = Products	H...	E...	K...

FIGURE 5.30 PRO/II entered reactions involved in the reaction menu for the chemical reactions presented in Example 5.4.

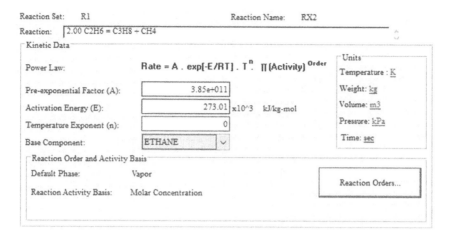

FIGURE 5.31 PRO/II required kinetic parameters of the first reaction for the case presented in Example 5.4.

AVEVA PROCESS SIMULATION

Start by placing the Source and PFR on the simulation environment (Canvas). Once the models are connected, the simulation is ready, squared, and solved, and so the Source is set up with a default fluid, and the PFR with the default reaction submodel that is both designed to work with one another in one first place on the canvas. The default reaction submodel set the reaction rate of all components to zero. To begin creating our reaction submodel, we need to make sure that we are in the model writing role.

The defKinetics is usually the default reaction set for the PFR (renamed here to Kin_E), and the PFR requires kinetic information to work correctly. Right click and choose Edit to start customizing the Kin_E submodel; in the description, set an

| Stream Name | | S1 | S2 |
Stream Description			
Phase		Vapor	Vapor
Temperature	C	800.0	800.1
Pressure	KPA	304.0	304.0
Flowrate	KG-MOL/HR	6703.0	9182.4
Total Molar Comp. Rates	KG-MOL/HR		
ETHANE		4788.0	2164.2
ETHYLENE		0.0	2171.3
PROPANE		0.0	0.2
ACETYLN		0.0	20.6
CYPR		0.0	21.3
12BD		0.0	122.9
METHANE		0.0	308.3
H2		0.0	2458.8
H2O		1915.0	1915.0

FIGURE 5.32 PRO/II generated process flowsheet and stream table for the case presented in Example 5.4.

1) C2H6 --> C2H4(MIXED) + H2(MIXED)   ▼

Reacting phase   **Vapor**   ▼   Rate basis   *Reac (vol)*   ▼

Power Law kinetic expression

If To is specified   Kinetic factor $=k(T/To)^n e^{-(E/R)[1/T-1/To]}$

If To is not specified   Kinetic factor $=kT^n e^{-E/RT}$ [Edit Reactions]

k   **4.65e+13**

n   *0*

E   **65210** **cal/mol**   ▼

To   *C*   ▼   [Solids]

[Ci] basis   *Molarity*   ▼

FIGURE 5.33 The first reaction's kinetic parameters entered in Aspen reaction's menu for the case presented in Example 5.4.

ethylene production set for easy reference for the explanation. Enter all parameters involved in the reaction. Parameters are variables that tend not to change with changing process conditions and other models. Provide the parameters name, what type of variable orders, and a default starting condition or value; depending on the type of variable, we may need to define the unit of measure or set the parameter to be an array by selecting the dim column.

The last section that we need to edit is the equation section. The equation section is where we will find all of our equations for the reaction we are trying to implement for this reaction set. Optionally, we can enter a description of the equation. It is not necessary to rearrange the equation in any fashion. In this manner, Aveva Process Simulation is exceptionally flexible. Also, the equations' writing falls elementary equation writing standards, similar to what would be done in a cell when writing the formula for ease of data entry.

	Units	S1 ▼	S2 ▼
− Mole Flows	kmol/hr	6703	9068.17
C2H6	kmol/hr	4788	2291.74
H2	kmol/hr	0	2343.82
C2H4	kmol/hr	0	2083
C3H8	kmol/hr	0	0.300247
CH4	kmol/hr	0	282.475
C3H6	kmol/hr	0	21.5099
C4H6	kmol/hr	0	108.978
C2H2	kmol/hr	0	21.355
H2O	kmol/hr	1915	1915

FIGURE 5.34 Process flowsheet and stream table property generated by Aspen Plus for the case presented in Example 5.4.

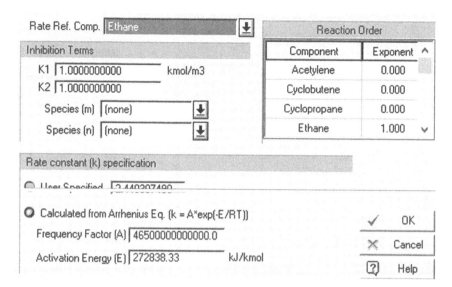

FIGURE 5.35 SuperPro required kinetic parameters of the first reaction entered in the reaction's menu for the case presented in Example 5.4.

Edit the custom fluid (DefFluid) and click on Methods, and select Peng–Robinson. For Phases, select Vapor/Liquid (VLE) from the pull-down menu. Drag and drop the fluid onto the Source icon to use our new reaction submodel; we can select it from the pulldown menu at the top of the reactors data entry window. Double click on the reactor icon and fill in the Reactor Data (number of tubes [Nt] = 1, length [L] = 1, volume [V] = 20 m³). Click on Calculation

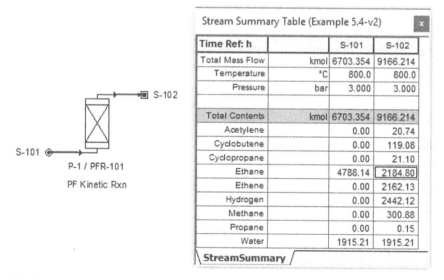

FIGURE 5.36 SuperPro generated the process flowsheet and stream table properties for the case described in Example 5.4.

FIGURE 5.37 Aveva Process Simulation generates the process flowsheet and list of equations and parameters for the case described in Example 5.4.

Options, set Operation to Isothermal, and click on Operations and set $T_2 = 800°C$. Figure 5.37 shows the list of equations of the process flowsheet of two streams conditions.

CONCLUSION

Comparison of simulation results and manual calculations are close to each other.

PROBLEMS

5.1 VOLUME OF CSTR REACTOR

The inlet molar feed rate to the CSTR reactor is 50 kmol/h ethanol, 50 kmol/h diethylamine, and 100 kmol/h water. The reaction is,

$$A(ethanol) + B(diethylamine) \xrightarrow{k} C(triethylamine) + D(water)$$

The reaction is in second-order concerning ethanol, and the rate constant, k

$$k = \left(4775 \frac{m^3}{kmol.h}\right) exp\left(\frac{-10,000 \frac{kJ}{kmol}}{RT}\right)$$

Find the reactor volume that achieves 90% conversion of ethanol,

- If the reactor is isothermal,
- If the reactor is adiabatic,

- Compare the results in a, and b.
- Use UniSim software or others to simulate the reaction process (NRTL-ideal is a suitable fluid package for liquids)

5.2 CONVERSION IN THE PFR REACTOR

A measure of 100 kgmol/h of acetone fed into an isothermal PFR. The inlet temperature and pressure of the feed stream are 750°C and 1.5 atm, respectively. The reaction is taking place in the vapor phase. Acetone (CH_3COCH_3) reacted to ketene (CH_2CO) and methane (CH_4) according to the following chemical reactions:

$$CH_3COCH_3 \rightarrow CH_2CO + CH_4$$

The reaction is in first-order concerning acetone, and the specific reaction rate can be expressed by

$$k = 8.2 \times 10^{14} \left(s^{-1}\right) exp \frac{-2.845 \times 10^5 \left(kJ/kmol\right)}{RT}$$

Calculate the reactor volume required to achieve 45% of limiting components. Use UniSim software or other available software packages to simulate the reaction process (PRSV is a proper fluid package in UniSim)

5.3 STYRENE PRODUCTION

Styrene is produced from the dehydrogenation of ethylbenzene in a PFR, according to the following reaction,

$$C_8H_{10} \rightarrow C_8H_8 + H_2$$

The feed consists of 780 kmol/h pure ethylbenzenes, the reaction is isothermal, and the inlet temperature and pressure are 600°C, and 1.5 atm, respectively. The reaction rate is in first order for ethylbenzene. Calculate ethylbenzene's percent conversion if the reaction takes place in a 150 m³, 3.0 m-long PFR. The reaction rate is:

$$r = -kP_{EB},$$

where P_{EB} is the partial pressure of ethylbenzene. The specific reaction rate constant,

$$k = 200 \frac{kmol}{m^3 s k P_a} exp\left[-90,000 \frac{kJ}{kmol} /RT\right]$$

Use UniSim software or other available software packages to simulate the reaction process (Peng–Robinson is an appropriate fluid package).

5.4 ETHYLENE PRODUCTION

Ethylene is produced by dehydrogenation of ethane in an isothermal PFR at 800°C and 3 atm. The reaction taking place is

$$C_2H_6 \xrightarrow{K_1} C_2H_4 + H_2$$

Reaction rate: $r_1 = k_1 C_{C2H6}$
The specific reaction rate constant:

$$k = 4.65 \times 10^{13} \left(s^{-1}\right) exp\left(-2.7 \times 10^5 \frac{kJ}{kmol} /RT\right)$$

Find the reactor volume that is required to achieve 65% conversion of ethane,

- If the reaction takes place isothermally,
- If the reaction takes place in an adiabatic reactor,
- Discuss results in parts a, and b.

Use UniSim software or other available software packages to simulate the reaction process (Peng–Robinson is a suitable fluid package).

5.5 CATALYTIC REACTION

Consider the multiple reactions taking place in a PBR. Table 5.8 shows the reaction parameters of the list of the chemical reactions taking place in the PBR.
The following reactions are taking place simultaneously

$$CH_4 + H_2O \Leftrightarrow 3H_2 + CO$$

$$CO + H_2O \Leftrightarrow H_2 + CO_2$$

TABLE 5.8
Reaction Rate Constant

ρ_b	1,200 kg/m^3
A_1	5.517×10^6 mol/kg.s.atm
Ea_1	1.849 × 10^8 J/mol
R	8.314 J/mol/K
P	30.0 atm
K_a	4.053 atm^{-1}
A_2	4.95 × 10^8 mol/kg/s
E_{a2}	1.163 × 10^5 J/mol
K_{eq2}	$e^{-4946+\frac{4897}{T}}$ (T in K)

The reaction rate

$$r_1 = \frac{\rho_b * A_1 * e^{(-E_{a1}/R*T)} P_{CH4}}{(1 + K_a * P_{H_2})}$$

The reaction rate of the second reaction, r_2

$$r_1 = \rho_b * A_1 * e^{-E_{a2}/R*T)} \left(y_{CH4} * y_{H2O} - \frac{y_{CO_2} * y_{H_2}}{K_{eq2}} \right)$$

where ρ_b is the bulk density, A_1 and A_2 are the pre-exponential factor or Arrhenius constant of the first and second reactions. E_{a1} and E_{a2} are the activation energies of the first and second reactions, respectively. k_a is an absorption parameter, and y_i is the mole fraction of the ith component. Feed enters the reactor at 350°C, 30 atm, and 2,110 mol/s with feed mole fractions: 0.098 CO, 0.307 H_2O, and 0.04 CO_2, 0.305 hydrogens, 0.1 methane, and 0.15 nitrogen.

Use UniSim software or other available software packages to simulate the reaction process (Peng–Robinson is a suitable fluid package).

5.6 PRODUCTION OF TRANS-BUTENE

Pure cis-2-butene reacts to produce trans-2-butene in a CSTR at 100 kgmol/h, 10 bar, and 25°C. Calculate the volume required to achieve 90% conversion. The reaction rate:

$$r_A = -kC_A, k = 0.004 \, s^{-1}$$

Use UniSim software or other available software packages to simulate the reaction process (Peng–Robinson is a suitable fluid package).

5.7 PRODUCTION OF PROPYLENE GLYCOL

Propylene glycol resulted from the reaction of propylene oxide and an excess amount of water in a 2 m^3 stirred-tank reactor (CSTR) operating isothermally at atmospheric pressure and 25°C. The combined water, ethylene oxide mixture, is fed at a rate of 500 kgmol/h consisting of 80% water, and the balance is ethylene oxide.

$$C_3H_6O + H_2O \rightarrow C_3H_8O_2$$

The reaction rate is first-order concerning propylene oxide, and the reaction rate constant is

$$k = 1.7 \times 10^{13} \exp\left(\frac{75330 \frac{kJ}{kmol}}{R} \right)$$

Use UniSim/Hysys (NRTL) to find the ethylene oxide conversion. Consider kinetic reaction type, the reaction phase is a combined liquid, and the base component is ethylene oxide.

5.8 ETHYLENE PRODUCTION

Determine the volume necessary to produce ethylene from cracking a feed stream of 100 kgmol/h of pure ethane. The reaction is irreversible and elementary. The PFR (50 m³) operates isothermally at 880°C and a pressure of 5 atm.

$$C_2H_6 \rightarrow C_2H_4 + H_2$$

The reaction rate is first-order concerning ethane. The Arrhenius constant, $A = 1.3 \times 10^{13}$ and activation energy, $E = 3.5 \times 10^5$ kJ/kmol. Using UniSim/Hysys, determine the fraction conversion of ethane to ethene. The appropriate fluid package is Peng–Robinson, which uses the kinetic reaction type, and the reaction is in the vapor phase.

5.9 KETENE PRODUCTION

Calculate the PFR reactor volume to achieve 15% conversion of acetone to ketene and methane. The reactor is operating in adiabatic conditions. The following reaction is taking place:

$$CH_3COCH_3 \rightarrow CH_2CO + CH_4$$

The reaction is irreversible and elementary; the reaction rate constant is

$$k\left(s^{-1}\right) = 8.2 \times 10^{14} \exp\left(\frac{-2.85 \times 10^5 \left(\frac{J}{mol}\right)}{RT}\right)$$

The feed rate is 138 kgmol/h acetone at 730°C and 150 kPa. Use UniSim/Hysys to find the reactor volume. The recommended fluid package is PRSV; consider kinetic reaction type and vapor phase.

5.10 PRODUCTION OF ISOBUTANE

Normal butane isomerized to is-butane in a CSTR reactor. This irreversible elementary reaction is carried out adiabatically in the liquid phase under high pressure using a liquid catalyst with a specific reaction rate of 0.5 s⁻¹. The feed enters at 330 K and 1 atm. Calculate the PFR volume necessary to process 160 kmol/h at 70% conversion of a mixture of 90 mol % n-butane and 10 mol % i-pentane, which is considered inert. Simulate the CSTR in UniSim/Hysys (NRTL); the reaction phase is overall.

REFERENCES

1. Schmidt, L. D., 1998. The Engineering of Chemical Reactions, Oxford University Press, New York, NY.
2. Fogler, H. S., 2020. Elements of Chemical Reaction Engineering, 6th edn, Prentice-Hall, NJ.
3. Levenspiel, O., 1998. Chemical Reaction Engineering, 3rd edn, John Wiley, New York, NY.
4. Sundaram, K. M. and G. F. Froment 1977. Modeling of thermal cracking kinetics. 1. Thermal cracking of ethane, propane, and their mixtures, Chemical Engineering Science, 32, 601.
5. Froment, G. F., B. O. Van de Steen, P. S. Van Damme, S. Narayanan and A. G. Goossens, 1976. Thermal cracking of ethane and ethane-propane mixtures, Journal of Industrial and Engineering Chemistry, Process Design and Development, 1, 495.

6 Distillation Column

At the end of this chapter, students should be able to:

1. Describe the purpose of the distillation column.
2. Determine the minimum reflux ratio.
3. Compute the minimum and the actual number of stages and optimum feed stage.
4. Calculate the tower diameter.
5. Verify manual calculations with simulation prediction obtained with UniSim/Hysys, PRO/II, Aspen Plus, SuperPro Designer, and Aveva Process Simulation.

6.1 INTRODUCTION

Distillation is a separation technique based on differences in boiling point, which results in a dissimilarity in vapor pressure. It is essential to know at least two factors in any distillation column design: the first is the minimum number of plates required for the separation if no product or practically no product exits the column (the condition of total reflux). The second point is the minimum reflux that to accomplish the design separation. Distillation columns consist of three main parts: condenser, column, and reboiler.

Condenser: The condenser temperature should be high enough to use cooling water in the condenser. If this requires very high pressure, then consider a refrigerated condenser.

Column: The bottom pressure must be higher than the pressure at the top of the tower so that the vapors can move from the bottom to the top. Specify a condenser and a reboiler pressure drop as zero.

Reboiler: The reboiler temperature should not be too high to enable using low-pressure steam in the reboiler because the steam cost is directly proportional to steam pressure [1].

The continuous distillation column has no moving parts, and the vapor generated in the reboiler moves the tower upward. Liquid moves down the column. Heat is added to the reboiler to evaporate the liquid mixture entering the reboiler; by contrast, heat is removed in the condenser to totally or partially condense the overhead vapor. The trays in the column represent an equilibrium achieving steady-state conditions.

6.2 SEPARATION OF BINARY COMPONENTS

Binary component distillation is the simplest form of separation. Commonly, feed to distillation columns consists of more than two components [2–4]. Figure 6.1 shows the schematic diagram of a typical one feed distillation column.

DOI: 10.1201/9781003167365-6

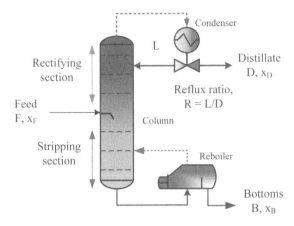

FIGURE 6.1 A schematic diagram of a typical distillation column consists of a total condenser, column, and reboiler.

6.2.1 MATERIAL BALANCE AND ENERGY BALANCE AROUND THE COLUMN

Total balance:

$$F = D + B \tag{6.1}$$

Component mass balance:

$$x_F F = x_D D + x_B B \tag{6.2}$$

Energy balance:
   The energy balance around the column

$$F h_F + Q_{reboiler} = h_D D + h_B B + Q_{condenser} \tag{6.3}$$

Figure 6.2 shows the top section (rectifier), bottom section (stripper), and feed section. Derive from these three material balances and column operating lines. R is the reflux (L) to distillate flow (D) and is called the reflux ratio.

6.2.2 MATERIAL BALANCE ON THE TOP SECTION OF THE COLUMN

Assuming constant molar overflow,

$$y_{n+1} \ V_{n+1} = L_n x_n + D x_D \tag{6.4}$$

The reflux ratio

$$R = \frac{L}{D} \tag{6.5}$$

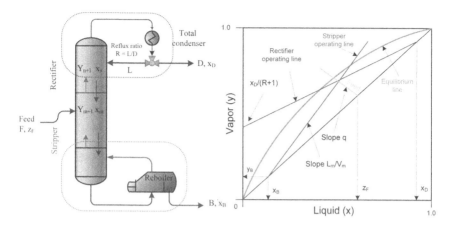

FIGURE 6.2   Schematic diagram of a distillation column material balance, feed zone, rectifying, and stripping sections.

The vapor leaving the top of the column equals recycled liquid and distillate,

$$V = L + D \tag{6.6}$$

Substituting L as a function of R,

$$V = (R + 1)D \tag{6.7}$$

The rectifying section operating line

$$y_{n+1} = \frac{R}{R+1} x_n + \frac{1}{R+1} x_D \tag{6.8}$$

6.2.3   Material Balance on the Bottom Section of the Column

Assuming constant molar flow of vapor and liquid,

$$L_m x_m = y_{m+1} V_{m+1} + B x_B \tag{6.9}$$

The bottom section operating line

$$y_{m+1} = \frac{L_m}{V_m} x_m - \frac{B}{V} x_B \tag{6.10}$$

If $x_B = x_m$, moreover, $V = L - B$, then $y_{m+1} = x_B$, and the stripping section's operating line crosses the diagonal at point (x_B, y_B); this is always true no matter the type of reboiler, as long as there is only one bottom product. Construct the lower operating line using the point (x_B, y_B), and the slope L/V is $L/(L-B)$.

6.2.4 MATERIAL BALANCES ON THE FEED TRAY

Material balance around the feed tray (Figure 6.3):

$$yV + L'x = Lx + V'y + Fz_F \tag{6.11}$$

Rearranging,

$$y(V - V') = x(L - L') + Fz_F \tag{6.12}$$

q is the fraction of liquid in the feed stream:

$$q = \frac{L - L'}{F} = 1 + \frac{V' - V}{F} \tag{6.13}$$

Rearranging Equation 6.12 by dividing both sides of the equation by F

$$y\left(\frac{V' - V}{F}\right) = x\left(\frac{L - L'}{F}\right) + z_F \tag{6.14}$$

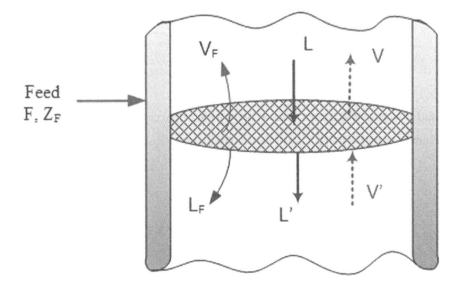

FIGURE 6.3 Schematic diagram of a distillation column portion represented the material balance around the feed tray.

Feed tray operating line, substituting q in equation 6.13

$$y = x\left(\frac{q}{q-1}\right) - z_F\left(\frac{1}{q-1}\right) \qquad (6.15)$$

where q is the number of moles of the saturated liquid on the feed tray per mole of feed. If the feed enters at its boiling point, then q = 1. If the feed enters as vapor at the dew point, then q = 0. For a cold liquid feed, q > 0; for superheated vapor, q < 1; and for the feed being part liquid and part vapor, q is the fraction of the feed that is liquid (0 < q < 1).

$$q = \frac{H_V - H_F}{H_V - H_L} = \frac{(H_V - H_L) + C_{pL}(T_B - T_F)}{H_V - H_L} = \frac{\lambda + C_{pL}(T_B - T_F)}{\lambda} \qquad (6.16)$$

where λ is the latent heat of vaporization (J/mol), C_{PL} is the heat capacity of the liquid feed (kJ/kg mol K), T_B is the feed's boiling point, and T_F is the feed stream temperature.

6.3 MULTICOMPONENT DISTILLATION

In multicomponent distillation, the equilibrium depends on all components. The complete composition of top and bottom products required trial-and-error calculations. Separation is between Light Key (LK) and Heavy Key (HK) components. Components lighter than LK are the main components in the head, and those heavier than HK are the main component in the bottom. Use the two methods in the design of multicomponent distillation: the shortcut method and the rigorous distillation method.

6.3.1 SHORTCUT DISTILLATION METHOD

The shortcut column performs Fenske–Underwood shortcut calculations for simple refluxed towers. Calculate the Fenske minimum number of trays and the Underwood minimum reflux ratio. Use a specified reflux ratio to calculate the vapor and liquid traffic rates in the enriching and stripping sections, the condenser duty and reboiler duty, the number of ideal trays, and the optimal feed location. The shortcut column estimates the column performance restricted to simple refluxed columns. This operation provides initial estimates for most simple columns. For more realistic results, use the rigorous distillation column operation.

6.3.2 MINIMUM NUMBER OF TRAYS AT TOTAL REFLUX RATIO, N_{MIN}

The Fenske equation determines the minimum number of equilibrium stages, N_{min}

$$N_{min} = \frac{\ln\left[(x_{LK}/x_{HK})_D (x_{HK}/x_{LK})_B\right]}{\log(\overline{\alpha}_{LK/HK})} \qquad (6.17)$$

where LK and HK, and $\bar{\alpha}_{LK/HK}$ is the average geometric relative volatility defined by

$$\bar{\alpha}_{LK/HK} = \left[(\alpha_{LK/HK})_F (\alpha_{LK/HK})_D (\alpha_{LK/HK})_B \right]^{1/3} \tag{6.18}$$

F is the feed, D is a distillate, and B is the bottom product [3].

6.3.3 MINIMUM REFLUX RATIO, R_{MIN}

An approximate method for calculating the minimum reflux ratio, R_{min}, is given by Underwood equations.

$$1 - q = \sum_{i=1}^{n} \frac{\alpha_i x_{F,i}}{\alpha_i - \phi} \tag{6.19}$$

where n is the number of individual components in the feed, α_i is the average geometric relative volatility of component (i) in the feed mixture to the HK component, $x_{F,i}$ is the mole fraction of component (i) in the feed, and q is the moles of saturated liquid on the feed tray per mole of feed.

$$R_{min} + 1 = \sum_{i=1}^{n} \frac{\alpha_i x_{Di}}{\alpha_i - \phi} \tag{6.20}$$

where α_i is equal to K_i/K_{HK}, where K_{HK} is the K value for the HK, x_{Di} is the distillate mole fraction. The correct value of ϕ lies between the values of (α) for the HKs and LKs; that is, solve for ϕ from Equation 6.3, then use this value to get R_{min} in Equation 6.4. Simultaneously, solving Equations 6.3 and 6.4 estimates the correct value.

6.3.4 NUMBER OF EQUILIBRIUM STAGES

Gilliland correlation calculates the number of equilibrium stages (N).
   Select a reflux ratio that is $R = (1.1 - 1.5)R_{min}$
   The number of theoretical stages, N

$$\frac{N - N_{min}}{N + 1} = 0.75 \left(1 - \left(\frac{R - R_{min}}{R + 1} \right)^{0.566} \right) \tag{6.21}$$

The Gilliland correlation is suitable for preliminary estimates but has some restrictions:

   1. The number of components must be between 2 through 11
   2. q value must be between 0.28 and 1.42

3. Pressure must be from vacuum to 600 Psig (41.83 atm)
4. α must be between 1.11 and 4.05
5. R_{min} must be between 0.53 and 9.0
6. N_{min} must be from 3 to 60

6.3.5 FEED STREAM LOCATION

The most popular method for determining the best feed locations is the Kirkbride equation given in Equation 6.22. Calculate the ratio of trays above the feed tray (N_D) and the ones below the feed tray (N_B) using the following equation:

$$\ln\left(\frac{N_D}{N_B}\right) = 0.206 \ln\left(\left(\frac{x_{HK}}{x_{LK}}\right)_F \left(\frac{x_{LK,inB}}{x_{HK,inD}}\right)^2 \left(\frac{B}{D}\right)\right) \tag{6.22}$$

where N_D is the number of equilibrium stages above the feed tray, while N_B is the number of equilibrium stages below the feed tray. B is the molar flow rate of the bottom; D is the molar flow rate of distillate. Since $N_D + N_B = N$, where N is the total number of equilibrium stages, we can easily calculate the feed plate's location.

6.3.6 COMPOSITION OF NON-KEY COMPONENTS

The Fenske equation helps estimate other components' splits if the split of one is specified.

$$\frac{x_{Di}}{x_{Bi}} = \left(\bar{\alpha}_{ij}^{N_{min}}\right)\frac{(x_{HK})_D}{(x_{HK})_B} \tag{6.23}$$

x_{Di} is the mole fraction of the ith component in the distillate; x_{Bi} is the mole fraction of ith component in the bottom. N_{min} is the minimum number of stages from the Fenske equation. The average geometric relative volatility of ith component relative to HK component ($\bar{\alpha}_{ij}$). The estimated splits are close to being correct, even though N is not at a minimum. Multiply both sides of Equation 6.23 by D/B,

$$x_{Di} \times D = d_i, \ x_{Bi} \times B = bi, \ (x_{HK})_D \times D = d_j, \ (x_{HK})_B \times B = b_j$$

Once N_{min} is known, the Fenske equation can calculate the molar flow rates d and b for non-key components

$$\frac{d_i}{b_i} = \left(\bar{\alpha}_{ij}^{N_{min}}\right)\frac{d_j}{b_j} \tag{6.24}$$

where i is the non-key component, and j is the reference component (or the heave key component, HK),

a material balance on the column.

$$f_i = b_i + d_i$$

$$d_i = \frac{f_i\left(\overline{\alpha}_{ij}^{\,N_{min}}\right)\left(d_j/b_j\right)}{1+\left(\overline{\alpha}_{ij}^{\,N_{min}}\right)\left(d_j/b_j\right)} \tag{6.25}$$

The distillate flow rate, D:

$$D = \sum d_i \tag{6.26}$$

6.4   COLUMN DIAMETER

Most of the factors that affect column operation are vapor flow conditions, either excessive or too low. Vapor flow velocity is dependent on column diameter. Weeping determined the minimum vapor flow required, while flooding determines the maximum vapor flow allowed. The incorrect ratio of vapor/liquid flow conditions can cause foaming, entrainment, flooding, and weeping. Foaming is the expansion of liquid due to the passage of vapor or gas, and high vapor flow rates cause foaming, entrainment by excessively high vapor flow rates, flooding by excessive vapor flow, and weeping dumping by low vapor flow. Excessively high liquid flow or a mismatch between the liquid flow rate and the downcomer area causes downcomer flooding. The tower inside cross-sectional area (A_T) computed at a fraction f (typically 0.75–0.85) of the vapor flooding velocity (U_f), the vapor mass flow rate (G):

$$G = f \times U_f\left(A_T - A_d\right)\rho_g \tag{6.27}$$

The A_d is the cross-sectional area of the downcomer. Rearranging the equation for tower inside cross-sectional area:

$$A_T = \frac{G}{f \times U_f \times\left(1 - A_d/A_T\right)\rho_g} \tag{6.28}$$

The ratio of the downcomer cross-sectional area to tray cross-sectional area:

$$\frac{A_d}{A_T} = 0.1,\ \text{for}\ F_{LG} \le 0.1,\ F_{LG} = \left(\frac{L}{G}\sqrt{\frac{\rho_g}{\rho_l}}\right)$$

$$\frac{A_d}{A_T} = 1 + \frac{(F_{LG} - 0.1)}{9}, \text{ for } 0.1 \le F_{LG} \le 0.1$$

$$\frac{A_d}{A_T} = 0.2, \text{ for } F_{LG} \ge 1.0$$

The tower inside diameter,

$$D_T = \sqrt{\frac{4 A_T}{\pi}} \qquad (6.29)$$

The flooding velocity (U_f) is computed based on a force balance on a suspended liquid droplet.

$$U_f = C \left(\frac{\rho_l - \rho_g}{\rho_g} \right)^{0.5} \qquad (6.30)$$

where C, the capacity parameter is given by

$$C = C_{SB} F_{ST} F_F F_{HA} \qquad (6.31)$$

Calculate the parameter C_{SB} from the correlation of Fair reference [5]. For 24 in tray spacing, Figure 6.4 shows C_{SB} as a function of the flow ratio parameter (F_{LG}):

$$F_{LG} = \left(\frac{L}{G} \sqrt{\frac{\rho_g}{\rho_l}} \right) \qquad (6.32)$$

where L and G are the mass flow rate of liquid and vapor, respectively. F_{ST} is the surface tension factor, $F_{ST} = (\sigma/20)^{0.2}$, and σ = liquid surface tension (dyne/cm). F_F is the foaming factor, $F_F = 1$ for non-foaming systems, most distillation applications. $F_F = 0.5 - 0.75$ for foaming systems, absorption with heavy oils. F_{HA} is the hole–area factor for valve and bubble cap:

For Sieve trays: $F_{HA} = 1$
For $A_h / A_a \ge 1$: $F_{HA} = 1$
For $0.06 \le A_h / A_a \le 0.1$, $F_{HA} = 5(A_h / A_a) + 0.5$

where A_h is the whole total area on a tray and A_a is the active area of the tray where bubbling area occurs: $A_a = (A_T - 2A_d)$

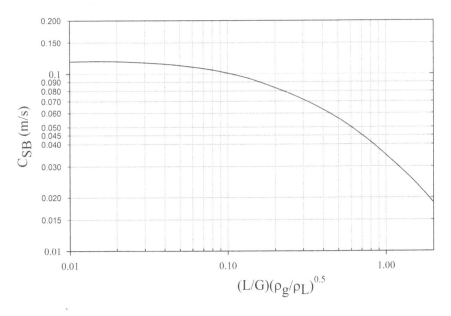

FIGURE 6.4   Capacity value, C_{SB} values for 24 in tray spacing [5].

Rectifying and stripping vapor and liquid flow rates:

1. Rectifying liquid: $L_R = RD$
2. Rectifying vapor: $V_R = (R + 1)D$
3. Stripping liquid: $L_S = L_R + qF$
4. Stripping vapor: $V_S = V_R - (1 - q)F$

where R is the reflux ratio, q is the feed index, 1 for boiling liquid, and 0 for the saturated vapor, and F is the feed flow rate.

Example 6.1: Tray Diameter Calculation

Compute the diameter of a distillation column consisting of valve trays, and the liquid phase flow rate is 2.71 kg/s, vapor phase flow rate is 0.307 kg/s, the density of the liquid is 519 kg/m³, gas-phase density is 17.54 kg/m³, liquid surface tension is 7.1 dyne/cm.

SOLUTION

First, the flooding velocity is determined:

$$U_f = C \left(\frac{\rho_L - \rho_g}{\rho_g} \right)^{0.5}$$

The capacity factor

$$C = C_{SB}F_{ST}F_{F}F_{HA}$$

C_{SB} is a function of F_{LG}:

$$F_{LG} = \left(\frac{L}{G}\sqrt{\frac{\rho_v}{\rho_l}}\right) = \left(\frac{2.71\text{ kg/s}}{0.307\text{ kg/s}}\sqrt{\frac{17.54\text{ kg/m}^3}{519\text{ kg/m}^3}}\right) = 1.62$$

From Figure 6.4, for 24 in tray spacing, $C_{SB} = 0.09$ m/s; $F_F = 1$, for distillation column where no foaming exists; $F_{HA} = 1$, for valve trays.

$$F_{ST} = (\sigma/20)^{0.2} = (7.1/20)^{0.2} = 0.813$$

The flooding velocity for 24 in tray spacing, U_f

$$U_f = C_{SB}F_{ST}F_{F}F_{HA}\left(\frac{\rho_L - \rho_g}{\rho_g}\right)^{0.5} = 0.09(0.813)(1)(1)\left(\frac{519\frac{\text{kg}}{\text{m}^3} - 17.54\frac{\text{kg}}{\text{m}^3}}{17.54\frac{\text{kg}}{\text{m}^3}}\right)^{0.5} = 0.39\frac{\text{m}}{\text{s}}$$

The ratio of downcomer cross-sectional area to tray cross-sectional area, A_d/A_T; since, $F_{LG} > 1$:

$$\frac{A_d}{A_T} = 0.2$$

Assuming the column is operating at 80% of flooding velocity, the cross-sectional column area

$$A_T = \frac{G}{f \times U_f \times (1 - A_d/A_T)\rho_g} = \frac{0.307\text{ kg/s}}{(0.8)\left(0.39\frac{\text{m}}{\text{s}}\right)(1-0.2)\left(17.54\frac{\text{kg}}{\text{m}^3}\right)} = 0.07\text{ m}^2$$

The tower inside diameter, D_T

$$D_T = \sqrt{\frac{4A_T}{\pi}} = \sqrt{\frac{4 \times 0.07}{\pi}} = 0.30\text{ m}$$

Example 6.2: Binary Distillation Column

A feed stream flow rate of 10 kmol/h of a saturated liquid consists of 40 mol% benzene (B) and 60% toluene (T). The desired distillate composition is 99.2 mol% of benzene and a bottom product composition with 98.6 mol% of toluene.

The relative volatility, benzene/toluene (α_{BT}), is 2.354. The reflux is saturated liquid, and the column has a total condenser and a partial reboiler. The feed is at 95°C and 1 bar.

1. Use the Fenske equation to determine a minimum number of trays (N_{min}).
2. Determine optimum feed stage for the minimum number of stages.
3. Find the minimum reflux ratio, R_{min} or $(L/D)_{min}$.
4. Calculate actual reflux ratio(R), if $R = 1.1\ R_{min}$.
5. Calculate the actual number of trays.
6. Define the optimum location of the feed tray.
7. Calculate the flow rates of the rectifying liquid and vapor, stripping liquid and vapor.
8. Compare the manual results with the available software packages, UniSim, PRO/II, Aspen Plus, SuperPro, and Aveva Process Simulation.

MANUAL SOLUTION

1. Using the Fenske equation to determine,

$$N_{min} = \frac{\ln\left(\left(\dfrac{x_{LK}}{x_{HK}}\right)_D \left(\dfrac{x_{HK}}{x_{LK}}\right)_B\right)}{\ln(\bar{\alpha}_{LK/HK})}$$

Substituting required values:

$$N_{min} = \frac{\ln\left(\left(\dfrac{0.992}{0.008}\right)_D \left(\dfrac{0.986}{0.014}\right)_B\right)}{\ln(2.354)} = \frac{9.075}{0.856} = 10.6$$

For a binary component distillation, the minimum number of theoretical trays can also be done, such as

$$N_{min} = \frac{\ln\left((x/1-x)_D / (x/1-x)_B\right)}{\ln \alpha_{BT}}$$

Substitute the essential parameters

$$N_{min} = \frac{\ln\left((0.992/(1-0.992))_D / (0.986/(1-0.986))_B\right)}{\ln(2.354)} = 10.59$$

2. Calculate the optimum feed stage for the minimum number of the stage, $N_{min,F}$

$$N_{min,F} = \frac{\ln\left((x_{LK}/x_{HK})_D / (x_{LK}/x_{HK})_F\right)}{\ln \alpha_{LK-HK}} = \frac{\ln\left((0.992/0.008)_D / (0.4/0.6)_F\right)}{\ln(2.4)} = 6.1$$

3. Underwood equations to find the minimum reflux ratio, R_{min} or $(L/D)_{min}$

$$1-q = \sum_{i=1}^{n} \frac{(\alpha_i x_i)_F}{\alpha_i - \phi} = 1-1 = \sum_{i=1}^{n} \frac{(\alpha_i x_i)_F}{\alpha_i - \phi}$$

Since feed is saturated liquid, $q = 1$,

$$0 = \sum \frac{\alpha_i x_{Fi}}{\alpha_i - \phi} = \frac{\alpha_B x_{B,F}}{\alpha_B - \phi} + \frac{\alpha_T x_{T,F}}{\alpha_T - \phi} = \frac{2.354 \times 0.4}{2.354 - \phi} + \frac{1 \times 0.6}{1 - \phi}$$

Solving for ϕ

$$0 = 0.94(1-\phi) + 0.6(2.354 - \phi) \Rightarrow \phi = 1.528$$

Substituting the value of $\phi = 1.528$ in the Underwood equation:

$$R_{min} + 1 = \sum \frac{(\alpha_i x_i)_D}{\alpha_i - \phi} = \frac{(\alpha_B x_B)_D}{\alpha_B - \phi} + \frac{(\alpha_T x_T)_D}{\alpha_T - \phi} = \frac{(2.354)(0.992)}{2.354 - 1.528} + \frac{(1)(0.008)}{1 - 1.528}$$

Simplify further,

$$R_{min} + 1 = 2.83 - 0.0152 = 2.82, \text{ accordingly, the minimum}$$
reflux ratio is $R_{min} = 1.82$

4. The actual reflux ratio, R,
   since $R = 1.1\, R_{min}$

$$R = 1.1 \times R_{min} = 1.1 \times 1.82 = 2.0$$

5. The actual number of trays N (or equilibrium number of trays) using Gilliland correlation to estimate the number of stages and optimum feed location:

$$\frac{N - N_{min}}{N+1} = 0.75\left(1 - \left(\frac{R - R_{min}}{R+1}\right)^{0.567}\right)$$

Substituting values of N_{min}, R_{min}, and R in Gilliland correlation:

$$\frac{N - 10.59}{N+1} = 0.75\left(1 - \left(\frac{2 - 1.82}{2+1}\right)^{0.567}\right)$$

$(N - 10.59)/N + 1 = 0.6$, solving for $N \Rightarrow N = 28$. Actual number of stages is 28 stages.

6. The optimum feed tray:

$$\frac{N_D}{N_B} = \left(\frac{x_{F,HK} \times x_{B,LK} \times B}{x_{F,LK} \times x_{D,HK} \times D} \right)^{0.206}$$

First, determine the values of B and D:

$$D = \left(\frac{z - x_B}{x_D - x_B} \right) F$$

Benzene balance:

$$D = \left(\frac{0.4 - 0.014}{0.992 - 0.014} \right) (10.0 \text{ kmol/h}) = 3.95 \text{ kmol/h}$$

The bottom molar flow rate, B

$$F = D + B \Rightarrow B = 10 - 3.95 = 6.05 \text{ kmol/h}$$

Then optimum feed tray location is calculated:

$$\frac{N_D}{N_B} = \left(\frac{0.6 \times 0.014 \times 6.05}{0.4 \times 0.008 \times 3.95} \right)^{0.206} = 1.33$$

Simplifying,

$$\frac{N_D}{N_B} = 1.33 \ \& \ N_D + N_B = 28$$

Substitute $N_D = 1.33 \times N_B$

$$N_B + 1.33 N_B = 28$$

Solving for N_B and N_D

$$N_B = 12 \text{ and } N_D = 16$$

Tray number 16 from the top is the feed tray. Using Kirkbride method:

$$\ln\left(\frac{N_D}{N_B} \right) = 0.206 \ \ln\left(\left(\frac{x_{HK}}{x_{LK}} \right)_F \left(\frac{x_{LK,inB}}{x_{LK,inD}} \right)^2 \left(\frac{B}{D} \right) \right)$$

Substituting values of mole fraction of LK and HK in feed stream distillate and bottom stream and mole flow rate of bottom and distillate in Kirkbride equation:

$$\ln\left(\frac{N_D}{N_B}\right) = 0.206 \ln\left(\left(\frac{0.6}{0.4}\right)\left(\frac{0.014}{0.008}\right)^2\left(\frac{6.05}{3.95}\right)\right) = 0.402$$

$$N_D/N_B = 1.5$$

The number of trays above the feeds tray (N_D) to those in the bottom trays (N_B) is 1.5, and $N_D + N_B = 28$. Solving for N_B and N_D, $N_B = 11$, $N_D = 17$.

The number of stages above the feed tray is 17 trays counted from the top down; this means that the feed tray is the tray number 17.

7. The flow rates of the rectifying liquid and vapor, stripping liquid and vapor.
   - Rectifying vapor, $V_R = D \times (R + 1) = 3.95\ (2 + 1) = 11.58$ kmol/h.
   - Rectifying liquid, $L_R = D \times R = 3.95 \times 2 = 7.9$ kgmol/h.
   - Stripping liquid, $L_S = L_R + F = 7.9 + 10 = 17.9$ kgmol/h.
   - Stripping vapor, $V_s = V_R = 11.58$ kgmol/h.
8. Compare manual calculation with simulation results, UniSim, PRO/II, Aspen Plus, SuperPro, and Aveva Process Simulation.

UniSim/Hysys Simulation

Perform distillation in Hysys using the shortcut column and continuous distillation column. The data obtained from the shortcut distillation column is an initial estimate of the rigorous distillation column. The specs needed are the number of stages, number of feed stage, reboiler and condenser pressure, reflux ratio, and distillate rate. In the shortcut column, there are no iterations or advanced specs. The rigorous model takes more effort to converge, but there are several ways to help the solver get to a solution. If the rigorous model does not converge with shortcut column's data, more easily achieved specs need to be used.

Shortcut Column Method

Using Hysys shortcut distillation method, select Peng-Robinson (PR) as the fluid package, connect feed and product streams, distillate and product streams, condenser and reboiler energy streams, and fully specifying feed stream. While on Design/Parameters page, the following data should be made available:

- LK in the bottom: Benzene with mole fraction 0.014
- HK in distillate: Toluene, mole fraction 0.008
- Condenser pressure: 1 atm
- Reboiler pressure: 1 atm
- External reflux ratio: 2

Specify light and heavy components, top and bottom pressure (in the design, parameter's page). Specify the reflux ratio to be 1.1 times the minimum reflux ratio, $R = 1.1\ R_{min}$. Obtain the minimum reflux ratio from Design/parameters menu ($R_{min} = 1.824$) and the column performance from the performance page by clicking

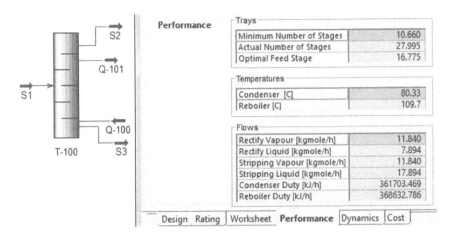

FIGURE 6.5 UniSim generated the performance of the shortcut column for the separation method presented in Example 6.2.

the Performance tab button. Figure 6.5 shows the simulated results. The minimum number of trays rounded up to 11 trays, the actual number of trays is 28, and the optimum feed tray is tray 17, counting from the top. Results are in good agreement with those obtained by hand calculations. The rectifying vapor is 11.840 kmol/h, and the rectifying liquid molar flow rate is 7.894 kgmol/h. Figure 6.5 shows the process flow diagram (PFD) page and the performance page of the shortcut distillation method generated by UniSim for the case described in Example 6.2 (double click on the column and click the Performance tab).

UniSim/Hysys Rigorous Column

After selecting the Distillation Column from the object palate or case (Main), select the Peng–Robinson (PD) as the appropriate fluid package. Connect feed and product streams, top and bottom product streams, and two energy streams, one for the condenser and one for the partial reboiler. Enter the following data:

- Fully specify the feed stream (i.e., providing feed flow rate and composition, and stream conditions, such as temperature and pressure or vapor/phase fraction and temperature or pressure)
- Feed stream inlet stage: 17
- Total number of trays: 28
- Condenser and reboiler pressure: 1 atm
- Reflux ratio: 2
- Distillate liquid rate: 3.947 kmol/h

The Monitor page is necessary to ensure that the degree of freedom is zero and uncheck the unknown variables' check buttons. Providing all the above-required data and running the system, the results should look like that shown in Figure 6.6. The Hysys predicted results show that top and bottom product streams' molar flow rates are close to the manual calculations.

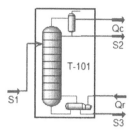

Streams		S1	S2	S3
Temperature	C	95.00	80.33	109.7
Pressure	kPa	101.3	101.3	101.3
Molar Flow	kgmole/h	10.00	3.947	6.053
Comp Mole Frac (Benzene)		0.400	0.992	0.014
Comp Mole Frac (Toluene)		0.600	0.008	0.986

FIGURE 6.6 UniSim generated a rigorous distillation process flow diagram and streams summary for the case existing in Example 6.2.

PRO/II SIMULATION

Two distillation methods are available in PRO/II; the shortcut distillation method and the continuous distillation method.

PRO/II Shortcut Distillation Method

The object-palette selects the shortcut column and places it in the PFD area (simulation environment). Select the reboiler and condenser in the popup window. Connect the feed stream and two product streams. Double click on the shortcut distillation block diagram and fill in the required data windows as shown in Figure 6.7. Click on the buttons with blue borders to provide the necessary data.

Minimum Reflux

Double click on the Minimum Reflux button and check the performed minimum reflux calculation; LK is benzene, HK is toluene, and reflux ratio is 1.1.

Specifications

The composition of benzene in the bottom stream (S3) is 0.014, and the toluene composition in the distillate is 0.008 (Figure 6.8).

Products

Set the molar flow rate of the product stream S2 to 4 kgmol/h. One needs to enable the minimum reflux calculations in the shortcut column to see the

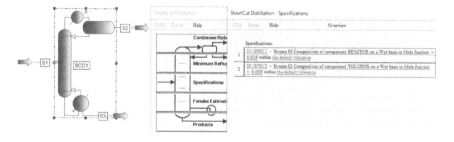

FIGURE 6.7 PRO/II shortcut distillation required data and column specifications for the case described in Example 6.2.

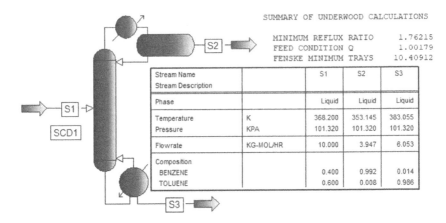

FIGURE 6.8 PRO/II generates the shortcut distillation, flow diagram and streams summary, and the Summary of Underwood Calculations for the case described in Example 6.2.

minimum number tray and feed location. Once this option is selected, then running the simulation generates a text report. The required results are under the Summary of Underwood Calculations extracted from PRO/II generated report, as shown in Figure 6.8, the figure depicts the exit streams flow rate and composition.

PRO/II Rigorous Distillation Method

The Distillation unit in the PRO/II object palette is selected and placed in the simulation area. Enter the proper components (Benzene and Toluene) and suitable thermodynamic package (PR). Specify the number of theoretical trays (28) and include a condenser and reboiler.

Connections

Press the Streams button to begin connecting streams. Connect the feed stream to the left side of the distillation column. Define the inlet stream name or keep the stream default name S1.

Condenser

The condenser has several port connections. The top port usually is a vapor product, while the bottom red port is a liquid product, and the bottom green port is water. For this case, we will be connecting the fluid product port from the condenser. Several factors need to define the distillation column in PRO/II; double click on the column to open the specification window. The following are required to be defined indicated by the red border:

- Pressure profile
- Feeds and products
- Performance specifications

The Condenser and Reboiler are defined by defaults, indicated by the blue border. Conversion Data, Thermodynamic Systems, and Initial Estimates are all user optional, they can be defined or redefined, but it is not necessary, indicated by the green border. Heaters, Coolers, and Pump rounds are optional additions to the distillation column. The column can be renamed by typing the desired name into the unit: field, currently T1. Clicking on Pressure Profile will open this window:

Pressure Profile

The column's pressure can be defined either by the overall or by individual trays. To change between these two systems, select the appropriate radio button. Generally, it requires the top tray pressure and the overall pressure drop, either per tray or across the whole column. Individual trays allow the user to specify the pressure on individual trays. The minimal trays to define are the top and bottom trays. Once the column's pressure profile is determined, press OK to save the changes and return to the Column window. If the pressure profile is appropriately defined, the box will now have a blue border.

Feeds and Products

In this window, the user defines the feed tray. The user can also set whether both vapor and liquid from the feed stream will go to the feed tray or the feed flashes when entering the column. The feed stream's liquid portion will go to the feed tray for flashed feed, while the vapor will go to the tray above the feed tray. Besides, an estimate of one of the two product stream flow rates is needed. Both stream flows have red borders, but once one is defined, the other becomes set. Setting both streams can lead to over specifications of the system; fixed the feed tray and the product flow rate and then click OK to save the changes and return to the column window. If the feed tray and product streams have correctly defined, that box will now have a blue border.

Performance Specifications

If column performance specifications are not defined, PRO/II will prompt the user to establish the specification's product flow rate. If the user does this, the flow rate will not be an estimate but a factor that PRO/II will try to fit. Clicking "Performance Specifications" will open a window.

The distillation column has a wide variety of possible parameters to fit. This example will use one parameter from the tower and one parameter from a product stream. Start by clicking on the parameter in the COL1SPEC1 line. Select either stream or column under the drop-down menu; only the defined column will appear under the unit name, even if there are multiple distillation units in the process flow diagram. Then click on the parameter to choose the column parameter to fit and open this window. All possible parameters appear in the window; each has multiple sub-choices and can be for individual trays. This example will be using a fixed reflux ratio as a fitting parameter. After selecting the parameter, click OK to save the changes. The new parameter will appear highlighted in blue and the units in green. Repeating this process for line COL1SPEC2, but instead selecting stream and the desired stream name in the first parameter window that appears will open this window. This example selects one of the product streams' compositions as the parameter to fit (Figure 6.9).

Now the user can simulate the process by pressing the Run button in the toolbar. The system will become blue colored if calculations are successfully converged (Figure 6.10).

☑ Add Specifications and Variables

		Specifications:	Active:
Cut	1	COL1SPEC1 - Stream S2 Flowrate of All Components on a Wet basis in kg-mol/hr = 4 within the default tolerance	☑
Insert			
Reset	2	COL1SPEC2 - Column T1 Reflux Ratio on a Mole basis = 1.938 within the default tolerance	☑

FIGURE 6.9   PRO/II performance specifications page for the case defined in Example 6.2.

ASPEN PLUS SIMULATION

In Aspen Plus, several distillation models can be used depending on the process method and properties. There are two types of common interest for distillation in Aspen:

- The Shortcut distillation column using the Winn-Underwood-Gilliland (DSTWU) provides basic calculation and simple input; it is the multi-component shortcut distillation method.
- RadFrac is the rigorous simulation of a single column.

In this example, we will use the DSTWU column, which performs a base calculation and provides simple input. From the DSTWU column, we will observe the molar and mass flows of input and output streams and the reflux ratio used in the rigorous distillation column (RadFrac).

1. First, create a blank simulation. Add benzene and toluene components in the individual component ID cells.
2. Next, we need to specify the method and parameters for our simulation. Switch to the Properties tab in the navigation pane. Select the methods

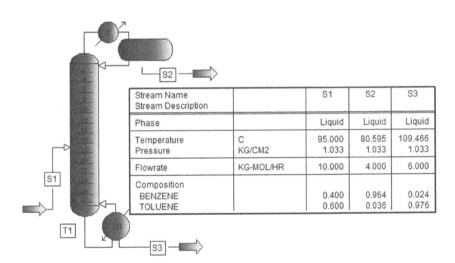

Stream Name Stream Description		S1	S2	S3
Phase		Liquid	Liquid	Liquid
Temperature Pressure	C KG/CM2	95.000 1.033	80.595 1.033	109.466 1.033
Flowrate	KG-MOL/HR	10.000	4.000	6.000
Composition BENZENE TOLUENE		0.400 0.600	0.964 0.036	0.024 0.976

FIGURE 6.10   Process flowsheet and stream properties table generated by PRO/II for the separation process defined in Example 6.2.

folder and specify the method name to NRTL-RK; in the binary interactions folder within the Parameters folder, select NRTL-1 (Peng–Robinson could also be used as both components are nonpolar). The data displayed are the interaction parameters between the components of benzene and toluene. Now it is time to specify our feed stream and the distillation column.

3. Click the Simulation tab (bottom left corner) to switch to the primary flowsheet.

4. In the model palette, click to the Columns tab, select the DSTWU column and drag it onto the primary flowsheet. Then, click the material icon in the model palette and create inlet and outlet streams for the distillation column. Red and blue arrows appear around the column. A red arrow signifies a required stream for the design specification; blue arrows signify an optional stream. Connect the feed and product streams where Aspen indicates that they are required and name the newly added streams.

5. Switch to the primary flowsheet (click Simulation). Then, click on the feed stream to access its specifications menu. Specify the temperature, pressure, and total molar flow. Then specify the composition to mole fraction.

6. Now, in the Block folder of the navigation pane, select the B1 folder in the column specifications box, specify the Reflux ratio (R) to –1.1 (the negative sign means that you want to use the 1.1 as a multiplier to the minimum reflux ratio, $R = 1.1\ R_{min}$), and in the pressure box, specify the condenser and reboiler pressure values (1 atm). Then, in the Key component recovered box, identify the Light key (benzene) with a recovery value (0.992). Specify the Heavy key (Toluene) with a recovery value (0.006).

7. Specify the feed stream conditions. The feed stream flow rate is 10.0 kmol/h, saturated liquid at 1 atm, and consists of 40 mol% benzene (B) and 60% toluene (T).

8. Now, open up the Data Browser window. Notice that we are only required to update our data input in the Blocks tab. For this simulation, we will be inputting:
   - The reflux ratio
   - The key component recoveries
   - The tower pressures

For our purposes, we will assume that the tower has no pressure drop throughout it. However, we will set the condenser and reboiler pressures to 1 atm to aid in our separation process. We will start with an input reflux ratio of $1.1 \times R_{min}$. The component recovery values that are input are equal to each component in the distillate divided by each component's amount in the feed. For this reason, a recovery of 99% for benzene and 0.5% for the toluene is reasonable if our distillation tower is operating well; the results should look like as shown in Figure 6.11. Using Aspen shortcut distillation column, the minimum reflux ratio is 3.499, the actual reflux ratio is 3.849, the minimum number of stages is 12, the number of actual stages is 28, and the feed stage is stage 15 (Figure 6.11).

Aspen Plus Rigorous Distillation Column

To start a new project, click on the "New" button that appears when Aspen Plus starts; then click Create. Alternatively, use the option, go to file > new. This example will show how to use the RadFrac, the most commonly used tower.

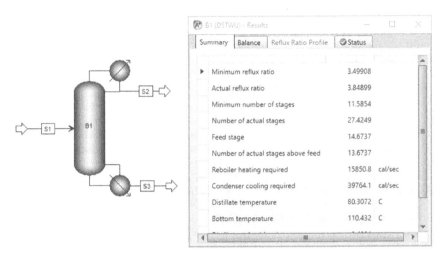

FIGURE 6.11 Aspen Plus generated results of the shortcut distillation for the case described in Example 6.2.

This example desires to separate benzene and toluene, with a 99.2% mass purity of benzene in our 4.0 kmol/h product stream.

1. The first step of creating a simulation is defining the components and the properties of the simulation. Start by telling Aspen Plus which compounds will be present in our system. Having benzene and toluene in our system are common chemicals; type them directly into the Component ID box. Type benzene and then press enter. Type toluene and then press enter.
2. The folder named Methods on the left side of the screen (in the red box) has a small red half-filled circle icon on the folder icon (designates the section that needs more input data).
3. Click on the little white arrow to the left of this folder and then click on Specifications. Here is where we will have to tell Aspen Plus which equations to use to estimate each of our components' properties. These components are nonpolar so that we will use the Peng–Robinson (PENG-ROB) equation of state (EOS). In Aspen Plus, click on the drop-down menu for Method Name and find PENG-ROB. Aspen filled in the rest of the sheet, and now the Methods folder on the left side of the screen has a blue checkmark on it, which means we have provided enough information.
4. The Properties environment is the background information that we just put in that Aspen Plus needs to calculate all the things we will want it to figure later. Now we are ready to start our simulation.
5. Click on "Simulation" right below Properties to go to the Simulation environment. We now have a blank sheet to make a flowsheet of our process. To draw our process, we will need to use blocks and streams to tell Aspen Plus what components are going where and what is happening to them between the entrance and the exit of our process.

6. The picture below is the Model Palette, and it has everything we will need to draw the flowsheet. Click on the Columns tab, and Aspen Plus gives various ways to model a distillation column and other separation unit operations. We will be using RadFrac, which is a rigorous method of solving a distillation column.

7. Click on the RadFrac icon and then click anywhere in the white space above to put a distillation column in the flowsheet. Then click on the material icon to put in our streams. Notice that the distillation column now has red and blue arrows going in and out. The red arrow on the left is where the feed stream connects. The arrows coming out to the right are product and water streams. Hover over them with the mouse; it will describe each arrow. When connecting the streams, each stream must be strictly connected where Aspen wants it to be. Otherwise, it will not be noticeable. The red arrows indicate required streams, and the blue arrows indicate optional streams.

8. Click on the red arrow going into the distillation column, and then click again somewhere to the left of the picture to generate the feed stream. The circle below and to the right of the column is the reboiler, where we will get our bottoms product, toluene, out. Click on this red arrow and again somewhere to the picture's right to get a product stream. The circle above and to the right of the column is a condenser. We can get either a vapor or a liquid distillate product, but we will just use a liquid distillate. Click on the lower red arrow on the condenser and click again somewhere to the column's right to generate the benzene product stream.

9. Our flowsheet is complete, but we still have to tell Aspen Plus the background information of each of the blocks and streams. To do this, we will go back to the folders on the left side of the screen. Click on the little white arrow next to Streams and then click on 1, which is our feed stream. We need two of either temperature, pressure, and vapor fraction. We will use temperature and pressure and set them to 369 K (96°C) and 1 atm, respectively. To change the units on the pressure, use the drop-down menu and find atm. Our total flow rate will be 10 kmol/h. To the right, there is a box for composition. To specify this, we want Mole-Flow in kmol/h, with 0.4 mole fraction benzene and the other being toluene.

10. Click on the white arrow next to Blocks and then click on B1, the distillation column's default name. Notice that several tabs on this page need information. We will start with Configuration and work our way through the rest of them. We want 25 stages with a total condenser. We want a distillate rate of 4.0 kmol/h, and we will start with a Reflux ratio on a mole basis of 2.0. Manipulate the reflux ratio to get the desired purity.

11. Now we will specify the streams, which is all of telling Aspen Plus where our feed stream connects to our tower. Let us put that at Stage 15 and leave the Above-Stage Convention. The product streams can be left where they are, and we do not have any pseudo streams. The last thing to specify for this tower is the pressure. To do this, we will set the stage 1/condenser pressure to 5 psi and leave the optional boxes blank to make this simulation simpler.

12. When we specified a total condenser, it became fully defined, and the reboiler established from the Kettle type we put in. Our simulation is

ready to run and calculate, and the run button is present under the home tab. There is always a run button that is just the triangle on the very top ribbon.

13. Click on the Run button and wait for a few seconds for Aspen Plus to calculate the simulation. If everything worked just fine, there would be no warning or error message to pop up, and the bottom left corner will read "Results Available." Now we will look at our results. There are several ways to do this, but the easiest way is to go back to the Main Flowsheet, click on a stream to select it, and then right click on it to choose Results.

14. To summarize, the following data obtained from the shortcut column (DSTWU) provides the data required for the rigorous column.
    • Fully specify the feed stream (feed stream flow rate, composition, and two of the three variables; temperature, pressure, and vapor fraction)
    • Number of stages: 27
    • Condenser type: Total
    • Distillate rate: 4 kmol/h
    • Reflux rate: 3.85
    • Location of feed stream: 15
    • Condenser pressure: 1 atm

Providing all required data and running the system, the stream's results summary should look like in Figure 6.12.

SuperPro Simulation

From the toolbar, select Unit Procedures, then distillation, and then Continuous (Shortcut). The shortcut distillation column is selected and added to the process flowsheet. Connect the feed stream and two product streams to the unit. Fully specify the feed stream, and in the component separation page (double click on the column icon and click on the second tap from top), near "set by the user," add relative volatiles of feed stream components (Benzene: 2.345 and Toluene: 1) and percentage in the distillate (Benzene: 99.2% and Toluene: 0.62%). The number of theoretical stages obtained with SuperPro Designer is 17, and the minimum reflux ratio is 2.245 (Figure 6.13).

The relative volatilities should be calculated using the Antoine equation or took from other software. Distillate stream composition is also needed, so pre-material

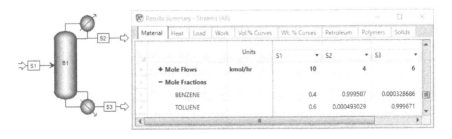

FIGURE 6.12 Aspen Plus generated the rigorous distillation column flowsheet and stream summary for the case presented in Example 6.2.

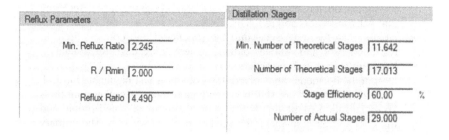

FIGURE 6.13   Shortcut distillation operating with SuperPro, the calculated number of theoretical stages is 17 generated for the case presented in Example 6.2.

balance calculations are required. The reflux ratio to minimum reflux ratio (R/R_{min}), the column pressure, and the feed quality (q) is required. Vapor linear velocity and stage efficiency are kept as SuperPro default values. Figure 6.14 shows the system process flow diagram and stream summary.

AVEVA PROCESS SIMULATION

The basic configuration of the distillation column in Aveva Process Simulation is as follows:

1. Start by dragging a distillation column from the Process model library onto the canvas (simulation area). Resize the icon, and connect the feed stream to the tower. The column model icon is in the process model library.
2. Double click on the column and change the number of stages to 28. Specify the feed stream to be at stage 17, and make the pressure at the top of the column 10 bar.
3. There are ways to determine these parameters in Process Simulation, but let us assume we have been given them for the primary design case. Now set both the condenser and Reboiler modes to internal. Furthermore, notice that two new unit operations have now appeared in

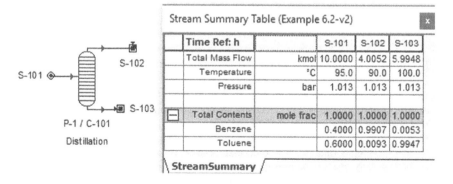

FIGURE 6.14   SuperPro generated process flow diagram and streams summary for the case presented in Example 6.2.

the simulation. It is always a good idea to start designing base cases using internal condensers and reboilers.

4. Finally, drag two new sinks into the simulation and connect them to the condenser and reboiler outlets. When doing this, the simulation becomes square and solved.

5. Despite being square and solved, the column is still highlighted in red. That is because we are still in configuration mode. Configure mode is an initialization mode that helps us build up an estimated profile and the column before invoking the solver's rigorous solution. The primary design case assumes no liquid or vapor contact is not a particularly good column model.

6. Let us refine our design by changing the column's setup mode to solve it by changing the liquid-vapor contact fraction to one.

7. This case is a standard procedure when designing columns in Aveva Process Simulation and will produce a detailed tray by tray calculation for the column, and just like that, the column is configured.

8. Let us drag variable references onto the sheet again to see the Benzene composition in the top product and the column reflux ratio.

9. Right click on the distillation column and select Show specification for both variables. We can now specify the Benzene concentration and unspecify the column's reflux ratio to have Process Simulation solved for the optimal reflux ratio.

10. If we set the benzene concentration to 99.9 percent more, then Process Simulation shows that we will need a reflux ratio of about 2.0.

11. Even though the Benzene concentration is an external variable from the column, Process Simulation still knows how all of our variables interact and can quickly solve all these variables in the simulation. Nevertheless, for now, the primary design case is complete.

12. To summarize, achieving an initial rigorous solution, perform the following steps:
    - Set the number of stages
    - Set feed tray location
    - Set Condenser type (typically internal)
    - Set the Reboiler type (typically internal)
    - Set Setup to Solve
    - Set Contact to 1 (or vapor and liquid will bypass each other)

Figure 6.15 summarizes the results.

Example 6.3: Multicomponent Distillation Column

Two thousand kilomoles per hour of saturated liquid is feed to a distillation column at atmospheric pressure. The feed stream composition mole fractions are 0.056 n-propane, 0.321 n-butane, 0.482 n-pentane, and 0.141 n-hexane. A distillation column separated the mixture. The column has a total condenser and a partial reboiler; the reflux ratio is 3.5, a fractional recovery of 99.4% of n-butane is desired in the distillate and 99.7% of n-pentane in the bottom stream. Calculate the flow rate of distillate and bottom products using a manual calculation. Verify the manual calculations with Hysys/UniSim, PRO/II, Aspen Plus, SuperPro Designer, and Aveva Process Simulation.

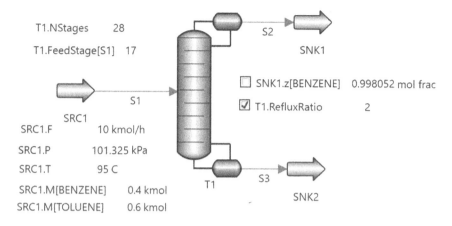

T1.NStages   28

T1.FeedStage[S1]   17

SRC1

SRC1.F   10 kmol/h

SRC1.P   101.325 kPa

SRC1.T   95 C

SRC1.M[BENZENE]   0.4 kmol

SRC1.M[TOLUENE]   0.6 kmol

S1

S2

SNK1

☐ SNK1.z[BENZENE]   0.998052 mol frac

☑ T1.RefluxRatio   2

T1

S3

SNK2

FIGURE 6.15   Aveva Process Simulation generated the process flow diagram of the distillation column as described in Example 6.2.

SOLUTION

MANUAL CALCULATIONS

The k-values of the feed stream components (saturated liquid, 10.41°C) are calculated using Antoine equations and shown in Tables 6.1–6.3 for feed stream, distillate, and bottom products, respectively.

The k-values of the components in distillate stream ($T_{bp} = -12$°C). The k-values of the components in the bottom stream ($T_{bp} = 41.6$°C). The order of the feed stream volatilities are: propane > n-butane > n-pentane > n-hexane. Since the

TABLE 6.1
Feed Stream k-Value for the Case Presented in Example 6.3

Components	k-Values	Key Components	Relative Volatility
Propane	6.35	LNK	16.28
n-butane	1.48	LK	3.8
n-pentane	0.39	HK	1
n-hexane	0.103	HNK	0.264

TABLE 6.2
Distillate Stream k-Value for the Case Described in Example 6.3

Components	k-Values	Key Components	Relative Volatility
Propane	3.18	LNK	23.21
n-butane	0.63	LK	4.6
n-pentane	0.137	HK	1
n-hexane	0.031	HNK	0.226

TABLE 6.3

Bottom Stream k-Value for the Case Described in Example 6.3

Components	k-Values	Key Components	Relative Volatility
Propane	13.88	LNK	11.81
n-butane	3.83	LK	3.26
n-pentane	1.175	HK	1
n-hexane	0.382	HNK	0.325

fractional recovery of n-butane in the distillate is stated, it is LK. The fractional recovery of n-pentane in the bottom is specified; hence, it is HK. Accordingly, propane = LNK (light, non-key) and n-hexane = HNK (heavy, non-key). Thus, we can assume that there is no propane in the bottom and no n-hexane in the distillate. From the given information regarding fractional recovery, we can write the equations given below.

Material Balance

The fractional recovery of n-butane in distillate:

$$0.994 = \frac{\text{Moles butane in distillate}}{\text{Moles butane in the feed}} = \frac{x_{c4,d} \times D}{z_{C4} \times F}$$

The number of moles of C_4 in the distillate,

$$x_{C4,d} \times D = 0.994 \times (F \times z_{C4})$$

The number of moles of C_4 in the feed,

$$x_{C4,b} \times B = (1 - 0.994) \times (F \times z_{C4})$$

Fractional recovery of n-pentane in the bottom

$$0.997 = \frac{\text{Moles pentane in bottom}}{\text{Moles pentane in the feed}} = \frac{x_{C5,b} \times B}{z_{C5} \times F}$$

The number of moles of C5 in the bottom,

$$x_{C5,b} \times B = 0.997 \times (F \times z_{C5})$$

The number of moles of C5 in the feed stream

$$x_{C5,d} \times D = (1 - 0.994) \times (F \times z_{C5})$$

Assuming all propane has gone to distillate and no n-hexane exists in distillate:

$$x_{C3,d} \times D = F \times z_{C3}$$

The number of moles of C_6 in the distillate,

$$x_{C6,d} \times D = 0$$

Substitute values of known concentration and flow rates (distillate).

$$x_{C3,d} \times D = 2000(0.056) = 112 \text{ kmol/h}$$

The molar flow rate of C_4 in distillate

$$x_{C4,d} \times D = (0.994)(20,000)(0.321) = 638.5 \text{ kmol/h}$$

The mole flow rate of C_5 in distillate

$$x_{C5,d} \times D = (1-0.994)(2000)(0.482) = 2.89 \text{ kmol/h}$$

The mole flow rate of C_6 in distillate

$$x_{C6,d} \times D = 0 \text{ kmol/h}$$

The total distillate mass flow rate:

$$D = \sum (D \times x_{i,d}) = 112 + 638.5 + 2.90 = 753.4 \text{ kmol/h}$$

Repeating the same procedure for the bottom stream:

$$x_{C3,d} \times B = 0 \text{ kmol/h}$$

The mole flow rate of C_4 in bottoms

$$x_{C4,b} \times B = (1-0.994)(20,000)(0.321) = 3.85 \text{ kmol/h}$$

The mole flow rate of C_5 in bottoms

$$x_{C5,b} \times B = (0.997)(2000)(0.482) = 961.1 \text{ kmol/h}$$

The mole flow rate of C_6 in bottoms

$$x_{C6,bot} \times B = Fz_{C6} = 282 \text{ kmol/h}$$

The total molar flow rate of the bottom:

$$B = \sum (B \times x_{i,dist}) = 3.85 + 961.1 + 282 = 1247 \text{ kmol/h}$$

Checking the mass balance:

$$D + B = 2000$$

Using the Fenske equation to determine the minimum number of trays, N_{min}

$$N_{min} = \frac{\ln\left((x_{LK}/x_{HK})_D (x_{HK}/x_{LK})_B\right)}{\ln(\bar{\alpha}_{LK/HK})}$$

The average relative volatility, $\bar{\alpha}_{LK/HK}$

$$\bar{\alpha}_{LK/HK} = \left(\alpha_{LK/HK,F} \times \alpha_{LK/HK,D} \times \alpha_{LK/HK,B}\right)^{1/3} = (3.8 \times 4.6 \times 3.26)^{1/3} = 2.514$$

The minimum number of theoretical trays, N_{min}

$$N_{min} = \frac{\ln\left((0.8475/0.0038)_D (0.771/0.0031)_B\right)}{\ln(2.514)} = 8.106$$

UNISIM/HYSYS SIMULATION

Two distillation methods are available in Hysys/UniSim: the Shortcut distillation method and the Continuous distillation method.

UniSim Shortcut Method

Select the shortcut distillation method, connect the feed stream, two product streams, and two energy streams. Select Peng–Robinson EOS as the thermodynamic fluid package. The feed flow rate is 2,000 kmol/h, the pressure is 1 atm, and the compositions were specified as given in the example since the feed is saturated liquid, vapor/phase fraction in Hysys sets to zero. While on the Design/Parameters page, specify the LK mole fractions in the bottom and HK in the distillate.

- The mole fraction of LK (n-butane) in bottom is 3.85/1,247 = 0.0031.
- The mole fraction of HK (n-pentane) in distillate is 2.89/753.4 = 0.0038.
- Click on the Performance tab (Figure 6.16) to find the minimum number of trays (8.373), actual number of trays (10.56), and the optimum feed stage (5.554).

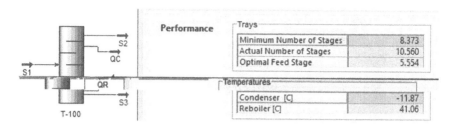

Performance	Trays		
	Minimum Number of Stages		8.373
	Actual Number of Stages		10.560
	Optimal Feed Stage		5.554
	Temperatures		
	Condenser [C]		-11.87
	Reboiler [C]		41.06

FIGURE 6.16 UniSim generated a process flow diagram of the shortcut distillation column and stream summary for the case presented in Example 6.3.

UniSim/Hysys Rigorous Distillation Column

Use the data from the shortcut distillation column as an initial estimate to the rigorous distillation column. After connecting and specifying the feed stream, connecting the top and bottom product streams and energy streams using a total condenser, and enter the following data:

- Add the component and select the appropriate fluid package (Peng–Robinson).
- Feed stream inlet stage: 6
- Total number of trays: 11
- Condenser and reboiler pressure: 1 atm
- Reflux ratio: 3.5
- Distillate liquid rate: 743 kmol/h

Providing all the above-required data and running the system, the results should look like that shown in Figure 6.17. On the Monitor page, the degree of freedom must be zero. The product streams conditions and molar flow rates are close to those obtained by the shortcut column and hand calculation.

PRO/II SIMULATION

Two distillation methods are available in PRO/II; the Shortcut distillation method and the Continuous distillation method.

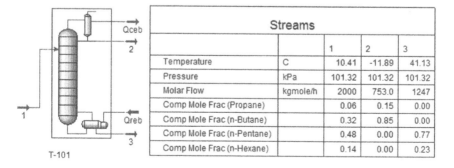

Streams				
		1	2	3
Temperature	C	10.41	-11.89	41.13
Pressure	kPa	101.32	101.32	101.32
Molar Flow	kgmole/h	2000	753.0	1247
Comp Mole Frac (Propane)		0.06	0.15	0.00
Comp Mole Frac (n-Butane)		0.32	0.85	0.00
Comp Mole Frac (n-Pentane)		0.48	0.00	0.77
Comp Mole Frac (n-Hexane)		0.14	0.00	0.23

FIGURE 6.17 UniSim generates the PFD of the rigorous distillation column and streams summary for the case presented in Example 6.3.

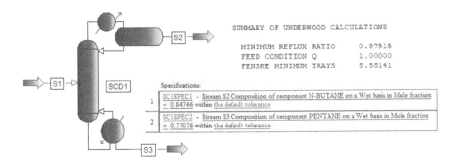

FIGURE 6.18 PRO/II shortcut distillation PFD and Summary of Underwood Calculations for the case presented in Example 6.3.

Shortcut Distillation Method

Using the shortcut distillation method in PRO/II, Peng–Robinson was selected as the thermodynamic fluid package. Perform minimum reflux calculations should be selected to get the Underwood results summary. The system required the following information: feed stream is fully specified, the LK is n-butane, and the HK is n-pentane. The ratio of reflux ratio to minimum reflux ratio is 0.87915 (i.e., 3.5/0.87915). The system converged successfully after providing the required data. Figure 6.18 shows the process flowsheet with stream summary and the Underwood calculations.

PRO/II Continuous Column

Use the Distillation column for continuous distillation. Select Peng–Robinson as the thermodynamic fluid package. The following information is required to run the system:

- The feed stream is fully specified
- Number of theoretical trays: 11
- Overall top tray pressure: 1 atm
- Pressure drop per tray: 0
- Feed tray: 7
- The distillate (Stream S2) molar flow rate is 746.662 kmol/h. The molar flow rate of stream S2 is not used as a specification but as an initial guess
- The reflux ratio is 3.5
- Mole fraction of n-propane in the distillate is 0.15

Once completing all required information, running the system should lead to the results shown in Figure 6.19. The rigorous column predictions are similar to the shortcut column.

Aspen Plus Simulation

There are two distillation columns in Aspen Plus; the shortcut (DSTWU) column and the rigorous distillation column (RadFrac). The DSTWU module provides a shortcut for distillation columns with a single feed and only distillate and bottoms products using the Winn–Underwood–Gilliland methods. The method helps make initial estimates on the stage number and feed stage for the process and provides an initial forecast for the rigorous techniques. The RadFrac model is the primary separation block in Aspen Plus. The RadFrac block can perform simulation, sizing, and rating of the tray and packed columns.

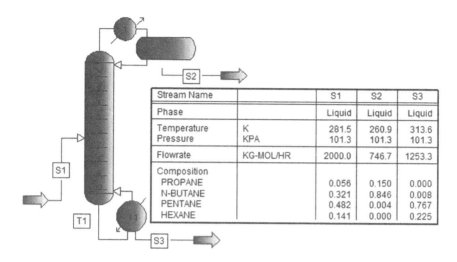

Stream Name		S1	S2	S3
Phase		Liquid	Liquid	Liquid
Temperature	K	281.5	260.9	313.6
Pressure	KPA	101.3	101.3	101.3
Flowrate	KG-MOL/HR	2000.0	746.7	1253.3
Composition				
PROPANE		0.056	0.150	0.000
N-BUTANE		0.321	0.846	0.008
PENTANE		0.482	0.004	0.767
HEXANE		0.141	0.000	0.225

FIGURE 6.19 PRO/II generated results of the continuous distillation column for the case presented in Example 6.3.

Aspen Plus Shortcut Method

Fully specify the feed stream conditions and component compositions: 2,000 kmol/h of saturated liquid is feed to a distillation column at atmospheric pressure. The feed stream composition mole fractions are 0.056 n-propane, 0.321 n-butane, 0.482 n-pentane, and 0.141 n-hexane. Select Peng–Robinson as the thermodynamic fluid package.

Figure 6.20 depicts the PFD and stream summary results using the Aspen Plus shortcut (DSTWU) distillation method. Provide the following data to the shortcut distillation method:

- Reflux ratio: 3.5
- LK: n-butane, fraction recovery in distillate: 0.994
- HK: n-pentane, fractional recovery in distillate: 0.003
- Condenser (total) and reboiler pressure: 1 atm

The column performance summary obtained using the Aspen Plus shortcut column is shown in Figure 6.20. The minimum reflux ratio is 0.873, the actual reflux ratio is 3.5, the minimum number of the stages is 9, the number of actual stages is 12, and the feed stage is stage number 7. The result is within the range of those obtained with hand calculations, Hysys, and PRO/II.

Aspen Plus Rigorous Distillation Method

The rigorous column in Aspen is RadFrac under the column tab in the model library. Enter the following data after building the process flowsheet:

- The feed is 2,000 kmol/h of saturated liquid (vapor fraction = 0) fed to a distillation column at atmospheric pressure
- The feed stream composition mole fractions are 0.056 n-propane, 0.321 n-butane, 0.482 n-pentane, and 0.141 n-hexane
- Number of trays: 11

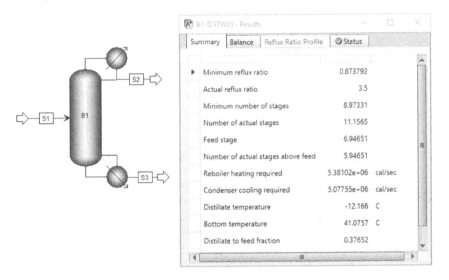

FIGURE 6.20 Aspen Plus generated results of the summary of the shortcut distillation method for the case described in Example 6.3.

- Condenser: Total
- Distillate rate: 753 kmol/h
- Reflux ratio: 3.5
- Feed stream: stage 6.947
- Top stage condenser pressure: 1 atm

Running the system leads to the results shown in Figure 6.21.

SUPERPRO SIMULATION

The shortcut distillation column is used and added to the PFD, from Unit Procedures in the toolbar, distillation, and then Continuous (Shortcut). Feed stream and two product streams are connected to the unit as shown in Figure 6.22. The feed stream

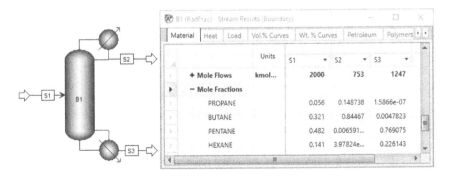

FIGURE 6.21 Aspen Plus generated a process flow diagram for rigorous distillation columns and stream conditions for the case described in Example 6.3.

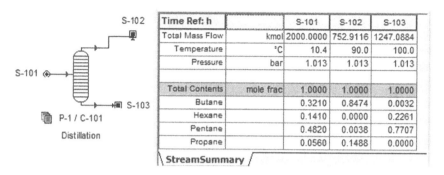

Time Ref: h			S-101	S-102	S-103
Total Mass Flow	kmol	2000.0000	752.9116	1247.0884
Temperature	°C	10.4	90.0	100.0
Pressure	bar	1.013	1.013	1.013
Total Contents	mole frac	1.0000	1.0000	1.0000
Butane		0.3210	0.8474	0.0032
Hexane		0.1410	0.0000	0.2261
Pentane		0.4820	0.0038	0.7707
Propane		0.0560	0.1488	0.0000

StreamSummary

FIGURE 6.22 SuperPro shortcut distillation PFD and stream summary estimated the theoretical number of stages rounded to 11 for the case described in Example 6.3.

is specified, and in the operating conditions, relative volatiles of feed stream components and percentage in distillate filled in as shown in Figure 6.23. The number of theoretical stages obtained with SuperPro (10.224) and minimum reflux 1.108 (i.e., 3.5/4.565).

AVEVA PROCESS SIMULATION

The basic configuration of the distillation column in Aveva Process Simulation is as follows:

1. Start by dragging a distillation column from the Process library onto the canvas (simulation area). Resize the icon, and rename it (e.g., Example 6.3).
2. Double click on the column and change the number of stages to 12. Specify the feed stream to be at stage 6, and make the pressure at the top of the column 1 atm.
3. Now set both the condenser and reboiler modes to internal. Furthermore, notice that two new unit operations have now appeared in the simulation. It is always a good idea to start designing base cases using internal condensers and reboilers.
4. Finally, drag two new sinks into the simulation and connect them to the condenser and reboiler outlets. When doing this, the simulation becomes square and solved.

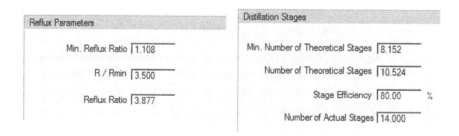

FIGURE 6.23 SuperPro required operating conditions, and the minimum reflux ratio is 1.108, reflux ratio is 3.877 for the case described in Example 6.3.

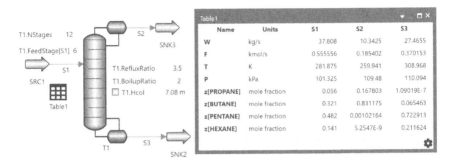

FIGURE 6.24 Aveva Process Simulation generated process flow diagram and streams summary absorption process described in Example 6.3.

5. Despite the system being square and solved, the column is still highlighted in red. That is because we are still in configuration mode. Configure mode is an initialization mode that helps us build up an estimated profile and the column before invoking the solver's rigorous solution. The primary design case assumes no liquid or vapor contact is not a particularly good column model.

6. Let us refine our design by changing the column's setup mode by changing the liquid-vapor contact fraction to one (if it does not work, gradually from 0 to 1).

7. This case is a standard procedure when designing columns in Aveva Process Simulation and will produce a detailed tray by tray calculation for the column; the column is configured (Figure 6.24).

8. Right click on the distillation column and select Show specification for both variables. We can now specify the benzene concentration and the column's reflux ratio unspecified, to have Process Simulation solved for the optimal reflux ratio.

9. To summarize, achieving an initial rigorous solution, perform the following steps:
   - Set the number of stages
   - Set feed tray location
   - Set Condenser type (typically internal)
   - Set Reboiler (usually internal)
   - Set Setup to Solve
   - Set Contact to 1 (or vapor and liquid will bypass each other)

Example 6.4: Multicomponent Separation

Design a distillation column to separate a mixture of 33% n-hexane, 37% n-heptane, and 30% n-octane. The feed stream is composed of 60% vapor (temperature is 105°C). The column operates at 1.2 atm. The feed rate is 100 kmol/h. The distillate product stream should contain 0.01 mole fraction n-heptane, and a bottom product stream should contain 0.01 n-hexane.

1. Calculate the complete product composition and a minimum number of ideal plates at infinite reflux.

2. Estimate the number of ideal plates required for the separation if the reflux ratio is 1.5 R_{min}.

3. Find the optimum feed tray.
4. Compare the manually calculated results with the available software packages: UniSim, PRO/II, Aspen Plus, SuperPro, and Aveva Process Simulation.

SOLUTION

MANUAL CALCULATIONS

1. The product composition and a minimum number of ideal plates at infinite reflux, first by estimating the vapor pressure using Antoine coefficients:

$$log(P) = \frac{A-B}{(T+C)}$$

where P is in mmHg and T is in degree Celsius. Table 6.4 lists the Antoine equation parameters for the feed stream components. For the feed temperature at 105°C, the vapor pressure of pure hexane,

$$log(P) = \frac{A-B}{(T+C)} = \frac{6.87024 - 1168.72}{(224.210 + 105)}$$

The vapor pressure in atm is $P_i = 2.75$ atm

$$K_i = \frac{P_i}{P_{tot}}$$

Using simple material balance to determine the top and bottom product stream components:

The boiling point of the bottom product is 115°C, the boiling point of the distillate is 75°C. Table 6.5 shows the k-values of feed, distillate, and bottoms stream.

Using the Fenske equation to determine the minimum number of trays, N_{min}

TABLE 6.4
Parameters Used in Antoine Equations for the Case Presented in Example 6.4

Formulae	Component	A	B	C
C_6H_{14}	Hexane	6.87024	1,168.720	224.210
C_7H_{16}	Heptane	6.89385	1,264.370	216.636
C_8H_{18}	Octane	6.90940	1,349.820	209.385

TABLE 6.5
k-Values of Feed, Distillate, and Bottom Streams (1.2 atm)

Component		Feed		Distillate			Bottom		
	Key	(mol/h)	105°C	(mol/h)	x	75°C	(mol/h)	x	75°C
LK	n-hexane	33	2.31	32.32	0.99	1.0	0.68	0.010	2.92
HK	n-heptane	37	1.01	0.33	0.01	0.396	36.67	0.544	1.33
HNK	n-octane	30	0.45	0	0	0.156	30	0.446	0.61

$$N_{min} = \frac{\ln\left(\left(x_{LK}/x_{HK}\right)_D \left(x_{HK}/x_{LK}\right)_B\right)}{\ln\left(\bar{\alpha}_{LK/HK}\right)}$$

$$\bar{\alpha}_{LK/HK} = \left(\alpha_{Fij} \times \alpha_{Dij} \times \alpha_{Bij}\right)^{1/3} = \left(2.29 \times 2.52 \times 2.2\right)^{1/3} = 2.3$$

Substituting wanted values in the Fenske equation:

$$N_{min} = \frac{\ln\left(\left(0.99/0.01\right)_D \left(0.544/0.01\right)_B\right)}{\ln\left(2.3\right)} = 9.5$$

2. The feed liquid fraction, q
   Since 60% of the feed is vapor, and q is the liquid fraction, hence, q = 0.4; also, q is calculated as follows:

$$q = \frac{H_V - H_F}{H_V - H_L} = \frac{-1.72 \times 10^5 - \left(-1.861 \times 10^5\right)}{-1.72 \times 10^5 - \left(-2.065 \times 10^5\right)} = 0.4$$

Calculate the known values to calculate φ

$$1 - q = \sum_{i=1}^{n} \frac{\left(\alpha_i x_i\right)_F}{\alpha_i - \phi}$$

Substitute relative volatilities and mole fractions,

$$1 - 0.4 = \sum \frac{\alpha_i x_{Fi}}{\alpha_i - \varphi} = \frac{2.28 \times 0.33}{2.28 - \varphi} + \frac{1 \times 0.37}{1 - \varphi} + \frac{0.45 \times 0.30}{0.45 - \varphi}$$

Simplifying further,

$$0.6 = \frac{0.75}{2.27 - \varphi} + \frac{0.37}{1 - \varphi} + \frac{0.134}{0.446 - \varphi}$$

Solve for the value of φ, and there may be more than one value, accordingly, select the value between the light and HK relative volatility. By trial and error, $\varphi = 1.68$

$$R_{min} + 1 = \sum \frac{(\alpha_i x_i)}{\alpha_i - \varphi} = \frac{(2.525)(0.99)}{5.525 - 1.68} + \frac{(1)(0.01)}{1 - 1.68}$$

Simplifying

$$R_{min} + 1 = 2.94$$

The calculated minimum reflux ratio,

$$R_{min} = 1.94$$

The reflux ratio, R

$$R = 1.5 R_{min} = 1.5 \times 1.94 = 2.92$$

The ideal number of plates, N

$$\frac{N - N_{min}}{N + 1} = 0.75 \left[1 - \left(\frac{R - R_{min}}{R + 1} \right)^{0.5668} \right]$$

Substitute the required parameters,

$$\frac{N - 9.5}{N + 1} = 0.75 \left[1 - \left(\frac{2.92 - 1.94}{2.92 + 1} \right)^{0.5668} \right]$$

Simplify further,

$$\frac{N - 9.5}{N + 1} = 0.41$$

$N = 16.8$, and hence the total I number of theoretical trays rounded up is 17 trays.

3. Using Kirkbride method to calculate optimum feed tray:

$$\ln \left(\frac{N_D}{N_B} \right) = 0.206 \ln \left[\left(\frac{x_{HK}}{x_{LK}} \right)_F \left(\frac{x_{LK,inB}}{x_{LK,inD}} \right)^2 \left(\frac{B}{D} \right) \right]$$

Substituting required values:

$$ln\left(\frac{N_D}{N_B}\right) = 0.206 \ ln\left(\left(\frac{0.37}{0.33}\right)_F\left(\frac{0.01}{0.01}\right)^2\left(\frac{67.35}{32.65}\right)\right) = 0.182$$

The ratio of rectifying (N_D) to stripping number of trays (N_B) is

$$\frac{N_D}{N_B} = 1.2$$

And $N = 16.8 = N_D + N_B$
   Substituting values and rearranging:

$$16.8 = N_D + N_B = 1.2N_B + N_B$$

Solving for N_B and N_D

$$N_B = 7.6, \ N_D = 9.2$$

The number of trays above the feed tray is 9, and that below the feed tray is 8.0 trays; the feed tray is tray number 10.

4. Compare the manually calculated results with the available software packages: UniSim, PRO/II, Aspen Plus, SuperPro, and Aveva Process Simulation.

UNISIM/HYSYS SIMULATION
UniSim Shortcut Method

Add the component and select the suitable property package (Antoine). Using shortcut distillation, perform the process PFD utilizing Antoine package for thermodynamic properties calculations. Consider the following values extracted for the given data of Example 6.4:

LK in the bottom: n-hexane, mole fraction is 0.01
HK in distillate: n-heptane, mole fraction is 0.01

Figure 6.25 shows the process flowsheet and stream summary. The external reflux ratio is $3.972 = 1.5 \times R_{min}$, and the value of the minimum reflux ratio is 2.648 (from Design/Parameters page). From the performance page, the minimum number of trays is 10, the actual number of trays is 18, and the optimum feed tray is tray number 10 (Figure 6.25).

UNISIM/HYSYS RIGOROUS DISTILLATION COLUMN

The following data were obtained from the UniSim shortcut column and fed to the UniSim rigorous distillation column:

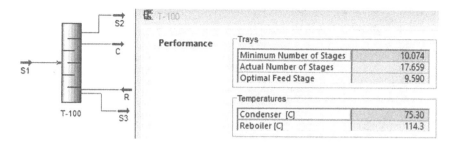

FIGURE 6.25 UniSim generates the performance of the shortcut distillation for the case presented in Example 6.4.

- Add the component and select the suitable property package (Antoine).
- Fully specify the feed stream by providing feed stream flow low rate, composition, and feed conditions (i.e., temperature and pressure or vapor/phase ratio with either temperature or pressure).
- Feed stream inlet stage: 10
- Total number of trays: 18
- Condenser and reboiler pressure: 121.6 kPa
- Reflux ratio: 3.972
- Distillate liquid rate: 32.65 kmol/h

Providing all the above-required data and running the system, the results should look like that shown in Figure 6.26.

SIMULATION WITH PRO/II

Consider the following two methods in designing the distillation column in PRO/II; the Shortcut method and the Rigorous distillation method.

PRO/II Shortcut Column

Figure 6.27 shows the process flow diagram using the PRO/II shortcut column. After building the process flowsheet, fill in the specifications page. The mole fraction of LK (n-hexane) in the bottom (S3) is 0.01. The mole fraction of HK (n-heptane) in distillate (S2) is 0.01. The Summary of Underwood Calculations

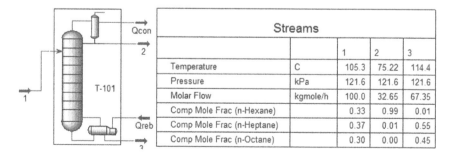

FIGURE 6.26 UniSim generates the process flow diagram and streams table of the continuous distillation column as presented in Example 6.4.

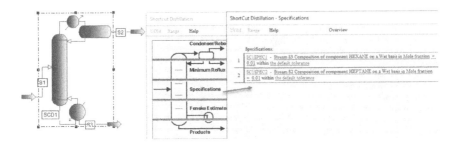

FIGURE 6.27 PRO/II shortcut distillation PFD, specifications page, and a Summary of Underwood Calculations for the case described in Example 6.4.

shows the minimum reflux ratio is 2.74, feed conditions; q = 0.388, the minimum number of trays rounded up to 12, and the total number of trays is 18 based on total reflux ratio to minimum reflux ratio (R/R$_{min}$ = 1.5).

PRO/II Rigorous Distillation Column

Employ PRO/II Distillation column for continuous distillation and the Peng–Robinson EOS for the property estimation. Fill in the following values obtained from the shortcut column in the PRO/II Distillation column:

- The feed stream is fully specified by providing flow rate, composition, pressure, and a liquid fraction (i.e., q = 0.4)
- Number of theoretical trays: 18
- Overall top tray pressure: 121.6 kPa
- Pressure drop per tray: 0
- Feed tray: 11
- The distillate (Stream S2) molar flow rate is 32.65 kmol/h. The molar flow rate of stream S2 is not used as a specification but as an initial guess
- Reflux ratio is 4.11 (i.e., 1.5 R$_{min}$)
- The mole fraction of n-propane in the distillate is 0.989

Once completing all required information, running the system should lead to the results shown in Figure 6.28.

ASPEN PLUS SIMULATIONS

Consider the two methods available in Aspen Plus, the Shortcut method using DSTWU block and the Rigorous distillation column using RadFrac column.

Aspen Plus Shortcut Simulations

The first step in the flowsheet simulation is to define process flowsheet connectivity by placing unit operations (blocks) and their connected streams. To determine a process flowsheet block, select a model from the Model Library (Column and then DSTWU) and insert it in the workspace. To define a process stream, select Streams from the Model Palette and click to establish each end of the stream connection on the available inlet and outlet locations of the existing blocks. To build process flowsheet using shortcut distillation from the model library, use Peng–Robinson EOS for the fluid package. Data needed for simulation are as follows:

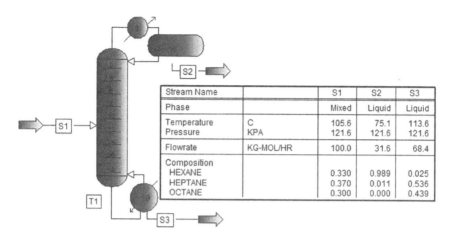

FIGURE 6.28 Process flowsheet of the distillation process and stream summary generated by PRO/II for the case existing in Example 6.4.

- The feed stream temperature is 105.3°C, and the pressure is 121.6 kPa, the total flow rate is 100 kmol/h, the mole fraction is 0.33 hexane, 0.370 heptane, and 0.3 octane.
- The reflux ratio is 3.972.
- The distillate is 0.98 (molar flow rate of n-hexane in distillate/molar flow rate of n-hexane in feed stream), and HK (n-heptane) in the distillate is 0.0088 (molar flow rate of n-heptane in distillate to molar flow rate of n-heptane in feed stream).
- Condenser and reboiler pressure is 121.59 kPa. Fill in all required data. The unit specifications, such as the minimum reflux ratio, are 2.72, and the actual reflux ratio is 3.972.
- Figure 6.29 depicts the Aspen Plus generated result with the shortcut method. The minimum number of stages is 10, the number of actual stages is 18, and the optimum feed stage is stage 11.

Aspen Rigorous Simulations

Provide the rigorous distillation column in Aspen Plus (RadFrac) from the model library. After building the process flowsheet, enter the following data:

- Number of trays: 18
- Condenser: total
- Distillate rate: 32.65 kmol/h
- Reflux ratio: 3.972
- Feed stream: stage 10
- Top stage condenser pressure: 121.6 kPa

Running the system leads to the results as shown in Figure 6.30. The stream's molar flow rate and conditions are similar to those obtained by the Aspen shortcut method.

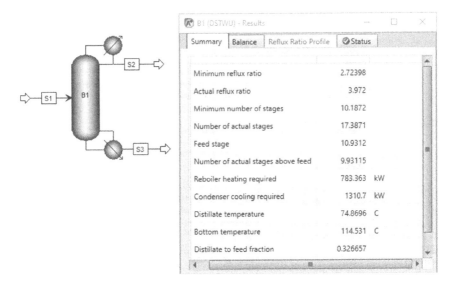

FIGURE 6.29 Aspen Plus generates the summary of shortcut results for the case presented in Example 6.4.

SuperPro Designer Simulations

The shortcut distillation column is used and added to the process flowsheet done in the previous examples. Feed stream and two product streams are connected to the unit, as shown in Figure 6.31. Specify the feed stream, and the operating conditions, relative volatilities of feed stream components, and percentage in distillate are specified. The number of theoretical stages obtained with SuperPro is 20, and minimum reflux is 1.588 (i.e., 2.382/1.5).

Aveva Process Simulation

The basic configuration of the distillation column in Aveva Process Simulation is as follows:

1. Start by dragging a distillation column from the Process library onto the canvas (simulation area). Resize the icon, and rename it (e.g., Example 6.4).

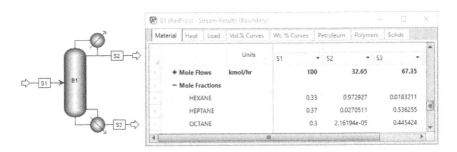

FIGURE 6.30 Aspen Plus generated the process flow diagram and streams summary for the distillation column described in Example 6.4.

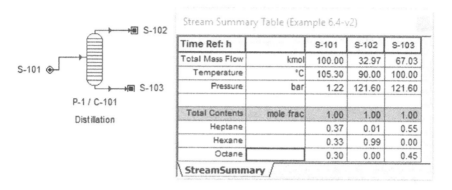

Time Ref: h		S-101	S-102	S-103
Total Mass Flow	kmol	100.00	32.97	67.03
Temperature	°C	105.30	90.00	100.00
Pressure	bar	1.22	121.60	121.60
Total Contents	mole frac	1.00	1.00	1.00
Heptane		0.37	0.01	0.55
Hexane		0.33	0.99	0.00
Octane		0.30	0.00	0.45

FIGURE 6.31 SuperPro predicted reflux ratio and the number of theoretical stages for the absorption system described in Example 6.4. The minimum reflux ratio 1.0, the theoretical number of trays is 21, R/R_{min} 1.5.

2. Double click on the column and change the number of stages to 17. Specify the feed stream to be at stage 8, and make the pressure at the top of the column 1 atm.

3. Now set both the condenser and Reboiler modes to internal. Furthermore, notice that two new unit operations have now appeared in the simulation. It is always a good idea to start designing base cases using internal condensers and reboilers.

4. Finally, drag two new sinks into the simulation and connect them to the condenser and reboiler outlets. When doing this, the simulation becomes square and solved.

5. Despite the system being square and solved, the column is still highlighted in red. That is because we are still in configuration mode. Configure mode is an initialization mode that helps us build up an estimated profile and the column before invoking the solver's rigorous solution. The primary design case assumes no liquid or vapor contact is not a particularly good column model.

6. Let us refine our design by changing the column's setup mode by changing the liquid-vapor contact fraction to one (if it does not work, gradually from 0 to 1).

7. This case is a standard procedure when designing columns in Aveva Process Simulation and will produce a detailed tray by tray calculation for the column; the column is configured (Figure 6.32).

8. Right click on the distillation column and select Show specification for both variables. We can now specify the benzene concentration and the column's reflux ratio unspecified, to have Process Simulation solved for the optimal reflux ratio.

To summarize, achieving an initial rigorous solution, perform the following steps:

- Set the number of stages
- Set feed tray location
- Set Condenser type (typically internal)
- Set Reboiler (usually internal)

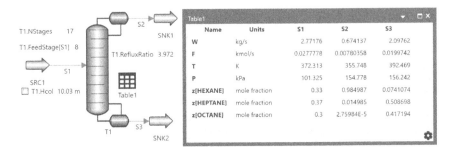

FIGURE 6.32 Aveva Process Simulation generates the process flow diagram and stream composition predicted for the separation process described in Example 6.4.

- Set Setup to Solve
- Set Contact to 1 (or vapor and liquid will bypass each other)

Example 6.5: Multicomponent Separation

A mixture of 100 kmol/h saturated liquid containing 30 mol% benzene, 25% toluene, and 45% ethylbenzene is to be separated by distillation column at atmospheric pressure, with 98% of the benzene and only 1% of the toluene to be recovered in the distillate stream. Consider a reflux ratio is 2:

1. Calculate the minimum number of ideal plates.
2. Calculate the approximate composition of the product streams.
3. Estimate the actual number of trays.
4. Determine the minimum number of trays if the distillation operates at 0.2 atm.
5. Compare the manually calculated results with those obtained from the available software packages, such as UniSim, PRO/II, Aspen Plus, SuperPro, and Aveva Process Simulation.

SOLUTION

MANUAL CALCULATIONS

Table 6.6 shows the relative volatilities of components at feed conditions obtained using the Antoine equation.

TABLE 6.6
List of the Relative Volatilities of Feed Components for the Case Described in Example 6.5

Key	Component	Feed Composition	Vapor Pressure (atm)	Relative Volatility (α)
LK	Benzene	0.3	2.37	2.44
HK	Toluene	0.25	0.97	1
	Ethyl benzene	0.45	0.47	0.49

MATERIAL BALANCE

The following data are available:

- The molar flow rate of benzene in distillate is $0.98 \times 30 = 29.4$ mol/h.
- Molar flow rate of benzene in bottom: $30 - 29.4 = 0.6$ mol/h.
- Molar flow rate of toluene in distillate: $0.01 (25) = 0.25$ mol/h.
- Molar flow rate of toluene in bottom: $25 - 0.25 = 24.75$ mol/h.

1. The minimum number of plates is obtained from the Fenske–Underwood equation using the relative volatility of the LK to the HK, which is the ratio of the K factors (Tables 6.7 and 6.8).
   At a first approximation, all ethylbenzene goes to the bottom.
   Using the Fenske equation to determine the minimum number of trays, N_{min}

$$N_{min} = \frac{\ln\left(\left(x_{LK}/x_{HK}\right)_D \left(x_{HK}/x_{LK}\right)_B\right)}{\ln\left(\bar{\alpha}_{LK/HK}\right)}$$

The relative volatilities,

$$\bar{\alpha}_{LK/HK} = \left(\alpha_{Fij} \times \alpha_{Dij} \times \alpha_{Bij}\right)^{1/3}$$

TABLE 6.7

Composition and k-Values of Distillate and Bottom Streams for the Case Described in Example 6.5

Component	Feed (mol/h)	Distillate (mol/h)	Bottom (mol/h)	x_D	k-Value	x_B	k-Value
Benzene	30	29.4	0.6	0.9916	1.005	0.0085	3.125
Toluene	25	0.25	24.75	0.0084	0.4146	0.3518	1.437
Ethyl benzene	45	0	45	0		0.639	0.7312

TABLE 6.8

Feed Stream Relative Volatilities for the Case Described in Example 6.5

	Component	Feed Composition	k-Value	Relative Volatility (α)
LK	Benzene	0.30	2.215	2.84
HK	Toluene	0.25	0.779	1.00
	Ethyl benzene	0.45	0.313	0.40

Substitute the available parameters,

$$N_{min} = \frac{\ln\left((0.9916/0.0084)_D (0.3518/0.0085)_B\right)}{\ln(2.44)} = 9.52$$

The number of ideal plates, including the partial reboiler, is 9.52. A more accurate estimate of N_{min} using the mean relative volatility based on top, feed, and bottom of the column

$$\bar{\alpha}_{LK/HK} = \left[\left(\alpha_{LK/HK}\right)_F \left(\alpha_{LK/HK}\right)_D \left(\alpha_{LK/HK}\right)_B\right]^{1/3} = [2.44 \times 2.42 \times 2.18]^{1/3} = 2.34$$

Substitute calculated relative volatility:

$$N_{min} = \frac{\ln\left((0.9916/0.0084)_D (0.3518/0.0085)_B\right)}{\ln(2.34)} = 10.0$$

Apply Fenske equation to benzene-ethyl benzene separation with $\alpha = 2.44/0.49 = 4.98$

$$N_{min} = \frac{\ln\left((x_{LK}/x_{HK})_D (x_{HK}/x_{LK})_B\right)}{\ln(\bar{\alpha}_{LK/HK})}$$

Substitute the values of the relative volatilities,

$$N_{min} = \frac{\ln\left((0.9916/x_{EB})_D (0.639/0.0085)_B\right)}{\ln(4.98)} = 9.52$$

Simplify further,

$$\ln\left(\left(\frac{0.9916}{x_{EB}}\right)_D \left(\frac{0.639}{0.0085}\right)_B\right) = \ln(5) \times 9.52$$

Simplify further

$$\left(\frac{0.9916}{x_{EB}}\right)\left(\frac{0.639}{0.0085}\right) = 4.5 \times 10^6$$

From which the mole fraction of ethylbenzene in the distillate, $x_{EB,D} = 1.7 \times 10^{-5}$, which is negligible,

2. Minimum reflux ratio, R_{min}. Since the feed is saturated liquid, $q = 1$

$$1-q = \sum_{i=1}^{n} \frac{\alpha_i x_{F,i}}{\alpha_i - \phi} = 1-1 = 0 = \frac{2.44 \times 0.3}{2.44 - \phi} + \frac{1 \times 0.25}{1 - \phi} + \frac{0.49 \times 0.45}{0.49 - \phi}$$

Solve for ϕ, the value should be between 2.44 and 1, by trial and error:
$\phi = 1.471$

$$R_{min} + 1 = \sum_{i=1}^{n} \frac{\alpha_i x_{Di}}{\alpha_i - \phi} = \frac{2.424 \times 0.9916}{2.424 - 1.471} + \frac{1 \times 0.0085}{1 - 1.471} = 2.5$$

The minimum reflux ratio: $R_{min} = 1.5$
3. Calculate the number of equilibrium stages using Gilliland correlation, N.
   Select a reflux ratio that is $R = (1.1$ to $1.5)$, R_{min}

$$\frac{N - N_{min}}{N + 1} = 0.75\left(1 - \left(\frac{R - R_{min}}{R + 1}\right)^{0.566}\right)$$

Substituting required data:

$$\frac{N - 9.52}{N + 1} = 0.75\left(1 - \left(\frac{2 - 1.5}{2 + 1}\right)^{0.566}\right) = \frac{N - 9.52}{N + 1} = 0.48$$

Simplify, and calculate the N value,

$0.522 \times N = 10 \Rightarrow N = 19$ trays. The actual number of stages is 19.

4. At 0.2 atm (the boiling point is 55°C), the relative volatilities are obtained
   using the Antoine equation. At a boiling point of 0.2 atm, the relative
   volatility of LK to HK is 2.84 compared to 2.44 at 1 atm.

$$N_{min} = \frac{\ln\left((0.9916/0.0084)_D (0.3518/0.0085)_B\right)}{\ln(2.84)} = 8.14$$

The heat of vaporization per mole is slightly higher at the low pressure,
but less reflux is required for a given number of plates because of the
higher relative volatility.

UniSim/Hysys Simulation

Consider the two distillation methods available in Hysys using the Shortcut method
and the Rigorous distillation method.

UniSim Shortcut Method

Select the shortcut distillation in Hysys after connecting and fully specifying feed
stream, connecting product and energy streams, selecting Peng–Robinson as the

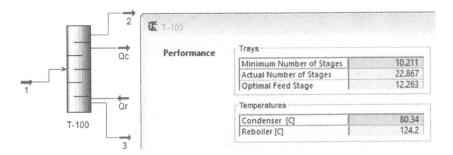

FIGURE 6.33 UniSim generated the performance of the shortcut distillation column for the case in Example 6.5.

thermodynamic fluid package. While on the Design/Parameters page, fill in the following data:

- LK in the bottom is benzene with mole fraction 0.0085
- HK in the distillate is toluene with a mole fraction is 0.0084
- Condenser and reboiler pressure are 1 atm
- The reflux ratio is 2

The minimum reflux ratio is 1.629. Figure 6.33 depicts the minimum number of trays rounded to 11 stages and the actual number of trays rounded to 23 stages; the optimum feed tray is tray number 13 from the top.

UniSim/Hysys Continuous Distillation Column

The following data are required and provided from results obtained in the shortcut column:

- The feed stream should be fully specified (i.e., fill in the feed stream conditions; the feed flow rate, compositions. The temperature and pressure, or the vapor/phase ratio with either temperature or pressure)
- Feed stream inlet stage: 12
- Total number of trays: 23
- Condenser and reboiler pressure: 101.3 kPa
- Reflux ratio: 2
- Distillate liquid rate: 29.65 kmol/h

Providing all the above required data and running the system, the results should look like as shown in Figure 6.34. Streams molar flow rates are the same as those obtained by the shortcut distillation method.

PRO/II Simulation

Use the following methods built-in PRO/II; the shortcut method and rigorous methods.

PRO/II Shortcut Column

Use the shortcut method in PRO/II, Peng–Robinson, as the suitable property package. Figure 6.35 predicts the Underwood calculation method's summary. The minimum

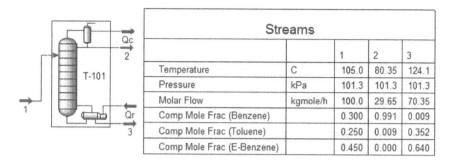

Streams				
		1	2	3
Temperature	C	105.0	80.35	124.1
Pressure	kPa	101.3	101.3	101.3
Molar Flow	kgmole/h	100.0	29.65	70.35
Comp Mole Frac (Benzene)		0.300	0.991	0.009
Comp Mole Frac (Toluene)		0.250	0.009	0.352
Comp Mole Frac (E-Benzene)		0.450	0.000	0.640

FIGURE 6.34 PFD and streams summary generated with Hysys continuous distillation column for the absorption case presented in Example 6.5.

reflux ratio is 1.567, the minimum number of trays is 10, and the actual number of stages is 18 trays. Values are close to the manual calculation results.

PRO/II Rigorous Distillation Column

PRO/II Distillation column is used for continuous distillation and select the Peng–Robinson EOS for the property estimation. Fill in the following values obtained from the shortcut column in the PRO/II Distillation column:

- The feed stream is fully specified (filling in feed flow rate, compositions, pressure, and feed at bubble point temperature)
- Number of theoretical trays: 18
- Overall top tray pressure: 1 atm
- Pressure drop per tray: 0
- Feed tray: 10
- The distillate (Stream S2) molar flow rate is 29.651 kmol/h. The molar flow rate of stream S2 is not used as a specification but as the initial guess
- Reflux ratio is 3.134 (i.e., 2 R_{min})
- Mole fraction of benzene in the distillate is 0.992

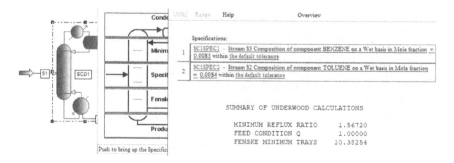

FIGURE 6.35 PRO/II calculated the Summary of Underwood Calculations for the case presented in Example 6.5.

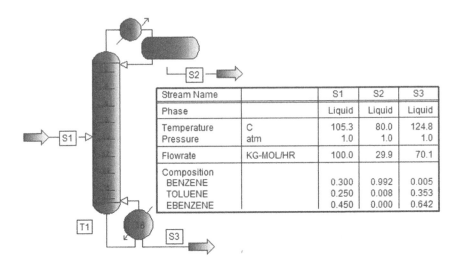

Stream Name		S1	S2	S3
Phase		Liquid	Liquid	Liquid
Temperature	C	105.3	80.0	124.8
Pressure	atm	1.0	1.0	1.0
Flowrate	KG-MOL/HR	100.0	29.9	70.1
Composition				
BENZENE		0.300	0.992	0.005
TOLUENE		0.250	0.008	0.353
EBENZENE		0.450	0.000	0.642

FIGURE 6.36 PRO/II generated the process flow diagram and streams table properties for the separation process described in Example 6.5.

Once completing all required information, running the system should lead to the results shown in Figure 6.36.

ASPEN PLUS SIMULATION

Consider the shortcut and rigorous methods in Aspen Plus.

Aspen Plus Shortcut Method

The first step in the flowsheet simulation is to define process flowsheet connectivity by placing unit operations (blocks) and their connected streams. To determine a process flowsheet block, select a model from the Model Library (Column and then DSTWU) and insert it in the workspace. To define a process stream, select Streams from the Model Palette and click to establish each end of the stream connection on the available inlet and outlet locations of the existing blocks. To build process flowsheet using shortcut distillation from the model library, use Peng–Robinson EOS for the fluid package. Data needed for simulation are as follows:

- The feed stream temperature is 105°C, and the pressure is 1 atm, the total flow rate is 100 kmol/h, the mole fraction is 0.30 benzene, 0.25 toluene, and 0.45 ethylbenzene.
- The reflux ratio is 2.
- The key component recoveries, 0.992 light key benzene, the heavy key is toluene 0.008.
- Condenser and reboiler pressure is 1 atm.
- Figure 6.37 depicts the Aspen Plus generated result with the shortcut method. The minimum number of stages is 10, the number of actual stages is 18, and the optimum feed stage is stage 11.

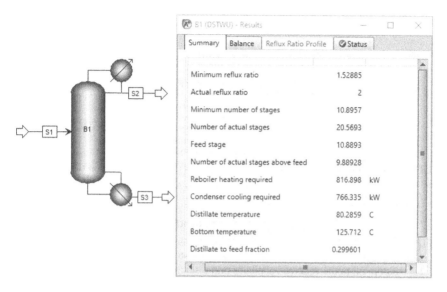

FIGURE 6.37 Aspen Plus predicted the minimum reflux ratio, the column minimum, and the actual number of trays for the case described in Example 6.5.

Aspen Plus Continuous Distillation Column Method

The most commonly used rigorous distillation column in Aspen Plus is RadFrac, and the model is available under the column in the Aspen Plus model library. After building the process flowsheet, enter the following data:

- Number of trays: 21
- Feed stream stage: 11
- Condenser: total
- Distillate rate: 29.56 kmol/h
- Reflux ratio: 2
- Condenser pressure: 1 atm

After providing the data from the shortcut method, Figure 6.38 shows the Aspen Plus generated process flow diagram and stream summary of the successful converged RadFrac distillation column.

FIGURE 6.38 Aspen Plus RadFrac column predicted the process flow diagram and the stream table of the case described in Example 6.5.

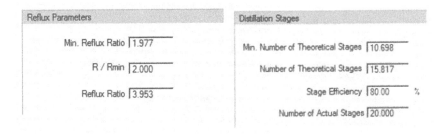

FIGURE 6.39   SuperPro predicted the reflux ratio and the number of theoretical trays for the case described in Example 6.5.

SuperPro Designer Simulation

Use the shortcut distillation column in SuperPro and add to the process flowsheet done in the previous examples. Connect the feed stream and two product streams to the column: fill in the relative volatilities and distillate stream composition. The feed stream is specified, and in the operating conditions, fill in the relative volatiles of feed stream components and percentage in the distillate. The number of theoretical stages obtained with SuperPro is 16, and the minimum reflux ratio is 1.977 (Figure 6.39), and Figure 6.40 depicts the column process flow diagram and summary.

Aveva Process Simulation

The basic configuration of the distillation column in Aveva Process Simulation is as follows:

1. Start by dragging a distillation column from the model library onto the canvas (Simulation page). Resize the icon, and connect the feed stream to the column.
2. Double click on the column and change the number of stages to 18. Specify the feed stream to be at stage 12, and make the pressure at the top of the column 1 atm.

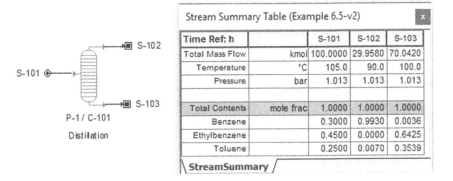

FIGURE 6.40   SuperPro predicted the process flow diagram and the stream summary table of the case described in Example 6.5.

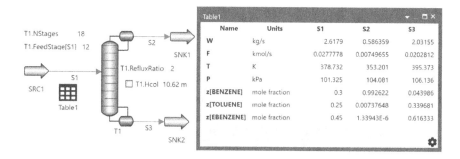

FIGURE 6.41   Aveva Process Simulation generates the process flow diagram and streams summary for the case presented in Example 6.5.

3. Now set both the condenser and reboiler modes to internal. Furthermore, notice that two new unit operations have now appeared in the simulation.
4. Drag two new sinks into the simulation and connect them to the condenser and reboiler outlets.
5. The column is highlighted in red because we are still in Configure mode. Configure mode is an initialization mode that helps us build up an estimated profile and the column before invoking the solver's rigorous solution.
6. Change the column's Setup mode to solve it by changing the liquid-vapor contact fraction to one (1), a standard procedure when designing columns in Aveva Process Simulation, and will produce a detailed tray-by-tray calculation for the column.
7. Drag variable references onto the sheet again to see the benzene composition in the top product and the column reflux ratio.
8. Right click on the distillation column and select Show specification for both variables. We can now specify the benzene concentration and the column's reflux ratio unspecified, to have Process Simulation solved for the optimal reflux ratio.
9. If we set the benzene concentration around 99.0 percent more, then Aveva Process Simulation shows that the column requires a reflux ratio of about 2.0.
10. Even though the benzene concentration is an external variable from the column, Aveva Process Simulation still knows how all of our variables interact and can quickly solve all these variables in the simulation. Nevertheless, for now, the primary design case is complete (Figure 6.41).

PROBLEMS

6.1   SHORTCUT DISTILLATION

The feed to a distillation column is at 207°C and 13.6 atm enters at 577 kmol/h with the following compositions in mole fraction: ethane 0.0148, propane 0.7315, i-butane 0.0681, n-butane 0.1462, i-pentane 0.0173, n-pentane 0.015, and n-hexane 0.0071. HK in the distillate is an i-butane 0.02 mole fraction; LK in the bottom is propane

0.025 mole fraction for reflux ratios of 1.5. Use UniSim (the suitable fluid package is PR) or any other software package to calculate the minimum reflux and column performance.

6.2 RIGOROUS DISTILLATION

The feed to a distillation column is at room conditions (T = 25°C, P = 1 atm). The concentration of the feed stream is 50% ethanol, 50% isopropanol in mass fractions. The feed is at a rate of 74 kg/h. We will assume a load of 35.169 kW for the reboiler heat duty, assume the number of trays to be 24, and the reflux ratio equals 3. Find the conditions of the exit streams. Use UniSim (the suitable fluid package is NRTL) or any other software package to calculate the minimum reflux and column performance.

6.3 CONTINUOUS DISTILLATION

Repeat Example 6.2; assume that the reboiler duty is unknown. The overhead ethanol concentration is 0.55. What is the amount of the reboiler load in kW?

6.4 CONTINUOUS DISTILLATION WITH KNOWN REBOILER DUTY

Repeat Example 6.2; calculate the reflux ratio that would give an ethanol concentration of 80% with a reboiler duty of 8.069×10^4 kcal/h.

6.5 SEPARATION OF BENZENE, TOLUENE, AND TRIMETHYL-BENZENE

Use a distillation column with a partial reboiler and a total condenser to separate a mixture of benzene, toluene, and trimethyl-benzene. The feed consists of 0.4-mole fraction benzene, 0.3 mole fraction toluene, and trimethyl-benzene balance. The feed enters the column as a saturated vapor and separated 95% of the distillate's toluene and 95% of the bottom's trimethyl-benzene. The column operates at 1 atm, the top and bottom temperatures are 390 and 450 K, respectively.

1. Find the number of equilibrium stages required at total reflux.
2. Find the minimum reflux ratio for the previous distillation problem using the Underwood method.
3. Estimate the total number of equilibrium stages and the optimum feed-stage location if the actual reflux ratio R equals 1.
4. Use UniSim software (the suitable fluid package is PRSV) or any other software package to calculate the minimum reflux ratio and column performance.

6.6 SEPARATION OF HYDROCARBON MIXTURES

A mixture with 4% n-C_5, 40% n-C_6, 50% n-C_7, and 6% n-C_8 distilled at 1 atm with 98% of n-C_6 and 1% of n-C_7 recovered in the distillate. Use UniSim (the suitable fluid package is PR) or any other software package to calculate the minimum reflux ratio for a liquid feed at its bubble point.

6.7 SEPARATION OF MULTICOMPONENT GAS MIXTURE

A feed at its bubble point temperature enters a distillation column. The feed contains propane (C_3), n-butane (n-C_4), i-butane (i-C_4), n-pentane (n-C_5) and i-pentane (i-C_5), the feed components' mole percent are 5%, 15%, 25%, 20%, and 35%, respectively. The operating pressure of the column is 405.3 kPa. Ninety-five percent of the i-C_4 is recovered in the distillate (top product) and 97% of the n-C_5 is recovered in the bottom product. Determine the top and bottom flow rats and composition, minimum reflux ratio, and the minimum number of trays at a reflux ratio of 2.25. Use UniSim (the suitable fluid package is PR) or any other software package to verify the manual calculations.

6.8 SEPARATION OF METHANOL, WATER, AND PHENOL

A distillation column is utilized to separate a mixture of methanol, water, and phenol. The feed stream is at 65°C and 1.7 bar, the inlet mass flow rate is 4,530 kg/h, the stream component mass fractions are 0.6, 0.39, and 0.01 for methanol, water, and phenol, respectively. Use the shortcut distillation method to determine the minimum reflux ratio, and for the fluid package, use NRTL-RK. The distillate's expected methanol recovery is 99% methanol (LK) and water 1% (HK). Assume the condenser pressure is at 1.1 bar and reboiler pressure at 1.7 bar.

6.9 SEPARATION OF DIMETHYL ETHER (DME)

Dimethyl ether at a feed rate of 135 kgmol/h (55°C and 1,040 kPa) containing 44 mol% DME, 46% H_2O, 8% methanol, and 2% ethanol is separated in five trays distillation column. It is desired to achieve 60 kmol/h as the top product containing a DME mole fraction of more than 98%. The reflux ratio is 2. Use UniSim/Hysys software package (PRSV) to simulate the distillation column and find the DME mole fraction in the distillate.

6.10 SEPARATION OF ETHANOL FROM METHANOL/ ETHANOL/WATER LIQUID MIXTURE

A feed to a distillation column is 75 kgmol/h at 50°C and 740 kPa, and contains 80 mol% water, 17% methanol, and 3% ethanol. Use the shortcut column in UniSim/Hysys (PRSV) to obtain 99% ethanol in distillate and 99% methanol in the bottom. Calculate the minimum reflux ratio. Using a reflux ratio of 1.5 times the minimum reflux ratio, find the actual number of trays.

REFERENCES

1. Wankat, P. C., 2011. Separation Process Engineering, 3rd edn, Prentice-Hall, Englewood Cliffs, NJ.
2. King, C. J., 1980. Separation Processes, 2nd edn, McGraw-Hill, New York, NY.

3. Seader, J. D. and E. J. Henley, 2019. Separation Process Principles with Applications Using Process Simulators, 4th edn, Wiley, New York, NY.
4. Humphrey, J. L. and G. E. Keller, 1997. Separation Process Technology, McGraw-Hill, New York, NY.
5. Fair, J. R. 1961. How to predict sieve tray entrainment and flooding, Petrochemical Engineering Journal 33, 45–52.
6. McCabe, W., J. Smith and P. Harriott, 2004. Unit Operations of Chemical Engineering, 7th edn, McGraw-Hill, New York, NY.

7 Gas Absorption

At the end of this chapter, students will be able to:

1. Define the factors considered in designing an absorber
2. Explain the importance of exhaust gas characteristics and liquid flow
3. Compute the minimum liquid flow rate required for separation
4. Determine the diameter and packing height of a packed bed column
5. Estimate the number of theoretical plates and the height of a plate tower

7.1 INTRODUCTION

Absorption is a process that refers to the transfer of a gaseous pollutant from the gas phase to the liquid phase. Absorbers are used extensively in the chemical industry to separate and purify gas streams, product recovery, and pollution control devices. The absorption processes can be physical or chemical; physical absorption occurs when the absorbed compounds dissolve in the solvent. Chemical absorption occurs when the absorbed compounds and the solvent react. Examples are separations of acid gases such as CO_2 and H_2S from natural gas using amine as solvents. Chemical engineers need to design gas absorbers that produce a treated gas of a desired purity with optimal size and liquid flow. The design depends on the existing correlations, and when required, laboratory and pilot plant data. For gas absorption, the two most frequently used devices are the packed tower and the plate tower. If designed and operated correctly, both these devices can achieve high collection efficiencies for a wide variety of gases. The design procedures' primary outcomes determine the column's diameter and the tower height [1–3].

7.2 PACKED BED ABSORBER

The design of an absorber used to reduce gaseous pollutants from process streams is affected by many factors such as the pollutant gathering efficiency, contaminant solubility in the absorbing liquid, liquid-to-gas ratio, stream flow rate, pressure drop, and construction details of the absorbers such as packing materials, plates, liquid distributors, entrainment separators, and corrosion-resistant materials. Solubility is an important factor affecting the amount of a pollutant or solute absorbed. Solubility is a function of both temperature and pressure. As temperature increases, the gas volume also increases; therefore, absorbed less gas due to the large gas volume. By increasing the pressure of a system, the amount of gas absorbed generally increases. Solubility data are obtained at equilibrium conditions, and it involves filling measured amounts of a gas and a liquid into a closed

DOI: 10.1201/9781003167365-7

vessel and allowing it to sit for some time. Henry's law expresses the equilibrium solubility of a gas-liquid system:

$$y = Hx \qquad (7.1)$$

where y is the mole fraction of gas in equilibrium with liquid; H is Henry's constant, mole fraction in vapor per mole fraction; and x is the mole fraction of the solute in equilibrium. Henry's law, which depends on total pressure, can predict solubility only when the equilibrium line is straight when the solute concentration is very dilute. Another form of Henry's law:

$$p = k_H c \qquad (7.2)$$

where p is the partial pressure of the solute in the gas above the solution, c is the solute concentration, and k_H is constant with the dimensions of pressure units divided by concentration units. The constant, known as Henry's constant, depends on the solute, the solvent, and the temperature. Some values for k_H for gases dissolved in water at 298 K include [4]:

- Oxygen (O_2): 769.2 L atm/mol
- Carbon dioxide (CO_2): 29.4 L atm/mol
- Hydrogen (H_2): 1282.1 L atm/mol
- Acetone (CH_3COCH_3): 28 L atm/mol

Example 7.1:   Henry's Law for Solubility Data

Calculate Henry's law constant (H) from the solubility of SO_2 in pure water at 30°C and 1 atm utilizing the data in Table 7.1.

SOLUTION

HAND CALCULATION

The mole fraction of SO_2 in the gas phase, y, is calculated by dividing the partial pressure of SO_2 by the total pressure of the system:

TABLE 7.1
Equilibrium Data of SO_2/H_2O at 30°C and 101.32 kPa

C_{SO2} (g of SO_2 per 100 g of H_2O)	P_{SO2} (kPa) (Partial Pressure of SO_2)
0.5	6
2.0	24.3

Source: Data from Peytavy, J. L. et al. (1990). Chemical Engineering and Processing, 27(3), 155–163.

$$y_1 = \frac{P_{SO_2}}{P_T} = \frac{6 \text{ kPa}}{101.3 \text{kPa}} = 0.06$$

The mole fraction of SO_2 in the gas phase

$$y_2 = \frac{P_{SO_2}}{P_T} = \frac{24.3 \text{ kPa}}{101.3 \text{ kPa}} = 0.239$$

The SO_2 mole fraction in the liquid phase, x, is calculated by dividing the moles of SO_2 dissolved in the solution by the liquid's total moles.

$$x = \frac{\text{Moles of } SO_2 \text{ in solution}}{\text{Moles of } SO_2 \text{ in solution} + \text{Moles of } H_2O}$$

The SO_2 mole fraction in the liquid phase, x_1

$$x_1 = \frac{(0.5 \text{ g } SO_2)/(64 \text{ g/mol})}{(0.5 \text{ g } SO_2)/(64 \text{g/mol}) + (100 \text{ g } H_2O)/(18 \text{ g/mol})} = 0.0014$$

The SO_2 mole fraction in the liquid phase, x_2

$$x_2 = \frac{(2.0 \text{ g } SO_2)/(64 \text{ g/mol})}{(2.0 \text{ g } SO_2)/(64 \text{ g/mol}) + (100 \text{ g } H_2O)/(18 \text{ g/mol})} = 0.0056$$

Henry's constant

$$H = \frac{y}{x} \Rightarrow \text{slope} = \frac{\Delta y}{\Delta x} = \frac{0.239 - 0.06}{0.0056 - 0.0014} = 42.62$$

Example 7.2: Minimum Liquid Flow Rate

Compute the minimum liquid mass flow rate of pure water required to cause a 90% reduction in the composition of SO_2 from a gas stream of 85 m³/min containing 3% SO_2 by volume. The temperature is 303 K, and the pressure is 101.32 kPa.

SOLUTION

MANUAL CALCULATIONS

The mole fraction of SO_2 in the gas phase:
  $Y_1 = 0.03$ mole fraction SO_2 (i.e., 3% SO_2 by volume equivalent to 03% gas mole percent and 0.03 mole fraction)

$y_2 = 0.003$ mole fraction of SO_2 in the exit stream (calcualted such that: 90% reduction of SO_2 from inlet concentration, what left in the exit stream is 10% of y_1 which is $0.1 \times 0.03 = 0.003$)

At the minimum liquid flow rate, the gas mole fraction of the pollutants moving into the absorber (y_1) will be in equilibrium with the liquid mole fraction of the solute pollutants released from the absorber (x_1); the liquid will be saturated with SO_2. At equilibrium: $y_1 = Hx_1$

The Henry's constant (H):

$$H = 42.62 \frac{\text{Mole fraction of } SO_2 \text{ in air}}{\text{Mole fraction of } SO_2 \text{ in water}}$$

The mole fraction of SO_2 in the liquid released from the absorber to achieve the required removal rate:

$$x_1 = \frac{y_1}{H} = \frac{0.03}{42.67} = 0.0007$$

The minimum liquid feed rate (L_m from the material balance):

$$y_1 - y_2 = \frac{L_m}{G_m}(x_1 - x_2)$$

Therefore,

$$\frac{L_m}{G_m} = \frac{y_1 - y_2}{x_1 - x_2} = \frac{0.03 - 0.003}{0.0007 - 0} = 38.4 \frac{\text{mol of water}}{\text{mol of air}}$$

Determine the inlet gas flow rate by converting the inlet stream volumetric flow rate to the gas molar flow rate. At standard conditions (0°C and 1 atm), there are 0.0224 m³/mol of an ideal gas, correction to 30°C and 1 atm, the gas molar volume (V_m) at the column operation conditions.

$$V_m = \left(0.0224 \frac{m^3}{gmol}\right)\left(\frac{303 \text{ K}}{273 \text{ K}}\right) = 0.025 \text{ m}^3/\text{mol}$$

Therefore, the gas molar feed rate (G_m):

$$G_m = Q_G\left(\frac{1 \text{ gmol of air}}{0.025 \text{ m}^3}\right)$$

The gas stream volumetric flow rate (Q_G) at 30°C is 85 m³/min, hence,

$$G_m = \left(\frac{85 \text{ m}^3}{\text{min}}\right)\left(\frac{1 \text{ gmol}}{0.025 \text{ m}^3}\right) = 3400 \text{ mol/min}$$

The minimum liquid feed rate (L_m) from the calculated minimum liquid-to-gas ratio,

$$\left(\frac{L_m}{G_m}\right)_{min} = 38.4 \frac{\text{mol of water}}{\text{mol of air}}$$

Therefore, the minimum liquid water feed rate (L_m):

$$\left(L_m\right)_{min} = G_m \left(\frac{L_m}{G_m}\right)_{min}$$

Substitution,

$$\left(L_m\right)_{min} = 3400 \frac{\text{mol}}{\text{min}} \left(38.4 \frac{\text{mole water}}{\text{mole air}}\right) = 1.3 \times 10^5 \frac{\text{mol}}{\text{min}}$$

Converting molar feed rate to the units of mass flow rate,

$$\left(L_m\right)_{min} = \left(130.0 \frac{\text{kgmol water}}{\text{min}}\right)\left(\frac{18 \text{ kg}}{\text{kgmol}}\right) = 2340 \text{ kg/min}$$

7.3 NUMBER OF THEORETICAL STAGES

The height of a transfer unit is a function of the type of packing, liquid and gas flow rates, pollutant concentration and solubility, liquid properties, and system temperature. Tower height is primarily a function of packing depth. For most practical applications, the height of the transfer unit of the packed bed columns is within the range of 0.3–1.2 m (1–4 ft). The required height of packing (H_{pack}) is determined from the theoretical number of overall transfer units (N_{tu}) needed to achieve a specific removal rate and the height of the transfer unit (H_{tu}) [2]:

$$H_{pack} = N_{tu} H_{tu} \tag{7.3}$$

The number of overall transfer units may be estimated graphically by stepping off stages on the equilibrium-operating line graph from inlet conditions to outlet conditions or by the following equation:

$$N_{tu} = \frac{\ln\left[(y_i - mx_i/y_o - mx_i)(1-(1/AF))+1/AF\right]}{1-(1/AF)} \tag{7.4}$$

where the absorption factor

$$AF = \frac{L_m}{mG_m} \tag{7.5}$$

where AF is the absorption factor, m is the slope of the equilibrium line on a mole fraction basis. The value of m is obtained from literature on vapor-liquid equilibrium data for a specific system. Since the equilibrium curve is typically linear in the concentration ranges usually encountered in air pollution control, the slope m would be constant for all applicable inlet and outlet liquid and gas streams. The slope may be calculated from mole fraction values using Equation 7.6 [5]:

$$m = \frac{y_o^* - y_i^*}{x_o - x_i} \tag{7.6}$$

The entering solute mole fraction (y_i^*) and leaving solute mole fractions (y_o^*) of the pollutant in the vapor phase are in equilibrium with the liquid phase inlet pollutant mole fractions (x_i) and exiting mole fraction (x_o), respectively. Based on the following assumptions [6, 7]:

- Henry's law applies to a dilute gas mixture
- The equilibrium curve is linear from x_i to x_o
- The pollutant concentration in the solvent is dilute enough such that the operating line can be considered a straight line

If $x_i = 0$ (i.e., a negligible amount of pollutant enters the absorber in the liquid stream) and $1/AF = 0$ (i.e., the slope of the equilibrium line is minimal and the L_m/G_m ratio is substantial), Equation 7.4 is simplified to

$$N_{tu} = \ln\left(\frac{y_i}{y_o}\right) \tag{7.7}$$

Several methods calculate the overall transfer unit's height; based on empirically determined packing constants [7–9]. One commonly used method involves determining the overall gas and liquid mass transfer coefficients; k_G and k_L. A significant difficulty in using this approach is that k_G and k_L values are frequently unavailable for specific pollutant-solvent systems of interest. For this purpose, the method used to calculate the overall transfer unit's height depends on the height of the gas and liquid film transfer units, H_L and H_G, respectively [8–13].

7.4 NUMBER OF THEORETICAL STAGES USING GRAPHICAL TECHNIQUE

The graphical technique estimates the theoretical stages using the material balance on a differential element from the absorbers.

Mass Balance:

$$-Vdy = K_y a.S.dz(y - y^*) \tag{7.8}$$

Rearranging,

$$\frac{K_y a.S}{V} \int dz = \int_{y_1}^{y_{N+1}} \frac{dy}{y - y^*} \tag{7.9}$$

The design equation for the absorption of a solute from a dilute solution:

$$Z = \left(\frac{V}{K_y a.S}\right)\left(\int_{y_1}^{y_{N+1}} \frac{dy}{y - y^*}\right) \tag{7.10}$$

The following expression can determine the packing height (Z):

$$Z = HTU_{OG} \times NTU_{OG}$$

Where NTU_{OG} is the overall number of gas-phase transfer units, and HTU_{OG} is the overall height of a gas-phase transfer unit, and a is the interfacial area (m²/m³), V is the total gas molar flow rate (kgmol/s), S is the cross-sectional area of the tower (m²), and K_y is the volumetric film mass transfer coefficient (kgmol/s). The integral part of the equation is the total number of trays, as shown in Figure 7.1.

Example 7.3:   Number of Theoretical Stages

A gas absorber removes 90% of the SO_2 from the 3.0 mol% SO_2/air gas mixture with pure water. The inlet gas flow rate is 206 kmol/h. The liquid water flow rate is 12,240 kmol/h, the temperature of the feed streams is 20°C, and the pressure is 1 atm.

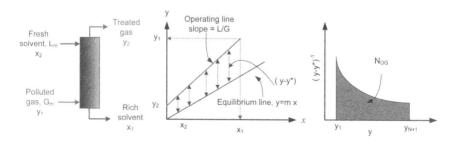

FIGURE 7.1 Schematic diagram of the operating and equilibrium lines in the absorption column and calculating the number of transfer units.

The packing material is 2 inches intalox saddles (Ceramic) randomly packed. Manually determine the number of theoretical stages required to achieve the necessary exit gas stream specifications. Henry's constant at the absorber operating conditions is 26. Compare manual calculations with the available commercial software packages (i.e., UniSim/Hysys, Aspen Plus, PRO/II, SuperPro, and Aveva Process Simulation).

SOLUTION

MANUAL CALCULATIONS

The number of theoretical stages in an absorber can be determined as follows:

$$N_{OG} = \frac{\ln\left[\left(y_1 - mx_2/y_2 - mx_2\right)\left(1 - \left(mG_m/L_m\right)\right) + mG_m/L_m\right]}{1 - mG_m/L_m}$$

The absorption factor ($AF = mG_m/L_m$):

$$AF = \frac{mG_m}{L_m} = \frac{26 \times 206 \text{ kmol/h}}{12240 \text{ kmol/h}} = 0.438$$

Substituting the absorption factor, the overall number of gas-phase transfer units (N_{OG}):

$$N_{OG} = \frac{\ln\left[\left(0.03 - 0/0.003 - 0\right)\left(1 - 0.438\right) + 0.438\right]}{1 - 0.438} = 3.2$$

HYSYS SIMULATION

After opening a new case in Hysys, select all components involved and the appropriate fluid package Peng-Robinson (PR), then in the simulation environment, choose the absorber from the object palette. Select the absorber icon from the object palette for this purpose. The feed streams are fully specified (i.e., air-in and water-in streams). The selection of the thermodynamic fluid package (non-random two liquids [NRTL]) is crucial. Using other fluid methods leads to different results. After trying various stages, three stages give the desired exit molar fraction of SO_2. Figure 7.2 shows the stream summary and the process flowsheet.

PRO/II SIMULATION

The selection of the appropriate thermodynamic model for the simulation of CO_2 or SO_2 in gas absorbers using water as a solvent is very important. Select the NRTL activity coefficient model to explain a liquid mixture's nonideal phase behavior between H_2O and SO_2. Henry's law option is also selected to calculate noncondensable supercritical gases such as H_2, CO, CO_2, CH_4, and N_2 in a liquid mixture. Click the Thermo tab, select Liquid Activity/NRTL, then click on the Add button. With a double click on the NRTL01, the thermodynamic data modification window pops up. Click on Enter Data of the NRTL vapor-liquid

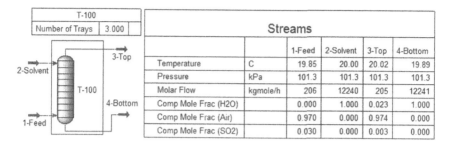

FIGURE 7.2 UniSim/Hysys generated the process flow diagram and streams summary for the SO₂ absorption case described in Example 7.3.

equilibria (VLE), k-values window pops up, and then click on Henry's law Enter Data as shown in Figure 7.3. Check the use of Henry's law for VLE of solute components. Figure 7.4 shows the stream summary result; the number of stages required to achieve the desired separation (i.e., the mole fraction of 0.003 in the exit air stream) is 4. The result obtained by PRO/II is closer to the hand-calculated results:

$$H = \exp\left(96.46 - \left(6706.144/293\right) - 12.3043\ln\left(293\right)\right) = 39.5$$

The value obtained by PRO/II is close to that obtained in Example 7.2 (H = 42.62).

ASPEN PLUS SIMULATION

In this example, we will be looking at ways to absorb gas impurities in Aspen Plus absorber module. The absorber is a device or a piece of equipment that removes one or more components from gas by absorbing it into a liquid solvent, and we

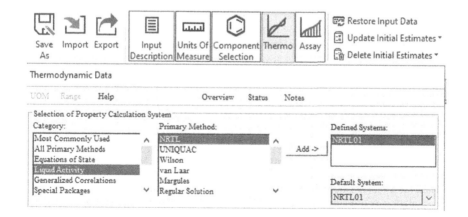

FIGURE 7.3 PRO/II selected the NRTL thermodynamic data to accommodate the absorption of CO₂ in the water system as described in Example 7.3.

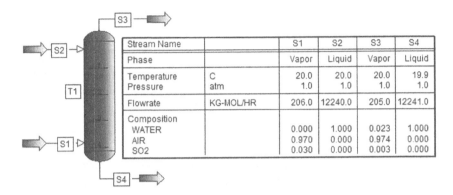

Stream Name		S1	S2	S3	S4
Phase		Vapor	Liquid	Vapor	Liquid
Temperature	C	20.0	20.0	20.0	19.9
Pressure	atm	1.0	1.0	1.0	1.0
Flowrate	KG-MOL/HR	206.0	12240.0	205.0	12241.0
Composition					
WATER		0.000	1.000	0.023	1.000
AIR		0.970	0.000	0.974	0.000
SO2		0.030	0.000	0.003	0.000

FIGURE 7.4 PRO/II generated the process flow diagram and streams summary for the SO_2 absorption system presented in Example 7.3.

define the stripper as one where we are going to remove one or more compounds from a liquid by mass transfer into the gas. The operation of both of these units is the same, and it is frequently convenient to call them both scrubbers. If you want to use a column for a scrubber, and when used for distillation, they would look the same. The only difference to the column itself would be that in a scrubber, you always have two different feed streams, one at the top and one at the bottom, and in a distillation tower, you can have any number of streams, and they could come in at any point in the column. The actual primary difference will be in the outlying equipment for distillation. The distillation column has to have a reboiler and a condenser and maybe pumps or other energy and flow management devices. In the case of scrubber, all needed is the tower itself, two feed streams, and two product streams.

Accordingly, we will be looking at how we can use RadFrac in Aspen Plus to model a scrubber. We will start our simulation by entering the components SO_2, air, and water. Next, when it comes to methods, we will enter NRTL as our method, but now, we do something a little different because I know that CO_2 and air should be noncondensable. I do not expect those to be found in the liquid phase, and they are only going to be in the vapor phase; hence specify them as Henry's component. Click on New and then type a name for the Henry compound, and you can name this anything you want, but it is convenient to use the name of your chemical itself. Make sure that all of the binary interaction parameters have now is established. In this case, it is Henry's law components. Then move on to the simulation page.

Choose the RadFrac block from the Columns subdirectory. Click on the down arrow next to the RadFrac block, a set of icons will pop up. These icons represent the same calculation procedure and are for various schematic purposes only. RadFrac is a rate-based nonequilibrium model for simulating all types of multistage vapor-liquid operations such as absorption, stripping, and distillation. RadFrac simulates actual tray and packed columns rather than the idealized representation of equilibrium stages. The next screen allows selecting the type of pressure specification to enter. Choose Top/Bottom and enter the pressure from the problem statement for segment 1. Segment 1 will refer to the first segment at the top of the tower and then click on Next. The packing material is 2 inch ceramic

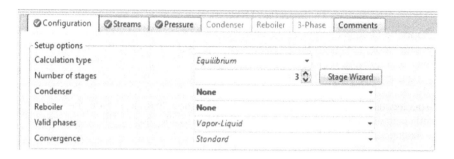

FIGURE 7.5 Aspen Plus required data in the configuration menu for the absorption case described in Example 7.3.

intalox saddles randomly packed. The packing height that gives the desired separation is the value found from hand calculation. The required size of packing necessary to achieve the split is 2 m. The column configuration is as shown in Figure 7.5. Click on Next to continue. The screen shown below asks to specify the location of feed inlets and outlets. Note that the gas stream should be outside the bottom stage (number of Stages + 1) because the stream location's convention is above the stage. Enter the following data (obtained from the manual calculation) in the column block specifications window:

- Property method: NRTL
- Number of stages: 3
- Diameter: 1 m (initial estimate)
- Height: 2 m
- Packing type: 2 inch ceramic intalox saddles
- Column pressure: 1 atm
- Condensor: None
- Reboiler: None

Enter the required inputs, and the simulation is ready to run. The result is obtained by clicking on the arrow next to the results heading. Figure 7.6 shows the Aspen Plus simulation result.

FIGURE 7.6 Aspen Plus generated a process flow diagram and stream summary using RadFrac columns for the case described in Example 7.3.

SuperPro Designer Simulation

SuperPro Designer simulates the absorber tower in two modes: design mode and rating mode. In Design Mode calculation, the user can specify the column diameter or the pressure drop/length. In User-Defined (Rating Mode), the user selects values for the number of units, column diameter, and height. In Design Mode, the material balances are specified by the user (% removed), and the separation specifications drive the equipment sizing calculation. In Rating Mode, the removal efficiency is either set by the user or calculated by the model. The basic steps for creating a design case:

1. Specify mode of operation (i.e., continuous or batch)
2. Register components and mixtures from:
   Tasks ≫ Register Components and Mixtures ≫ Pure Components.
   To register pure components available in the database: select chemical compounds needed for the simulation either by typing the name of a chemical in the entry box or by scrolling up and down in the Pure Component Database and selecting a chemical
3. Click Register
4. Unit procedures. To add a unit procedure, do the following: Unit Procedures ≫ [Type of Procedure] ≫ [Procedure], for example, to add and absorber, select unit Procedures ≫ Absorption/Stripping ≫ Absorber
5. Input and output streams

There are two modes in which the user can draw streams, Connect Mode and Temporary Connect Mode. Connect Mode is more convenient for drawing several streams at a time. To enter the connect mode, click on Connect Mode on the main toolbar; while in the Connect mode, the cursor will change to the Connect Mode Cursor. The simulation will remain in the Connect mode after drawing the stream(s). To leave the connect mode and return to select mode, click on the arrow next to Connect Mode. In the Connect Mode:

- Click once on an open area to begin drawing the stream.
- To change the direction of the stream, click once on the open area. The stream will bend at a 90° angle. Click once more to change the direction again. These direction changes are called stream elbows.
- Click once on an input port to connect the stream to the unit operation.

Enter Henry's constant in the Miscellaneous menu (Figure 7.7).

Tasks ≫ Edit Pure Components ≫ double click on the solute component. The inlet and exit streams are connected, as shown in Figure 7.8. Figure 7.9 shows the design components calculated with SuperPro. The estimated number of transfer units (NTUs) rounded off to 4.

Aveva Process Simulation

Perform the absorption column simulation in Aveva Process Simulation utilizing the Column icon in process library; a multistage fractionation of vapor and liquid mixtures. The Column is a single model used to simulate all possible column types such as distillater, absorber, and stripper. A typical workflow for column building

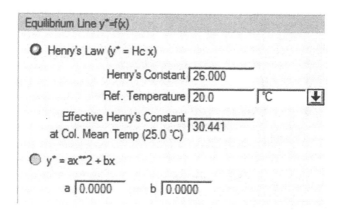

FIGURE 7.7 SuperPro required sulfur dioxide solute Henry's constant in the Miscellaneous menu for the case described in Example 7.3.

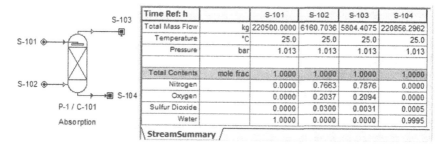

FIGURE 7.8 SuperPro generates the process flowsheet and streams summary for the case of absorption described in Example 7.3.

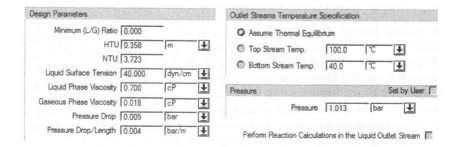

FIGURE 7.9 SuperPro designed the parameters menu, and the outlet streams temperature specification data required to solve the case absorption case presented in Example 7.3.

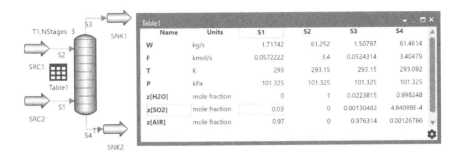

FIGURE 7.10   Aveva Process Simulation generated the process flow diagram and the stream conditions for the SO_2 absorption into the water (Example 7.3).

and preparing the simulation and achieving a squared/solved state, and building accurate absorption column is to certify that the feed information is accurate, and NRTL is the method of fluid property. We are making the base Column model a starting point to achieve initial conceptual design solution, and attain extended design for detailed engineering in process mode. Edit configuration and enter the following informations:

- Set the number of stages to 3.
- Set feed tray location: 1 for liquid solvent (top) and 3 for gas feed stream (bottom).
- Set Condenser type (None for Absorber).
- Set Reboiler (None for absorber).
- Set Setup to Solve.
- Set Contact between 0 to 1 (or vapor and liquid will bypass each other).

Figure 7.10 shows the squared and solved simulation of the absorption column to capture SO_2 from the air stream into the freshwater stream (solvent). The result shows the successful removal of SO_2 from the air with three stages absorption column.

7.5   PACKED BED COLUMN DIAMETER

The main parameter affecting the size of a packed column is the gas velocity. When the column's gas flow rate (at a fixed column diameter) reached a point where the liquid flowing down over the packing begins to be held in the void spaces between the filling, the gas-to-liquid ratio is termed the loading point. In other words, the loading point of a column is when the gas velocity is high enough to restrict the flow of liquid, after this point, the pressure drops at a much faster rate till another point, known as the flooding point, when all the liquid is carried away by the gas. It marks the start of the entrainment regime in columns. The column's pressure drop begins to increase, and the degree of mixing between the phases decreases. A further increase in gas velocity will cause the

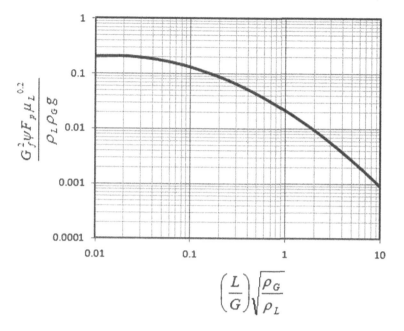

FIGURE 7.11   Eckert's correlation plotted at the flooding rate.

liquid to fill the void spaces in the packing. The liquid forms a layer over the top of the filling, and no more liquid can flow down through the tower. The pressure drop increases substantially, and mixing between the phases is minimal. This condition is referred to as flooding, and the gas velocity at which it occurs is the flooding velocity [9]. A typical operating range for the gas velocity through the columns is 50–75% of the flooding velocity; operating in this range; the gas velocity will be below the loading point. A standard and relatively simple procedure for estimating flooding velocity is to use a generalized flooding and pressure drop correlation as shown in Figure 7.11 [10]. Another factor influencing the tower diameter is the packing factor of the packing material used.

The packing factor is inversely proportional to the packing size, and the tower diameter requirements decrease as the packing material's size increases for the same inlet gas flow rate. The design is based on the gas flow rate under immersion conditions, the surface gas velocity (G_f), and the liquid-to-gas feed rate (L/G). The calculated surface gas velocity (G_f) can then determine the cross-column area (A) and the diameter (Dt). Figure 7.11 presents the relationship between G_f and the L/G ratio at the tower flood point. The abscissa value (x-axis) in the graph is stated as:

$$X_{axis} = \left(\frac{L}{G}\right)\sqrt{\frac{\rho_G}{\rho_L}} \tag{7.11}$$

where L and G are the liquid stream and gas stream's mass flow rate, respectively; ρ_G is the gas stream's density, and ρ_L is the absorbing liquid's density. The ordinate value (Y-axis) in the graph is expressed as

$$\gamma_{axis} = \frac{G_f^2 \psi \; F_p \mu_L^{0.2}}{\rho_L \rho_G g} \tag{7.12}$$

where G_f is the mass flow rate of the gas (kg/s m^2), ψ is the ratio of the density of the scrubbing liquid to water; F_p is the packing factor (m^2/m^3), μ_L is the viscosity of the solvent (cP), ρ_L is the liquid density (kg/m^3), ρ_G is the gas density (kg/m^3), and g is the gravitational constant, (9.82 m/s^2, 32.2 ft/s^2). The value of the packing factor, F_p, is obtained from Table 7.2 or vendors.

After calculating the Abscissa (X_{axis}) value, determine the corresponding coordinate (Y_{axis}) value from the flooding curve. The Y_{axis} ordinate calculated using the following equation:

$$Y_{axis} = 10^{\xi} \tag{7.13}$$

where $\xi = -1.668 - 1.085 \left(\log X_{axis} \right) - 0.297 \left(\log X_{axis} \right)^2$

Rearrange Equation 7.2 to solve for G_f:

$$G_f = \left[\frac{\rho_L \rho_G \; g \left(Y_{axis} \right)}{F_p \psi \mu_L^{0.2}} \right]^{0.5} \tag{7.14}$$

TABLE 7.2
Design Data for Various Random Packing

Packing Type	Size Inch (mm)	Weight (kg/m^3)	Surface Area (m^2/m^3)	Packing Factor F_pm^{-1}
Ceramic Raschig Rings	1/2 (13)	881	368	2,100
	1 (25)	673	190	525
	11/2 (38)	689	128	310
	2 (51)	651	95	210
	3 (76)	561	69	120
Ceramic Intalox Saddles	1/2 (13)	737	480	660
	1 (25)	673	253	300
	11/2 (38)	625	194	170
	2 (51)	609	108	130

Source: Data from Sinnott, R. K. (1999). Coulson & Richardson's Chemical Engineering. Vol. 6, 3rd edn., Butterworth Heinemann, Oxford.

where ρ_G and ρ_L in kg/m³, g = 9.82 m/s², F_p packing factor in m²/m³, and μ_L in cP; ψ is the ratio of specific gravity of the scrubbing liquid to that of water. Calculate the cross-sectional area of the tower as follows:

$$A = \frac{G_m}{f\,G_f} \tag{7.15}$$

f is a flooding factor. To prevent flooding of the column, operate at a fraction of G_f. The value of f typically ranges from 0.60 to 0.85. Calculate the diameter of the column from the cross-sectional area [9, 10]:

$$D_t = \sqrt{\frac{4A}{\pi}} \tag{7.16}$$

As a rule of thumb, the column's diameter should be at least 15 times the size of the packing used in the column. If this is not the case, the column diameter is recalculated using a smaller-diameter packing [10]. The superficial liquid flow rate entering the Absorber, L_f (lb/h-ft²), based on the cross-sectional, is

$$L_f = L/A \tag{7.17}$$

For the absorber to operate correctly, the liquid flow rate entering the column must be high enough to effectively wet the packing so that mass transfer between the gas and liquid can occur. Calculate the minimum value of $(L_f)_{min}$ required to wet the packing effectively using Equation 7.13:

$$\left(L_f\right)_{min} = MWR\ \rho_L a \tag{7.18}$$

where the minimum wetting rate (MWR) is in the units (m²/h), and a is the surface area-to-volume ratio of packing (m²/m³). The appropriate value of MWR is 0.079 m²/h, for ring packing larger than 3.0 inches and for structured grid packing. For other types of packing, an MWR value of 0.12 m²/h is recommended [7, 13].

Example 7.4: Absorption of SO_2 from Air with Pure Water

A gas absorber is employed to capture 90% of the SO_2 from a gas stream with pure water. The mass flow rate of the feed gas is 1.72 kg/s. The gas stream contains 3.0 mol% SO_2 in the air. The minimum liquid flow rate was 41 kg/s. The feed stream temperature is 20°C, and the pressure is 1 atm. The gas velocity should be no greater than 70% of the flooding velocity, and the randomly packed material is 2 inches intalox saddles (Ceramic). The liquid mass flow rate is 1.5 times the minimum fluid mass flow rate. Manually determine the packed column diameter. Compare manually obtained diameter with those predicted by the available software packages (UniSim, Aspen Plus, PRO/II, SuperPro, and Aveva Process Simulation).

SOLUTION

MANUAL CALCULATIONS

The liquid mass flow rate is 1.5 times the minimum liquid flow rate:

$$L = 1.5 \times L_m = 1.5 \times 2450 = 3675 \text{ kg/min}$$

The superficial flooding velocity is the flow rate per unit of the cross-sectional area of the tower.

$$\text{Abscissa} = \left(\frac{L}{G}\right)\left(\frac{\rho_g}{\rho_l}\right)^{0.5} = \left(\frac{3,675\text{kg/min}}{103 \text{ kg/min}}\right)\left(\frac{1.17 \text{ kg/m}^3}{1000 \text{ kg/m}^3}\right)^{0.5} = 1.22$$

Figure 7.11, with the abscissa of 1.22, moves up to the flooding line; the Y-axis ordinate value is 0.02.

$$G_f = \left[\frac{\rho_L \rho_G \, g\left(Y_{\text{axis}}\right)}{F_p \psi \mu_L^{0.2}}\right]^{0.5}$$

where $\rho_G = 1.17$ kg/m^3, density of air at 30°C; $\rho_L = 1,000$ kg/m^3, density of water at 30°C; $g = 9.82$ m/s^2, the gravitational constant; $F_p = 131$ m^2/m^3, the packing factor for 2 inch; Ceramic intalox saddles; $\psi = 1.0$, the ratio of specific gravity of the scrubbing liquid to that of water; and $\mu_L = 0.8$ cP, the viscosity of the liquid. The superficial flooding velocity at flooding:

$$G_f = \left[\frac{\left(1.17 \text{ kg/m}^3\right)\left(1000 \text{ kg/m}^3\right)\left(9.82 \text{ m/s}^2\right)\left(0.02\right)}{\left(131 \text{ m}^2/\text{m}^3\right)(1)\left(0.8 \text{ cP}\right)^{0.2}}\right]^{0.5} = 1.354 \frac{\text{kg}}{\text{s.m}^2}$$

The superficial gas velocity at operating conditions (G_{op}), where the absorber works at 70% of the flooding velocity, hence

$$G_{op} = f \times G_f, \text{ where } f = 0.70$$

Substituting the values,

$$G_{op} = (0.7)\left(1.354 \frac{\text{kg}}{\text{s.m}^2}\right) = 0.95 \frac{\text{kg}}{\text{s.m}^2}$$

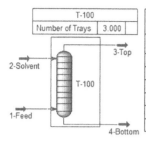

T-100			Streams				
Number of Trays	3.000			1-Feed	2-Solvent	3-Top	4-Bottom
Temperature		C		19.85	20.00	20.02	19.89
Pressure		kPa		101.3	101.3	101.3	101.3
Molar Flow		kgmole/h		206	12240	205	12241
Comp Mole Frac (H2O)				0.000	1.000	0.023	1.000
Comp Mole Frac (Air)				0.970	0.000	0.974	0.000
Comp Mole Frac (SO2)				0.030	0.000	0.003	0.000

FIGURE 7.12 UniSim generated process flow diagram and streams summary for the SO_2 absorption case described in Example 7.4.

The cross-sectional area of the packed tower is

$$A = \frac{G}{G_{op}} = \frac{(103 \text{ kg/min})(1 \text{ min/60s})}{0.95 \frac{\text{kg}}{\text{s.m}^2}} = 1.81 \text{ m}^2$$

The manually calculated tower diameter

$$D_t = \left(\frac{4A}{\pi}\right)^{0.5} = \left(\frac{4 \times 1.81 \text{ m}^2}{3.14}\right)^{0.5} = 1.52 \text{ m}$$

UNISIM SIMULATION

Select the Absorber icon from the object palette for this purpose. The feed streams are fully specified (i.e., air-in and water-in streams). Set the number of stages to three. NRTL is the appropriate fluid package. Figure 7.12 depicts the process flowsheet and the stream summary. The diameter of the tower is determined using Tools/utilities/trays sizing. The diameter obtained by UniSim tray sizing is 1.676 m, slightly different from that obtained by manual calculations.

PRO/II SIMULATION

I am using the data obtained in the manual calculations found in Example 7.3. The thermodynamic fluid package NRTL defines the system. Figure 7.13 represents the process flow diagram and stream summary. Figure 7.14 displays the column diameter (2.0 m) calculated with PRO/II, as shown in the captured part from the generated text report. The text report is generated from: Output ≫ Generate a text report.

ASPEN PLUS SIMULATION

Select the Absorber column from RadFrac in the Column subdirectory. After specifying the feed streams using NRTL for the property method, enter the following

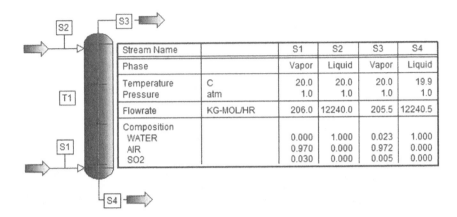

Stream Name		S1	S2	S3	S4
Phase		Vapor	Liquid	Vapor	Liquid
Temperature	C	20.0	20.0	20.0	19.9
Pressure	atm	1.0	1.0	1.0	1.0
Flowrate	KG-MOL/HR	206.0	12240.0	205.5	12240.5
Composition					
WATER		0.000	1.000	0.023	1.000
AIR		0.970	0.000	0.972	0.000
SO2		0.030	0.000	0.005	0.000

FIGURE 7.13 PRO/II generates the process flow diagram and stream summary for the absorption column described in Example 7.4.

data (obtained from the manual calculation for validation) in the column block specifications window:

- Number of stages: 3
- Diameter: 1 m (initial estimate)
- Height: 2 m
- The Packing type is Ceramic intalox saddles, 2 inches.
- Column pressure: 1 atm

After entering the required data, the system is ready to be run. Figure 7.15 depicts the Aspen Plus simulation results.

The packed bed column diameter is obtained from the column internals design specifications. While in the browser menu, click on Column Internals under Block, then click on Add new, double click on INT-1, and click on Add New again to generate a window named CS-1 as a default name. Fill in the following information:

- Start stages: 1 and the end-stage: 3.
- Internal type: Packing

```
ID                        =   COLSECT-1
PACKED HEIGHT             =   2.67 M
COLUMN ID                =   2.000 M
PACKING TYPE (NORTON)    =   Super INTALOX saddles, ceramic
PACKING FACTOR           =   130.00 M2/M3
NOMINAL PACKING DIAMETER =   51.000 MM
CAPACITY METHOD          =   fraction of flood
HETP METHOD              =   Norton
DP METHOD                =   Norton
```

FIGURE 7.14 PRO/II estimated the packed column height and diameter for the absorption system described in Example 7.4.

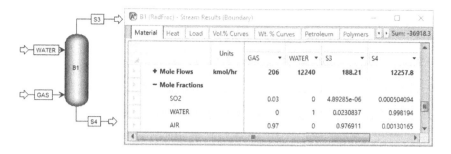

FIGURE 7.15 Aspen Plus generated a process flow diagram and streams summary using RadFrac columns for the case described in Example 7.4.

- Tray/packing type: INTX (intalox saddles).
- Material: Ceramic.
- Dimension: 2 inches.
- Packing height: 2 meters.

Click on View under details; the popup menu should look like as shown in Figure 7.16.

SuperPro Designer Simulation

Following the same procedure of Example 7.3, Section 7.3.5 provides the value of Henry's law constant, i.e., 6.34×10^{-4} atm m³/mol. This value obtained from the literature is entered in the Physical (constants) and then double click on the solute component name in the miscellaneous section: Tasks ≫ Edit Pure Components ≫ double click on the solute component (e.g., sulfur dioxide) (Figure 7.17). Figure 7.18 shows the design components calculated with SuperPro. The packing material is 2-inch ceramic intalox saddles [10]. Figure 7.19 shows the column specifications.

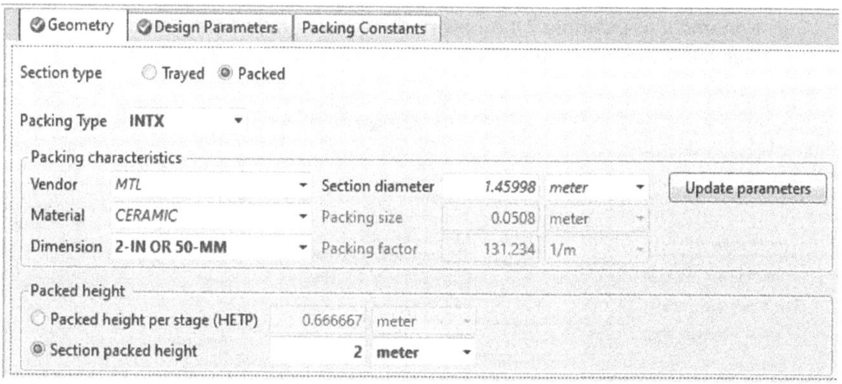

FIGURE 7.16 Aspen Plus generated the packed bed absorber diameter for the absorption case described in Example 7.4.

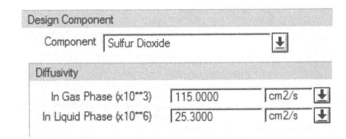

FIGURE 7.17 SuperPro generated the pure component properties for the case described in Example 7.4.

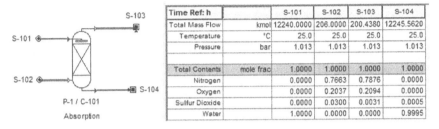

FIGURE 7.18 SuperPro generated process flowsheet and stream summary for the absorption case presented in Example 7.4.

AVEVA PROCESS SIMULATION

Start Aveva Process Simulation software, create a new simulation, rename it (e.g., Example 7.4). Select the DefFluid from Fluid's menu and choose the NRTL fluid package. Fully specify the gas stream and freshwater feed stream (W, T, P, and composition). Connect feed streams to the column inlet. Edit the configuration page:

1. Drag a Column object from the Process library onto the Canvas, next to feed.
2. Rename it to Example 7.4 by overwriting its default name directly on the Canvas.

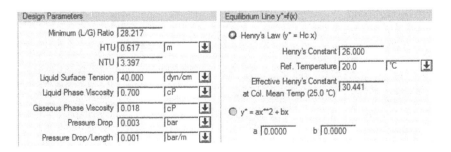

FIGURE 7.19 SuperPro specified the column design options and the equilibrium line for the absorption case described in Example 7.4.

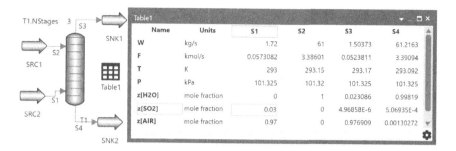

FIGURE 7.20 Aveva Process Simulation generated a process flow diagram to absorb SO_2 from a polluted air stream into the freshwater stream, the case described in Example 7.4.

3. Connect feed streams (S1 and S2).
4. Expand the column icon to an appropriate size by acting on its corner blocks.
5. Set the number of stages to 3.
6. Set the two feeds' location stages; 1 for liquid solvent (top) and 3 for gas feed stream (bottom).
7. Setup Condenser and Reboiler to "None" for Absorbers.
8. Set Setup to Solve.
9. Set Contact between 0 to 1; here, it is 0.6 (otherwise, vapor and liquid will bypass each other).
10. The packing material is available in the module "Stichlmair." Copy that model from the Process library to Example 7.4 model library. Select the packing material, as shown in Figure 7.20. The figure shows a squared and solved simulation of the absorption column to capture SO_2 from the air stream into the propylene carbonate stream (solvent). The result shows the successful removal of SO_2 from the air with three stages column.

Figure 7.20 shows the squared and solved simulation of the absorption column to capture SO_2 from the air stream into the freshwater stream (solvent). The result shows the successful removal of SO_2 from the air with three stages column.

7.6 PACKED-TOWER HEIGHT

The height of a packed column refers to the depth of packing material needed to accomplish the required removal efficiency; the more complicated the separation, the larger the height of packing necessary. Determining the proper size of packing is essential since it affects both the rate and absorption efficiency. Based on diffusion principles, several theoretical equations predict the required packing height [4].

7.6.1 ESTIMATION OF H_{OG} USING ONDE'S METHOD

The film mass-transfer coefficients k_G and k_L and effective wetted area of packing a_w are calculated using the correlation developed by Onda et al [11], which estimates H_G and H_L.

$$a_w = a_p \left\{ 1 - \exp \left[(-1.45) \left(\frac{\sigma_c}{\sigma} \right)^{0.75} \left(\frac{L}{a_p \, \mu_L} \right)^{0.1} \left(\frac{L^2 a_p}{\rho_L^2 g} \right)^{-0.05} \left(\frac{L^2}{\rho_L \sigma \, a_p} \right)^{0.2} \right] \right\} \quad (7.19)$$

k_G is calculated from the following correlations:

$$k_G \left(\frac{RT}{a_p D_G} \right) = 5.23 \left(\frac{G}{a_p \mu_G} \right)^{0.70} \left(Sc_G \right)^{1/3} \left(a_p d_p \right)^{-2} \quad (7.20)$$

k_L is calculated from the following correlations:

$$k_L \left(\frac{\rho_L}{g \mu_L} \right)^{1/3} = 0.0051 \left(\frac{L}{a_w \mu_L} \right)^{2/3} \left(Sc_L \right)^{-1/2} \left(a_p d_p \right)^{0.40} \quad (7.21)$$

where k_G is the gas mass transfer coefficient, kmol/m²s atm, and k_L is the liquid film mass transfer coefficient, kmol/m²s. a_p is the total packing surface area per packed bed volume (m²/m³), d_p is the packing size (m), L and G are the superficial mass velocity of liquid and gas (kg/m²s), μ_L is the liquid phase viscosity (kg/m.s), ρ_L is the liquid phase density (kg/m³), σ_L is the water surface tension (N/m), $\sigma_c = 61$ (dynes/cm) for ceramic packing, 75 dyne/cm for steel packing, and 33 dyne/cm for plastic packing; D_G and D_L are diffusivities in the gas and liquid phase, respectively, m²/s; R = 0.08314 m³ bar/kmol K; and g = 9.81 m/s². The d_p is the equivalent diameter of the packing calculated using the following equations:

$$d_p \, (m) = \frac{6}{a_p} (1 - \varepsilon) \quad (7.22)$$

The diffusion coefficient for the gas,

$$D_{AB,G} \left(\frac{cm^2}{s} \right) = \frac{0.001 T^{1.75} \left[M_{WA} + M_{WB} / M_{WA} . M_{WB} \right]}{P \left(a^{1/3} . b^{1/3} \right)} \quad (7.23)$$

where a and b are the atomic diffusion volumes of solute A in inert gas phase B, cm³/mol; M_{WA} and M_{WB} are the molecular weights of A and B, respectively; and P is the pressure, atm.

$$D_{AB,L} \left(\frac{cm^2}{s} \right) = \frac{9.89 \times 10^{-8} . V_B^{0.265} . T}{V_A^{0.45} \mu_L^{0.907}} \quad (7.24)$$

where V_A and V_B are the molar volume at the standard boiling point of solute A in liquid phase B, cm³/mol; T is the liquid stream temperature, K; μ_L is the liquid viscosity, kg/m s; and μ_G is the gas viscosity, kg/m s; the heights of transfer unit for the gas phase (H_G) and the liquid phase (H_L) are

$$H_G = \frac{G_m}{k_G.a_w.P}[=]\frac{(kmol/m^2s)}{(kmol/m^2s\ bar)(1/m)(bar)}[=]m \tag{7.25}$$

H_G is calculated from the following correlations:

$$H_L = \frac{L_m}{k_L.a_w.C_T}[=]\frac{(kmol/m^2s)}{(m/s)(1/m)(kmol/m^3)}[=]m \tag{7.26}$$

where P is the column pressure, atm; C_T is the total concentration, kgmol/m³ (ρ_L/molecular weight of solvent); G_m is the molar gas flow rate per cross-sectional area, kgmol/m² s; and L_m is the molar liquid flow rate per unit cross-sectional area, kgmol/m²s. The overall height of transfer units (HTUs),

$$H_{OG} = H_G + \frac{mG_m}{L_m}H_L \tag{7.27}$$

7.6.2 Estimation of H_{OG} Using Cornell's Method

The empirical equations for predicting the height of the gas and liquid film transfer units [9] is as follows:

The height of the liquid film transfer units, H_L

$$H_L = 0.305 \times \phi_h (Sc_L)^{0.5} \times K_3 \left(\frac{Z}{3.05}\right)^{0.15} \tag{7.28}$$

The height of the gas film transfer units, H_G

$$H_G = \frac{0.011 \times \psi_h (Sc_G)^{0.5}\left(\dfrac{D_c}{3.05}\right)^{1.11}\left(\dfrac{Z}{3.05}\right)^{033}}{(Lf_1f_2f_3)^{0.5}} \tag{7.29}$$

The term of diameter correction is 2.3 for a diameter greater than 0.6 m.

$$H_G = \frac{0.011 \times \psi_h (Sc_G)^{0.5}(2.3)\left(\dfrac{Z}{3.05}\right)^{033}}{(Lf_1f_2f_3)^{0.5}} \tag{7.30}$$

where K_3, ϕ_h, and ψ_h can be found elsewhere [9, 10], the following correlation is used, the percentage flooding correction factor (K_3) equals to one for the flooding percentage (F) less than 45, and for a higher percentage of flooding:

$$K_3 = -0.014(F) + 1.685 \tag{7.31}$$

Factors for H_G using Berl saddles, particle size $1\frac{1}{2}$ inch (38 mm), ψ_h

$$\psi_h = 9 \times 10^{-4} F^3 - 0.12F^2 + 5.29F + 0.834 \tag{7.32}$$

The factor for H_L using particle size $1\frac{1}{2}$ inch (38 mm), ϕ_h

$$\phi_h = 0.034 L_f^{0.4} \tag{7.33}$$

where L_f is the liquid mass velocity; kg/m^2 s.
Schmidt numbers for gas and liquid phases

$$Sc_G = \frac{\mu_G}{\rho_g D_G}, \ Sc_L = \frac{\mu_L}{\rho_L D_L}$$

Liquid properties correction factor:

$$f_1 = \left(\frac{\mu_L}{\mu_w}\right)^{0.16} \tag{7.34}$$

$$f_2 = \left(\frac{\rho_w}{\rho_L}\right)^{1.25} \tag{7.35}$$

$$f_3 = \left(\frac{\sigma_w}{\sigma_L}\right)^{0.8} \tag{7.36}$$

where $\mu_w = 1.0$ mPa s(1 cP), $\rho_w = 1000 \text{kg/m}^3$, $\sigma_w = 72.8$ mN/m. The height of the overall transfer unit

$$H_{OG} = H_G + \frac{mG_m}{L_m} \times H_L \tag{7.37}$$

Multiplying both sides of the equation by N_{OG},

$$N_{OG} \times H_{OG} = N_{OG} \times \left[H_G + \frac{mG_m}{L_m} \times H_L \right] \qquad (7.38)$$

Since

$$Z = H_{pack} = N_{OG} \times H_{OG}$$

The total height of the column is calculated from the following correlation:

$$H_{tower} = 1.4 \ H_{pack} + 1.02 \ D + 2.81 \qquad (7.39)$$

Equation 7.34 was developed from the information reported by gas absorber vendors and is applicable for column diameters from 2 to 12 feet (0.6–3.6576 m) and packing depths from 4 to 12 ft (1.2–3.6576 m). The surface area (S) of the gas absorber is calculated using the equation,

$$S = \pi D \left(H_{tower} + \frac{D}{2} \right) \qquad (7.40)$$

Example 7.5: Absorber Total Packing Height

The calculated value of the H_{OG} of the SO_2/H_2O system is 0.6 m. The packing material was Raschig ring ceramic, 2 inches. The process operating conditions are 20°C and 1 atm. Calculate the total height of packing required to achieve a 90% reduction in SO_2 the inlet concentration. Consider the following data (from Example 7.4):

H: The Henry's constant = 26
G_m: The molar flow rate of the inlet gas stream = 206 kmol/h
L_m: The molar flow rate of liquid stream = 12,240 kmol/h
x_2: The mole fraction of solute in entering liquid stream = 0 (pure water)
y_1: The mole fraction of solute in entering gas stream = 0.03
y_2: The mole fraction of solute in the exiting gas stream = 0.003

SOLUTION

MANUAL CALCULATIONS

The number of theoretical transfer units

$$N_{OG} = \frac{\ln \left[\left(y_1 - \frac{m.x_2}{y_2} - mx_2 \right) \left(1 - \frac{m.G_m}{L_m} \right) + \frac{m.G_m}{L_m} \right]}{1 - \left(\frac{m.G_m}{L_m} \right)}$$

The value of the inverse of the slope of the operating line to the equilibrium line:

$$\frac{m.G_m}{L_m} = \frac{26 \times 206}{12,240} = 0.438$$

Substituting values:

$$N_{OG} = \frac{\ln\left[\left(0.03 - \frac{0}{0.003} - 0\right)(1 - 0.438) + 0.438\right]}{1 - 0.438} = 3.2$$

Rounding off the number of transfer stages to 4
The total packing height, Z,

$$Z = H_{OG} \times N_{OG}$$

Given; $H_{OG} = 0.60$ m is the overall height of a transfer unit.
$Z = (0.6 \text{ m})(3.2) = 1.922$ m of packing height.

HYSYS/UNISIM SIMULATION

Following the same procedure in Example 7.3, employing tray sizing utilities in Hysys, the simulated packing height is 1.8 m (Figure 7.21). The column with three stages, achieving a mole fraction in the exit air, is 0.003.

PRO/II SIMULATION

Using the converged file in Example 7.3 with three stages, PRO/II estimates the packed column's height by double clicking on the column icon, and then the Tray Hydraulics/Packing and entering the packing materials (Figure 7.22a). Click on Enter Data and enter the type of packing, packing size, and packing factor (Figure 7.22b). Figure 7.22c shows the captured section from the Output

Section Results
○ Trayed   ● Packed   Export Pressures   View Warnings...

Packing Results

Section	Section_1		
Internals	Packed		
Section Diameter [m]	1.829		
Max Flooding [%]	58.01		
X-Sectional Area [m2]	2.627		
Section Height [m]	1.800		

FIGURE 7.21   UniSim/Hysys generates the packed bed absorber diameter and height for the case presented in Example 7.5.

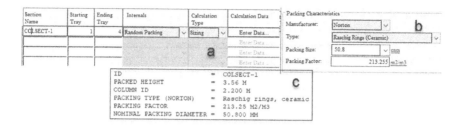

FIGURE 7.22   PRO/II tray hydraulics page (a), packing characterization menu (b), and generated height of the packed column (c).

generated text report that includes the simulated column packed height and column diameter. Figure 7.23 depicts the PFD and stream summary generated with the PRO/II software package.

ASPEN PLUS SIMULATION

In this case, Select RadFrac absorber from the Column subdirectory. Specify feed streams and connect outlet streams. Select NRTL for the property method. The number of stages is 3 trays. The height of the packed column is the manipulated variable (Figure 7.24). The simulated packed bed column with three stages achieves a 0.003 mol fraction of SO_2 in the exit air stream, as shown in Figure 7.24 (the conditions are the same as Example 7.3). The simulation is converged and the results displayed a packing height of 1.39 m.

SUPERPRO DESIGNER

Following the same procedure in Example 7.3. Right click on the column and select equipment data. The SuperPro predicted diameter is 1.5 m, and the column height is 3.063 (Figure 7.25).

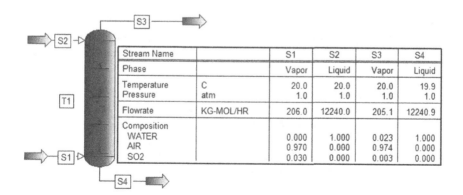

Stream Name		S1	S2	S3	S4
Phase		Vapor	Liquid	Vapor	Liquid
Temperature	C	20.0	20.0	20.0	19.9
Pressure	atm	1.0	1.0	1.0	1.0
Flowrate	KG-MOL/HR	206.0	12240.0	205.1	12240.9
Composition					
WATER		0.000	1.000	0.023	1.000
AIR		0.970	0.000	0.974	0.000
SO2		0.030	0.000	0.003	0.000

FIGURE 7.23   PRO/II generated the process flow diagram and streams summary for the absorption process case presented in Example 7.5.

Sections									
	Start Stage	End Stage	Diameter	Section Height	Internals Type	Tray Type or Packing Type	Section Pressure Drop	% Approach to Flood	Limiting Stage
CS-1	1	3	1.4777 meter	1.8 meter	PACKING	RASCHIG	0.0120346 bar	80	3

FIGURE 7.24 Aspen Plus predicted the packed bed height of the absorber and diameter for the case in Example 7.5.

AVEVA PROCESS SIMULATION

Select the DefFluid from Fluid's menu and for the method, choose the NRTL fluid package. Fully specify the gas stream and freshwater feed stream (W, T, P, and composition). Connect feed streams to the column inlet, and then process as follows:

1. Drag a Column object from the Process library onto the Canvas (simulation area), place next to feed steam.
2. Rename it to Example 7.5 by overwriting its default name directly on the Canvas.
3. Connect feed streams (S1 and S2).
4. Expand the column icon to an appropriate size by acting on its corner blocks.
5. Set the number of stages to 4.
6. Set the two feeds' location stages; 1 for liquid solvent (top) and 3 for gas feed stream (bottom).
7. Setup Condenser and Reboiler to "None" for Absorbers.
8. Set Setup to Solve.
9. Set Contact between 0 to 1; here, it is 0.6 (otherwise, vapor and liquid will bypass each other).

Figure 7.26 shows the squared and solved simulation of the absorption column to capture SO_2 from the air stream into the freshwater stream (solvent). The result shows the successful removal of SO_2 from the air with a 3.0 stages column.

Example 7.6:   Number of Theoretical Stages

Pure water absorbs SO_2 present in the air at 1 atm and 20°C. The gas feed stream is 5,000 kg/h; it contains 8 mol% SO_2 in the air. A 95% recovery of the sulfur dioxide is required. The absorbent is pure water with a feed rate of 29.5 kg/s.

Max Column Diameter	4.000	m	
Column Diameter	0.067	m	
Column Height	1.534	m	
Column Volume	5.382	L	

FIGURE 7.25 SuperPro estimated packed tower diameter and height for the case presented in Example 7.5.

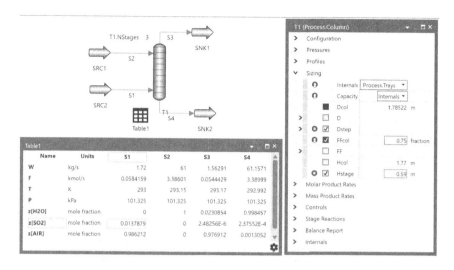

FIGURE 7.26 Aveva Process Simulation generated the process flow diagram of the packed bed absorber column, the process described in Example 7.5.

The packing material is Ceramic intalox saddles (2 inches). Design an absorber column for this purpose [9]. Compare manual calculation with those obtained from UniSim/Hysys, PRO/II, Aspen Plus, and Aveva Process Simulation. Table 7.3 shows the solubility of SO_2 in water.

SOLUTION

MANUAL CALCULATIONS

The equilibrium mole fraction of SO_2 in liquid

$$x = \frac{(0.05/64)}{0.05/64 + (100 - 0.05)/18} = 0.00014$$

TABLE 7.3
Solubility Data of SO_2

Mass Fraction SO_2 in Water	SO_2 Partial Pressure in Air (mmHg)
0.05	1.2
0.15	5.8
0.30	14
0.50	26
0.70	39
1.00	59

Source: Data from Perry, J.H. (1973). *Chemical Engineers' Handbook*, 5th edn, McGraw-Hill, New York, NY.

TABLE 7.4

Mole Fraction of SO₂ in Liquid Phase and Gas-Phase at Equilibrium

Mole Fraction of SO₂ in Water (x)	Mole Fraction of SO₂ in Air (y)
0.00014	0.002
0.00042	0.008
0.00085	0.018
0.00141	0.034
0.00198	0.051
0.00283	0.077

The equilibrium mole fraction of SO₂ in the gas phase

$$y = \frac{1.2 \text{ mmHg}}{760 \text{ mmHg}} = 0.002$$

Other values are calculated in the same manner, as shown in Table 7.4. The plot of the equilibrium line is shown in Figure 7.27.

The slope of the equilibrium line in Henry's law constant is 26 (i.e., m = 26). In this section, the number of theoretical stages, the column diameter, and column height are to be calculated as follows:

The gas molar flow rate, G_m

$$(G_m) = \frac{5000}{3600} = 1.39 \text{ kg/s} = \frac{1.39}{29} = 0.048 \text{ kmol/s}$$

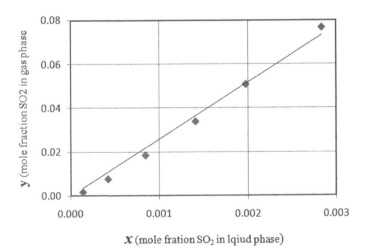

FIGURE 7.27 The equilibrium data of mole fraction of SO₂ in the gas phase versus SO₂ in the liquid phase.

The liquid molar flow rate, L_m

$$(L_m) = (29.5 \text{ kg/s}) = \left(\frac{\text{kmol}}{18 \text{ kg}}\right) = 1.64 \text{ kmol/s}$$

The number of theoretical stages:

$$N_{OG} = \frac{\ln\left[(y_1 - mx_2/y_2 - mx_2)(1 - mG_m/L_m) + mG_m/Lm\right]}{1 - mG_m/L_m}$$

Substituting values in the above equation:

$$N_{OG} = \frac{\ln\left[(0.08 - 0/0.004 - 0)(1 - (26)(0.048)/1.64) + (26)(0.048)/1.64\right]}{1 - (26)(0.048)/1.64} = 6.15$$

The calculated number of theoretical palettes is 6.15, rounded up to 7 theoretical trays.

Column Diameter

The gas' physical properties can be taken as those for air, as the concentration of SO_2 is low. The packing material: 38 mm (1 1/2 inches) ceramic intalox saddles, $F_p = 170 \text{ m}^{-1}$

The gas density

$$(\rho_g) \text{ at } (20°C) = \frac{29}{22.4} \times \frac{273}{293} = 1.21 \text{ kg/m}^3$$

The liquid density (ρ_L):

$$\rho_L \cong 1000 \text{ kg/m}^3$$

The liquid viscosity (μ_L):

$$\mu_L = 10^{-3} \text{Pas} = \text{Ns/m}^2$$

The x-axis of Figure 7.27 [9]:

$$\frac{L}{G}\sqrt{\frac{P_g}{P_L}} = \frac{29.5 \text{ kg/s}}{1.39 \text{ kg/s}}\sqrt{\frac{1.21 \text{ kg/m}^3}{10^3 \text{ kg/m}^3}} = 0.74$$

Design for a pressure drop of 20 mm H_2O/m packing from Figure 7.27 [9], Y-axis = 0.34

At flooding, Y-axis = 0.8

Percentage of flooding = $\sqrt{\dfrac{0.34}{0.8}} \times 100 = 66\%$, satisfactory.

The gas mass velocity at flooding (G_f):

$$G_f = \left[\frac{Y_{axis}\rho_G\left(\rho_L - \rho_G\right)}{1.31 F_P\left(\mu_L/\rho_L\right)} \right]^{1/2} = \left[\frac{0.34 \times 1.21\left(1000 - 1.21\right)}{1.31 \times 170\left(10^{-3}/10^3\right)^{0.1}} \right]^{1/2} = 0.87 \text{ kg/m}^2\text{s}$$

For 66% of flooding:

$$\text{Column area required} = A_c = \frac{G}{G_f} = \frac{1.39 \text{ kg/s}}{1.39 \text{ kg/m}^2 \text{ s}} = 1.6 \text{ m}^2$$

The diameter of the packed bed tower (D_t)

$$D_t = \sqrt{\frac{4}{\pi} \times A_c} = \sqrt{\frac{4}{\pi} \times 1.6} = 1.41 \text{ m}$$

The manually calculated bed diameter is 1.41 m, rounded to 1.5 m. The corrected column cross-sectional area ($A_{c,c}$).

$$A_{c,c} = \frac{4}{\pi} \times 1.5^2 = 1.77 \text{ m}^2$$

The ratio of column diameter to packing size $\left(\dfrac{D_t}{D_P}\right)$:

$$\frac{D_t}{D_P} = \frac{1.5 \text{ m}}{38 \times 10^{-3} \text{m}} = 39$$

The corrected percentage flooding at selected diameter = $66\% \times \dfrac{A_c}{A_{c,c}}$
Using Figure 7.1, the value of the Y-axis, $Y_{axis} = 0.01$

$$G_f = \left[\frac{\rho_L\rho_G g\left(Y_{axis}\right)}{F_P \psi \mu_L^{0.2}} \right]^{0.5} = \left[\frac{10^3 \text{ kg/m}^3 \times 1.21 \text{ kg/m}^3 \times 9.82 \text{ m/s}^2 \times 0.01}{170 \text{ m}^{-1} \times 1 \times \left(1 \text{ cP}\right)^{0.2}} \right]^{0.2} = 0.83$$

The required column cross-section area (A_c)

$$A_c = \frac{G}{f G_f} = \frac{1.39 \text{ kg/s}}{0.66 \times 0.87 \text{ kg/m}^2\text{s}} = 1.76 \text{ m}^2$$

The diamter of the packed bed tower (D_t)

$$D_t = \sqrt{\frac{4}{\pi} \times A_c} = \sqrt{\frac{4}{\pi} \times 1.76} = 1.498 \text{ m}$$

The calculated diameter is 1.498 meters, rounded up to 1.5 m.

OVERALL HEIGHT OF TRANSFER UNITS, H_{OG}

The overall HTUs are calculated using Cornell's method and Onda's method [9–11] and considered the highest value among the design's two methods.

CORNELL'S METHOD

The diffusivities of solute in the liquid phase, D_L

$$D_L \left(\frac{m^2}{s} \right) = 1.7 \times 10^{-9}$$

The diffusivities of solute in the gas phase, D_G

$$D_G \left(\frac{m^2}{s} \right) = 1.45 \times 10^{-5}$$

The gas (ρ_g) and liquid (ρ_L) densities are, $\rho_g = 1.21 \text{ kg/m}^3$, $\rho_L = 1,000 \text{ kg/m}^3$, the gas viscosity, $\mu_G = 0.018 \times 10^{-3} \text{ Ns/m}^2$. Substituting required values in Schmidt number for gas Sc_G and liquid Sc_L,

$$Sc_G = \frac{\mu_G}{\rho_s D_G} = \frac{0.018 \times 10^{-3}}{1.21 \times 1.45 \times 10^{-5}} = 1.04$$

Liquid-phase Schmidt number (Sc_L):

$$Sc_L = \frac{\mu_L}{\rho_L D_L} = \frac{1 \times 10^{-3}}{1000 \times 1.7 \times 10^{-9}} = 588$$

The gas velocity (G):

$$G = \frac{29.5}{1.77} = 16.7 \text{ kg/s.m}^2$$

Using Figure 7.27 [9] or Equations 7.26 through 7.28, the following values are obtained:

For 60% flooding, $K_3 = 0.85$ and $\psi_h = 80$, for $G = 16.7$, $\phi_h = 0.1$

H_L is calculated as follows:

$$H_L = 0.305 \times \phi_h \left(Sc\right)_L^{0.5} \times K_3 \left(\frac{Z}{3.05}\right)^{0.15}$$

Substituting values into the above equation, H_L

$$H_L = 0.305 \times 0.1 \left(588\right)^{0.5} \times 0.85 \left(\frac{Z}{3.05}\right)^{0.15} = 0.53Z^{0.15}$$

Since the column diameter is more significant than 0.6 m, the diameter correction term will be taken as 2.3.

$$H_G = \frac{0.011 \times \psi_h \left(Sc\right)_G^{0.5} \left(2.3\right) \left(\frac{Z}{3.05}\right)^{0.33}}{\left(Lf_1f_2f_3\right)^{0.5}}$$

At liquid water temperature 20°C,

$$f_1 = f_2 = f_3 = 1$$

Substituting values into H_G,

$$H_G = \frac{0.011 \times 80 \left(1.04\right)^{0.5} \left(2.3\right) \left(\frac{Z}{3.05}\right)^{0.33}}{\left(16.7\right)^{0.5}} = 0.35Z^{0.33}$$

The overall HTUs in the gas phase, H_{OG}

$$H_{OG} = H_G + \frac{mG_m}{L_m} \times H_L$$

Multiplying both sides of the equation by N_{OG}

$$N_{OG} \times H_{OG} = N_{OG} \times \left[H_G + \frac{mG_m}{L_m} \times H_L\right]$$

Since $Z = N_{OG} \times H_{OG}$

$$\frac{mG_m}{L_m} = \frac{26 \times 0.048 \text{ kmol/s}}{1.64 \text{ kmol/s}} = 0.8$$

Solving for Z,

$$Z = 7 \times \left[0.35 \ Z^{0.33} + 0.8 \times 0.53 \ Z^{0.15} \right]$$

The value of $Z = 9.25$, rounding off the packing height to $Z = 10$ m.

ONDA'S METHOD

The wetted area a_w is calculated from a correlation developed by Onda et al. [11].

$$a_w = a_p \left\{ 1 - \exp\left[(-1.45)\left(\frac{\sigma_c}{\sigma}\right)^{0.75}\left(\frac{L}{a_p \mu_L}\right)^{0.1}\left(\frac{L^2 a_p}{\rho_L^2 g}\right)^{-0.05}\left(\frac{L^2}{\rho_L \sigma a_p}\right)^{0.2} \right] \right\} [=] \frac{m^2}{m^3}$$

The value of σ_c is 61 dyne/cm for ceramic packing, 75 dyne/cm for steel packing, and 33 dyne/cm for plastic packing, where σ_L is the water surface tension (N/m); a_p is the total packing surface area per packed bed volume L (m²/m³), G represents the superficial mass velocity of liquid and gas (kg/m².s); μ_L is the liquid-phase viscosity (kg/m.s); ρ_L is the liquid-phase density (kg/m³), R = 0.08314 (bar m³/kmol K); σ_L is the surface tension of the liquid, for water at 20°C; $\sigma_L = 70 \times 10^{-3}$ N/m (70 dyne/cm); = g = 9.81 m/s²; and d_p is the particle diameter 38×10^{-3} m. From Table 7.2, for 38-mm intalox saddles: $a_p = 194$ m²/m³, σ_c for ceramics = 61×10^{-3} N/m (61 dyne/cm).

$$a_w \left(\frac{m^2}{m^3}\right) = a_p \left\{ 1 - \exp\left[(-1.45)\left(\frac{\sigma_c}{\sigma}\right)^{0.75}\left(\frac{L}{a_p \mu_L}\right)^{0.1}\left(\frac{L^2 a_p}{\rho_L^2 g}\right)^{-0.05}\left(\frac{L^2}{\rho_L \sigma a_p}\right)^{0.2} \right] \right\}$$

Substituting values in the above equation:

$$a_w = 194 \left\{ 1 - \exp\left[(-1.45)\left(\frac{0.06}{0.07}\right)^{0.75}\left(\frac{16.7}{0.194}\right)^{0.1}\left(\frac{(16.7^2)(194)}{(1000^2)(9.81)}\right)^{-0.05}\left(\frac{(17.6)^2}{(1000)(0.070)(194)}\right)^{0.2} \right] \right\}$$

The calculated wetted area, a_w

$$a_w = 194 \times 0.71 = 138 \ m^2/m^3$$

The value of k_L is determined using the following equation:

$$k_L \left(\frac{\rho_L}{g \mu_L}\right)^{1/3} = 0.0051\left(\frac{L}{a_w \mu_L}\right)^{2/3}\left(\frac{\mu_L}{\rho_L D_L}\right)^{-1/2}\left(a_p d_p\right)^{0.4}$$

Rearranging and substituting of given values:

$$k_L = \dfrac{0.0051\left(\dfrac{16.7}{(138)(0.001)}\right)^{2/3}\left(\dfrac{0.001}{(1000)(1.7\times10^{-9})}\right)^{-1/2}\left((194)(0.038)\right)^{0.4}}{k_L\left(\dfrac{1000}{(9.81)(0.001)}\right)^{1/3}}$$

The calculated mass transfer coefficient in the liquid phase (k_L):

$$k_L = 2.5\times10^{-4}\,\text{m/s}$$

The gas mass velocity (G_f) based on actual column diameter = $G_f/A = (1.39/1.77) = 0.79$ kg/m²s

$$k_G\left(\dfrac{RT}{a_p D_G}\right) = 5.23\left(\dfrac{G_f}{a_p\mu_G}\right)^{0.70}\left(\dfrac{\mu_G}{\rho_G D_G}\right)^{1/3}\left(a_p d_p\right)^{-2}$$

Rearranging and substituting given data:

$$k_G = \dfrac{5.23\left(\dfrac{0.79}{(194)(0.018\times10^{-3})}\right)^{0.7}\left(\dfrac{0.018\times10^{-3}}{(1.21)(1.45\times10^{-5})}\right)^{\frac{1}{3}}\left((194)(0.038)\right)^{-2}}{\left(\dfrac{(0.08314)(293)}{(194)(1.45\times10^{-5})}\right)}$$

The calculated mass transfer coefficient of the gas phase, k_G

$$k_G = 5.0\times10^{-4}\,\dfrac{\text{kmol}}{\text{s.m}^2.\text{bar}}$$

The molar gas velocity, G_m

$$G_m = \dfrac{G_f}{M_w} = \dfrac{0.79\ \text{kg/m}^2\text{s}}{29\ \text{kg/kmol}} = 0.027\ \text{kmol/m}^2\text{s}$$

The molar liquid velocity, L_m

$$L_m = \dfrac{L}{M_w} = \dfrac{16.7\ \text{kg/m}^2\text{s}}{18\ \text{kg/kmol}} = 0.93\ \text{kmol/m}^2\text{s}$$

The gas height transfer units, H_G

$$H_G = \frac{G_m}{k_G \times a_P \times P} = \frac{0.027}{5.0 \times 10^{-4} \times 138 \times 1.013} = 0.39 \text{ m}$$

C_T is the total liquid concentration water = $(\rho_L/M_{W,\,L})(1,000/18) = 55.6 \text{ kmol/m}^3$
The height of the transfer unit in the liquid phase (H_L):

$$H_L = \frac{L_m}{k_L \times a_P \times C_T} = \frac{0.93}{2.5 \times 10^{-4} \times 138 \times 55.6} = 0.49 \text{ m}$$

The height of the transfer unit in the gas phase, H_{OG}

$$H_{OG} = H_G \frac{mG_m}{L_m} \times H_L$$

Substituting the required values in the above equation,

$$H_{OG} = 0.39 + \frac{26 \times 0.027}{0.93} \times 0.49 = 0.8 \text{ m}$$

The packed bed height, Z

$$Z = N_{OG}H_{OG} = 7 \times 0.8 = 5.6 \text{ m}$$

Consider the higher value as the packed column height; accordingly, consider the estimated value determined using Cornell's method. Round up the packed bed size to 10 m.

HYSYS/UNISIM SIMULATION

Drag and drop the absorber from the object palette. Connect the inlet SO_2/air gas stream (Feed) to the column's bottom and the liquid (solvent) stream to the column's top. The exit gas stream is connected to the column's top, and the liquid stream is connected to the column's bottom. Feed streams are fully specified. Set the required number of trays so that the exit SO_2 mole fraction is 0.003 as required. In this example, the number of theoretical plates needed to achieve a mole fraction of 0.003 SO_2 in the exit air stream is more significant than four and less than five since the mole fraction in the air-out stream is 0.0038 (Figure 7.28). The Hysys theoretical number of trays is close to the number of trays obtained with manual calculations.

For tray sizing, follow the following steps: Tools ≫ Utilities ≫ Tray sizing ≫ Add Utilities

Then click on Tray sizing 1 ≫ Select TS ≫ TS-1 ≫ Auto section ≫ Packed Select the type of packing material; in this example, the packing material is:

Intalox Ceramic Saddles 1.5 in Random Packing. Click on the Performance tab to view the results shown in Figure 7.29.

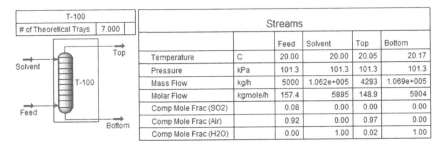

T-100				
# of Theoretical Trays	7.000			

Streams					
		Feed	Solvent	Top	Bottom
Temperature	C	20.00	20.00	20.05	20.17
Pressure	kPa	101.3	101.3	101.3	101.3
Mass Flow	kg/h	5000	1.062e+005	4293	1.069e+005
Molar Flow	kgmole/h	157.4	5895	148.9	5904
Comp Mole Frac (SO2)		0.08	0.00	0.00	0.00
Comp Mole Frac (Air)		0.92	0.00	0.97	0.00
Comp Mole Frac (H2O)		0.00	1.00	0.02	1.00

FIGURE 7.28 UniSim/Hysys generated the process flow diagram and stream summary of the absorber column for the process described in Example 7.6.

PRO/II SIMULATION

Following the same procedure in previous examples, Figure 7.30 shows the process flowsheet and stream summary. Figure 7.31 shows the PRO/II predicted column-packed height and the inside diameter.

Performance Results

Trayed
Table
Plot

Section Results
○ Trayed   ● Packed   [Export Pressures]   [View Warnings...]

Packing Results

Section	Section_1		
Internals	Packed		
Section Diameter [m]	1.219		
Max Flooding [%]	67.30		
X-Sectional Area [m2]	1.167		
Section Height [m]	3.714		
Section DeltaP [kPa]	1.181		

FIGURE 7.29 UniSim/Hysys generated the height and diameter of the packed bed absorber for the case described in Example 7.6.

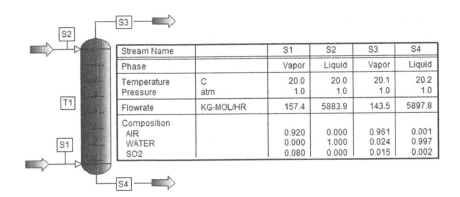

Stream Name		S1	S2	S3	S4
Phase		Vapor	Liquid	Vapor	Liquid
Temperature	C	20.0	20.0	20.1	20.2
Pressure	atm	1.0	1.0	1.0	1.0
Flowrate	KG-MOL/HR	157.4	5883.9	143.5	5897.8
Composition					
AIR		0.920	0.000	0.961	0.001
WATER		0.000	1.000	0.024	0.997
SO2		0.080	0.000	0.015	0.002

FIGURE 7.30 PRO/II generated PFD and stream summary for the case of absorption presented in Example 7.6.

```
ID                         =   COLSECT-1
PACKED HEIGHT              =   4.62 M
COLUMN ID                 =   1.600 M
PACKING TYPE (NORTON)     =   Super INTALOX saddles, ceramic
PACKING FACTOR            =   164.04 M2/M3
NOMINAL PACKING DIAMETER  =   38.100 MM
CAPACITY METHOD           =   fraction of flood
HETP METHOD               =   Norton
DP METHOD                 =   Norton
```

FIGURE 7.31 PRO/II calculated column packed height and inside diameter for the absorption case described in Example 7.6.

ASPEN PLUS SIMULATION

In this example, the number of theoretical stages required to achieve a 0.004 mole fraction of sulfur dioxide in the exit stream is set to 7 trays. Use Block Absbr2 in Aspen for this purpose. Choose Absrb2 under the column subdirectory by pressing the down arrow to the RadFrac icon's right. Air inlet stream and liquid inlet streams are fully specified (i.e., temperature, pressure, flow rate, and compositions to be entered). NRTL is the suitable property method. In the block setup options, the number of trays is 7; there is no condenser or reboiler for the absorber (Figure 7.32a). The pressure of the column is 1 atm. Figure 7.32b shows the inlet air entered below stage 7 (above stage 8).

Figure 7.33 shows the process flow diagram and stream summary after supplying all necessary information and running the system.

They are using the spreadsheet option to measure the height required to achieve sulfur dioxide concentration in stream three as 0.004. The manipulated variable, which is the height of the packed tower, is shown in Figure 7.34a. The converged value of the diameter is 1.92 m (Figure 7.34b).

SUPERPRO DESIGNER SIMULATION

Following the same procedure in the previous examples, the absorber is selected. Make sure that inlet streams are fully specified. The inlet conditions are as shown in Figure 7.35. The packed-material data are shown in Figure 7.36. The dimensionless packing constant (C_f) is taken from Treybal [10]. The resultant diameter and height are shown in Figure 7.36.

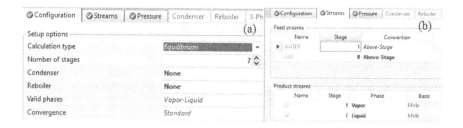

FIGURE 7.32 (a) Block setup configuration menu. (b) Feed and product stream inlet stages.

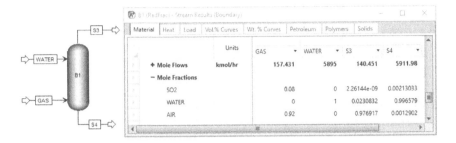

FIGURE 7.33 Aspen Plus generated a process flow diagram and streams summary for the case presented in Example 7.6.

FIGURE 7.34 (a) Aspen Plus design specification menu of packed bed height. (b) Limits of the manipulated variable.

AVEVA PROCESS SIMULATION

Select the DefFluid from Fluid's menu and choose the NRTL fluid package. Fully specify the gas stream and freshwater feed stream (W, T, P, and composition). Connect feed streams to the column inlet. Edit the configuration page:

1. Drag a Column object from the Process library onto the Canvas, next to feed.
2. Rename it to Example 7.6 by overwriting its default name directly on the Canvas.
3. Connect feed streams (S1 and S2).

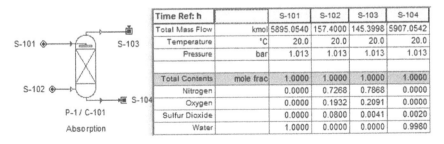

Time Ref: h		S-101	S-102	S-103	S-104
Total Mass Flow	kmol	5895.0540	157.4000	145.3998	5907.0542
Temperature	°C	20.0	20.0	20.0	20.0
Pressure	bar	1.013	1.013	1.013	1.013
Total Contents	mole frac	1.0000	1.0000	1.0000	1.0000
Nitrogen		0.0000	0.7268	0.7868	0.0000
Oxygen		0.0000	0.1932	0.2091	0.0000
Sulfur Dioxide		0.0000	0.0800	0.0041	0.0020
Water		1.0000	0.0000	0.0000	0.9980

FIGURE 7.35 SuperPro generated the process flow diagram and streams summary for the absorption case described in Example 7.6.

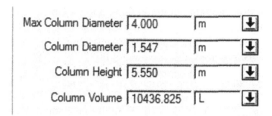

FIGURE 7.36   SuperPro calculated packed-tower height and diameter. Packing constant (c_f), 96.2, total specific area, 195 m^2/m^3, nominal diameter 38 mm, critical surface tension is 61 dyne/cm.

4. Expand the column icon to an appropriate size by acting on its corner blocks.
5. Set the number of stages to 3.
6. Set the two feeds' location stages; 1 for liquid solvent (top) and 3 for gas feed stream (bottom).
7. Setup Condenser and Reboiler to "None" for Absorbers.
8. Set Setup to Solve.
9. Set Contact between 0 and 1; here, it is 0.6 (otherwise, vapor and liquid will bypass each other).

Figure 7.37 shows the squared and solved simulation of the absorption column to capture SO_2 from the air stream into the freshwater stream (solvent). The result shows the successful removal of SO_2 from the air with three stages column.

Example 7.7:   Removal of CO_2 from Natural Gas

The absorption of CO_2 from CO_2/CH_4 natural gas stream into the physical solvent (propylene carbonate) took place in a packed column absorber. The inlet gas stream is 20 mol% CO_2 and 80 mol% methane. The gas stream flows at a

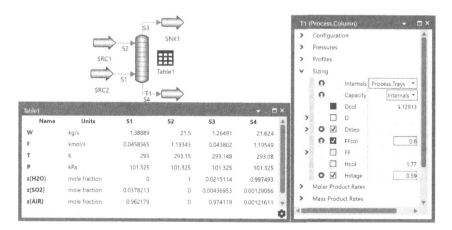

FIGURE 7.37   Aveva Process Simulation software generated the process flow diagram and the stream summary for the case described in Example 7.6.

7,200 m³/h, and the column operates at 60°C and 60 atm. The inlet solvent flow rate is 2,000 kmol/h. Use available software simulators to determine the concentration of CO_2 (mol%) in the exit gas stream, the column height (m), and the column diameter (m). Compare the simulators' predictions (results generated with UniSim/Hysys, PRO/II, Aspen Plus, SuperPro, and Aveva Process Simulation) to verify manual calculations. The selected packing material is a random ceramic raschig ring (2 inches). The carbon dioxide in the exit stream should not exceed 0.4 mol%, assume 60% flooding. The equilibrium data is: $y = m\,x$, where m is 27.34.

SOLUTION

MANUAL CALCULATIONS

In this section, manually calculate the NTUs, packed bed diameter, and height. Calculate the number of transfer units (NTUs) using the following equation:

$$NTU = \frac{\ln\left[\left(y_1 - \frac{m.x_2}{y_2} - mx_2\right)\left(1 - \left(\frac{1}{AF}\right)\right) + \frac{1}{AF}\right]}{1 - (1/AF)}$$

The absorption factor, AF

$$AF = \frac{L_m}{mG_m} = \frac{2000\ \text{kmol/h}}{27.34 \times 304.5\ \text{kmol/h}} = 0.24$$

Substituting values to calculate the NTUs:

$$NTU = \frac{\ln\left[(0.2 - 0/0.04 - 0)\left(1 - (1/0.24)\right) + 1/0.24\right]}{1 - (1/0.24)} = 5.12$$

The NTUs are 5.12 and rounded off into six stages. Calculate the diameter of the column using the following correlation and Figure 7.1.

$$X_{axis} = \left(\frac{L(\text{kg/h})}{G(\text{kg/h})}\right)\sqrt{\frac{\rho_G}{\rho_L}}$$

Substituting liquid and gas mass flow rate at the bottom of the absorber with corresponding densities into the above equation:

$$X_{axis} = \left(\frac{207,300\ \text{kg/h}}{6588\ \text{kg/h}}\right)\sqrt{\frac{52.64\ \text{kg/m}^3}{1,166\ \text{kg/m}^3}} = 6.66$$

From Figure 7.1 or using the following correlation:

$$Y_{axis} = 10^{\left(\left[-1.668-1.085\ (\log 6.66)\ -\ 0.297\left(\log 6.66\right)2\right]\right)} = 0.00176$$

The gas mass flow rate per cross-sectional area of the bed:

$$G_f = \left[\frac{\rho_L \rho_G g_c \left(Y_{axis}\right)}{F_p \psi \left(\mu_L\right)}\right]^{0.5}$$

For packing use, Raschig ring (ceramic, random), 2 inches.

$$G_f = \left[\frac{\left(1165.8\ kg/m^3\right)\left(94.83\ kg/m^3\right)\left(9.81\ m/s^2\right)\left(0.00176\right)}{\left(213.29\ 1/m\right)\left(1.17\right)\left(0.89\ cp\right)^{0.2}}\right]^{0.5} = 2.8\ kg/m^2 s$$

The bed cross-sectional area:

$$A = \frac{G}{G_f} = \frac{7.828\ kg/s}{2.8\ kg/m^2 s} = 0.652\ m^2$$

Determine the column diameter from the calculated area:

$$D = \sqrt{\frac{4A}{\pi}} = \sqrt{\frac{4 \times 0.652\ m^2}{\pi}} = 0.97\ m$$

The operating liquid mass flow rate per cross-sectional area of the tower:

$$L_f = \frac{207110.28\left(kg/h\right)\left(h/3600\ s\right)}{0.774\ m^2} = 74.33\ kg/m^2 s$$

The height of the gas transfer units:

$$H_G = \frac{0.011 \times \psi_h \left(Sc\right)_G^{0.5}\left(2.3\right)\left(\dfrac{Z}{3.05}\right)^{0.33}}{\left(L f_1 f_2 f_3\right)^{0.5}}$$

Values of f1

$$f_1 = \left(\frac{\mu_L}{\mu_w}\right)^{0.16} = \left(\frac{8.9 \times 10^{-4}\ \dfrac{kg}{m.s}}{9.4 \times 10^{-4}\ \dfrac{kg}{m.s}}\right)^{0.16} = 1.0$$

The value of f2

$$f_2 = \left(\frac{\rho_w}{\rho_L}\right)^{1.25} = \left(\frac{999.55 \text{ kg/m}^3}{1165.82 \text{ kg/m}^3}\right)^{0.16} = 0.83$$

Assume $f_3 = 1$.

Calculate the Schmidt number in the gas phase (Sc_G):

$$Sc_G = \frac{\mu_G}{\rho_G D_G} = \frac{\left(1.82 \times 10^{-5} \dfrac{\text{kg}}{\text{m.s}}\right)}{\left(1.12 \dfrac{\text{kg}}{\text{m}^3}\right)\left(2.36 \times 10^{-5} \dfrac{\text{m}^2}{\text{s}}\right)} = 0.69$$

Substituting the necessary values to calculate (H_G):

$$H_G = \frac{0.011 \times 80 \times 0.69^{0.5}\,(2.3)\left(\dfrac{Z}{3.05}\right)^{0.33}}{\left(74.33 \text{ kg/ft}^2\text{s} \times 0.83\right)^{0.5}} = 0.15\ Z^{0.33}$$

The height of the liquid phase transfer units (H_L):

$$H_L = 0.305 \times \phi_h\,(Sc)_L^{0.5} \times K_3 \left(\frac{Z}{3.05}\right)^{0.15}$$

The Schmidt number in the liquid phase (Sc_L):

$$Sc_L = \frac{\mu_L}{\rho_L D_L} = \frac{8.92 \times 10^{-4} \text{ kg/m s}}{\left(1165.82 \text{ kg/m}^3\right)\left(1.07 \times 10^{-9}\text{m}^2/\text{s}\right)} = 715.07$$

The liquid mass velocity (L_f): $L_f = 74.33 \text{ kg/m}^2\text{s}$

At $L_f = 74.33 \text{ kg/m}^2$ s, $\phi_h = 0.11$, and 60% flooding, $K_3 = 0.85$

$$H_L = 0.305 \times 0.11 \times (715.07)^{0.5} \times 0.85 \left(\frac{Z}{3.05}\right)^{0.15} = 0.65\ Z^{0.15}$$

The liquid phase HTU (H_L):

$$H_L = 0.65\ Z^{0.15}$$

The absorption factor (AF):

$$AF = \frac{L_m}{mG_m} = \frac{2000 \text{ kmol/h}}{27.34 \times 304.5 \text{ kmol/h}} = 0.24$$

The total HTUs (Z):

$$Z = N_{TU} \times \left(H_G + \frac{1}{AF} H_L \right) = 6 \times \left(0.15 \ Z^{0.33} + \frac{1}{0.24} \times 0.65 \ Z^{0.15} \right)$$

Solving for Z using the nonlinear algebraic solver in Polymath software package as shown below:

z(min) = 10
z(max) = 35
f(z) = 6 × (0.15 × z ^ 0.33 + (1/0.24) × 0.65 × z ^ 0.15)-z
z(0) = 25

or using goal seek in excel, the total height (Z) is 29.8 m.

HYSYS/UNISIM SIMULATION

Use UniSim/Hysys absorber object and consider the following information:

- Fluid package: Sour PR
- Number of trays: 6
- Packing material: Random, Ceramic Raschig ring, 1 1/2 inch

Figure 7.38 shows the feed stream condition, and Figure 7.39 displays the packed column diameter and height.

PRO/II SIMULATION

Select the absorber (Column) from the object library, and the number of stages is six stages. Use the PR equation of state to estimate system physical properties. Figure 7.40 shows the PRO/II predicted results after running the system. Figure 7.41 indicates the estimated column height and column inside diameter.

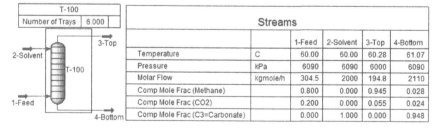

Streams		1-Feed	2-Solvent	3-Top	4-Bottom
Temperature	C	60.00	60.00	60.28	61.07
Pressure	kPa	6090	6090	6000	6090
Molar Flow	kgmole/h	304.5	2000	194.8	2110
Comp Mole Frac (Methane)		0.800	0.000	0.945	0.028
Comp Mole Frac (CO2)		0.200	0.000	0.055	0.024
Comp Mole Frac (C3=Carbonate)		0.000	1.000	0.000	0.948

T-100 | Number of Trays 6.000 — 3-Top — 2-Solvent — T-100 — 1-Feed — 4-Bottom

FIGURE 7.38 UniSim/Hysys generated process flow diagram (number of trays is 6) and streams summary for the case presented in Example 7.7.

Performance Results	Section Results ○ Trayed ● Packed			Export Pressures		View Warnings...
Trayed	Packing Results					
Table	Section		Section_1			
	Internals		Packed			
Plot	Section Diameter [m]		1.219			
	Max Flooding [%]		36.90			
	X-Sectional Area [m2]		1.167			
	Section Height [m]		3.184			

FIGURE 7.39   UniSim/Hysys utility tray sizing utility for the case described in Example 7.7.

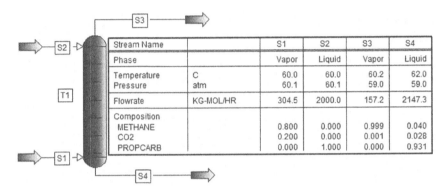

Stream Name			S1	S2	S3	S4
Phase			Vapor	Liquid	Vapor	Liquid
Temperature	C		60.0	60.0	60.2	62.0
Pressure	atm		60.1	60.1	59.0	59.0
Flowrate	KG-MOL/HR		304.5	2000.0	157.2	2147.3
Composition						
METHANE			0.800	0.000	0.999	0.040
CO2			0.200	0.000	0.001	0.028
PROPCARB			0.000	1.000	0.000	0.931

FIGURE 7.40   PRO/II generated process flow diagram and streams summary generated by PRO/II for the case described in Example 7.7.

ASPEN PLUS SIMULATION

The following are used to simulate packed bed absorber in Aspen Plus:

- Absorber type: RadFrac block/absorber.
- Fluid package: SRK.
- Configuration: Number of trays; 6, Condenser; None, Reboiler; None.

```
ID                          =   COLSECT-1
PACKED HEIGHT               =   5.33 M
COLUMN ID                   =   1.400 M
PACKING TYPE (NORTON)       =   Raschig rings, ceramic
PACKING FACTOR              =   213.25 M2/M3
NOMINAL PACKING DIAMETER    =   50.800 MM
CAPACITY METHOD             =   fraction of flood
HETP METHOD                 =   Frank rule of thumb
DP METHOD                   =   Norton
```

FIGURE 7.41   Column diameter and height generated by PRO/II for the case described in Example 7.7.

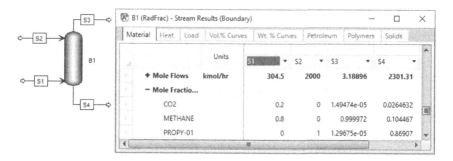

FIGURE 7.42 Aspen Plus generates the process flowsheet and streams summary using RadFrac for the case described in Example 7.7.

- Streams: S2 enters stage 1, convention; above the stage, S1 joins stage 6, convention; on stage.
- Packing material: Random, ceramic Raschig ring, 2 inches.

Figure 7.42 shows the feed stream conditions and streams summary. Aspen Plus calculated the packed bed diameter and height by clicking on the Column Internals folder, then clicking on Add New button, then double clicking on INT-1, and again clicking on Add New button. In the open menu, fill in the packing material specifications. Click on the View button to see that menu in Figure 7.43. Do not forget to select the packing option from the internal type pull-down menu.

SuperPro Designer Simulation

The absorber column is used, and the following data are supplied:

Henry's law constant, $H = 2.95 \times 10^{-2}$ atm m³/mol. Solute diffusivity in the gas phase = 0.233 cm²/s. Consider the following pieces of information:

- Solute diffusivity in the liquid phase = 1.07×10^{-5} cm²/s
- Solute diffusivity gas = 0.015 cP

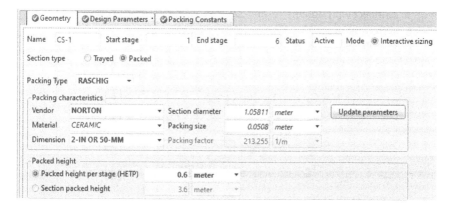

FIGURE 7.43 Aspen Plus required geometry for the packed column of the case described in Example 7.7.

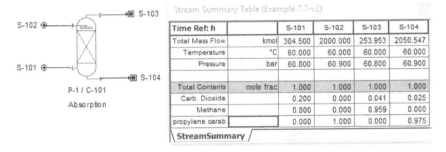

Stream Summary Table (Example 7.7-v2)					
Time Ref: h			S-101	S-102	S-103	S-104
Total Mass Flow	kmol	304.500	2000.000	253.953	2050.547
Temperature	°C	60.000	60.000	60.000	60.000
Pressure	bar	60.800	60.900	60.800	60.900
Total Contents	mole frac	1.000	1.000	1.000	1.000
Carb. Dioxide		0.200	0.000	0.041	0.025
Methane		0.800	0.000	0.959	0.000
propylene carab		0.000	1.000	0.000	0.975

StreamSummary

FIGURE 7.44 SuperPro generated a process flow diagram and stream summary for the case presented in Example 7.7.

- Liquid-phase viscosity = 1 cP
- Percent carbon dioxide removal = 83%
- Liquid surface tension = 52.3 dyne/cm

Packing material: Raschig ring ceramic, random packing, 2 inches (Cf = 65, a_p = 92 m²/m³). The process flowsheet and stream summary are shown in Figure 7.44. Packed bed specifications, HTUs, and NTUs are shown in Figure 7.44. Figure 7.45 shows the packed bed height and diameter.

AVEVA PROCESS SIMULATION

Select the DefFluid from Fluid's menu and choose the NRTL fluid package. Fully specify the gas stream and freshwater feed stream (W, T, P, and composition). Connect feed streams to the column inlet. Edit the configuration page as follows:

1. Drag a Column object from the Process library onto the Canvas, next to feed.
2. Rename it to Example 7.8 by overwriting its default name directly on the Canvas.
3. Connect feed streams (S1 and S2).

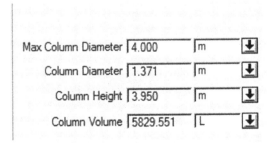

Max Column Diameter 4.000   m

Column Diameter 1.371   m

Column Height 3.950   m

Column Volume 5829.551   L

FIGURE 7.45 SuperPro software calculated the specification of the packed bed column for the case in Example 7.7.

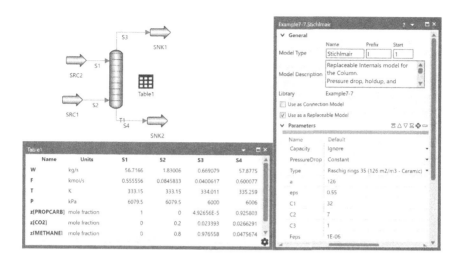

FIGURE 7.46 Aveva Process Simulation generated process flow diagram and estimated column diameter and height predicted for the case described in Example 7.7.

4. Expand the column icon to an appropriate size by acting on its corner blocks.
5. Set the number of stages to 3.
6. Set the two feeds' location stages; 1 for liquid solvent (top) and 3 for gas feed stream (bottom).
7. Setup Condenser and Reboiler to "None" for Absorbers.
8. Set Setup to Solve.
9. Set Contact between 0 to 1; here, it is 0.6 (otherwise, vapor and liquid will bypass each other)
10. Select the Stichlmair icon where packed material is available. Select the packing material, as shown in Figure 7.46. It shows the squared and solved simulation of the absorption column to capture SO_2 from the air stream into the propylene carbonate stream (solvent). The result shows the successful removal of SO_2 from the air with three stages column.

7.7 NUMBER OF THEORETICAL TRAYS

There are two methods that are most commonly used to determine the number of ideal trays required for a given removal method. One method used is a graphical technique. The number of ideal plates is obtained by drawing steps on a functional diagram. Figure 7.47 illustrates the procedure.

The second method is a simplified one used to estimate the number of plates. This equation can only be used if both the equilibrium and operating lines for the system are straight. This case is a valid assumption for most air pollution control systems [3]. The HETP in a tray absorber is referred to as height equivalent to a theoretical plate instead of the HTUs in a packed tower. The equation predicts the number of theoretical plates required to achieve a given removal efficiency.

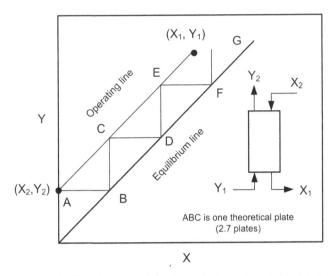

FIGURE 7.47 Graphically estimation of the theoretical number of the plate in a gas-liquid absorber column.

The operating conditions for a theoretical plate assume that the gas and liquid streams released from the plate are in equilibrium with each other. Three types of efficiencies are used to describe absorption efficiency for a plate tower; the overall efficiency concerns the entire Column, Murphree efficiency applies to a single plate, and local efficiency pertains to a specific location on a plate. The overall efficiency is the ratio of the number of theoretical plates to the number of actual plates. The overall tray efficiencies of absorbers operating with low-viscosity liquid fall in the range of 65–80% [13]. The height Z of the Column is calculated from the transfer unit HTU and the NTUs N_{OG}. The NTUs are given by the Kremser Method for theoretical trays (Absorber):

$$N_{OG} = \frac{\ln\left[y_1 - \dfrac{mx_1}{y_2} - mx_1\left(1-\left(\dfrac{1}{AF}\right)\right)+\left(\dfrac{1}{AF}\right)\right]}{\log(AF)} \quad (7.41)$$

where AF is the absorption factor, $L_m/m\, G_m$; L_m and G_m are the liquid and gas molar flow rates, kmol/h; m is the slope of equilibrium curve; y_1 *and* y_2 are the molar fractions of the entering and departing gas; and x_1 and x_2 are the mole fractions of leaving and entering liquid; Kremser method for theoretical trays (Stripper):

$$N = \frac{\log\left[\left(\dfrac{x_o - (y_{N+1}/m)}{x_N - (y_{N+1}/m)}\right)(1-(1/S)) + \dfrac{1}{S}\right]}{\log(S)} \quad (7.42)$$

where $S = m\, G/L$.

7.8 SIZING A PLATE TOWER ABSORBER

Another absorber used extensively for gas absorption is a plate tower. Here, absorption occurs on each plate or stage.

7.8.1 PLATE TOWER DIAMETER

Using the gas velocity through the tower, determine the minimum diameter of a single plate tower. If the gas velocity is too fast, liquid droplets are entrained, causing priming that causes liquid on one tray to foam and rises to the tray above. Priming reduces absorber efficiency by inhibiting gas and fluid Contact. Priming is a plate tower analogous to the flooding point in a packed column. The minor allowable diameter for a palette tower is

$$d_t = \psi \left(Q_G \sqrt{\rho_G} \right)^{0.5} \qquad (7.43)$$

where:
  Q_G is the volumetric gas flow rate m³/h;
  ψ is the empirical correlation, $m^{0.25} h^{0.5}/kg^{\circ 2S}$; and
  ρ_g is the gas density kg/m³.

Table 7.5 lists the empirical constant (ψ) for different types of trays.

Example 7.8: Design of the Plate Tower

A gas-liquid absorber tray tower is employed to capture SO_2 from a gas mixture (3% SO_2, 97% air) with pure water. The gas feed rate is 206 kmol/h, and the water feed rate is 12,240 kmol/h. The feed stream's temperature is 293 K, and the pressure is 101.32 kPa. The column tray type is a bubble cap. The concentration of SO_2 in the exit air should be less than 500 ppm. Manually determine the number of bubble cap trays and tower diameter. Compare manual results with

---

TABLE 7.5
Empirical Constant (ψ)

Tray	ψ (Metric units)	ψ (English Units)
Bubble cap	0.0162	0.1386
Sieve	0.0140	0.1198
Valve	0.0125	0.1069

Source: Data from Calvert, S. et al. (1972). Wet Scrubber System Study, Vol. 1, Scrubber Handbook, EPA-R2-72-118a, U.S. Environmental Protection Agency, Washington, DC.

---

the predicted results from the available software packages (UniSim/Hysys, PRO/II, Aspen Plus, SurperPro, and Aveva Process Simulation).

SOLUTION

HAND CALCULATIONS

The minimum acceptable diameter of the plate tower is determined using the following equation:

$$d_t = \psi\left(Q_G\sqrt{\rho_G}\right)^{0.5} = 0.0162\frac{m^{0.25}h^{0.25}}{kg^{0.25}}\left[\left(85\frac{m^3}{min}\left|\frac{60\ min}{h}\right.\right)\times\sqrt{1.17\frac{kg}{m^3}}\right] = 12\ m^{0.5}$$

The absorption factor (AF):

$$AF = \frac{L_m}{mG_m} = \frac{12,240\ kmol/h}{42.7\times206\ kmol/h} = 1.4$$

The total number of theoretical trays (N_{OG}):

$$N_{OG} = \left[\frac{0.03 - \dfrac{0}{0.0005} - 0\left(1-\left(\dfrac{1}{1.4}\right)\right)+\dfrac{1}{1.4}}{\ln(1.4)}\right] = 8.6$$

The number of trays rounded up to nine trays. The height of the tower: $Z = 9 \times 0.6096 = 5.5$ m.

UNISIM/HYSYS SIMULATION

In this example, the packed material of Example 7.3 is replaced by bubble cap trays using tray utilities. Figure 7.48 shows the stream summary and Figure 7.49 shows the column performance.

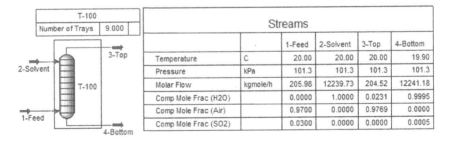

T-100		
Number of Trays	9.000	

Streams				
		1-Feed	2-Solvent	3-Top	4-Bottom
Temperature	C	20.00	20.00	20.00	19.90
Pressure	kPa	101.3	101.3	101.3	101.3
Molar Flow	kgmole/h	205.98	12239.73	204.52	12241.18
Comp Mole Frac (H2O)		0.0000	1.0000	0.0231	0.9995
Comp Mole Frac (Air)		0.9700	0.0000	0.9769	0.0000
Comp Mole Frac (SO2)		0.0300	0.0000	0.0000	0.0005

FIGURE 7.48 UniSim/Hysys generated the process flow diagram and stream summary of the bubble cap tray column for the case described in Example 7.8.

Performance	Section Results			
	⦿ Trayed ◯ Packed		Export Pressures	View Warnings...
Results				
Trayed	Tray Results			
Table	Section	Section_1		⌃
	Internals	Bubble Cap		
Plot	Section Diameter [m]	1.676		
	Max Flooding [%]	72.66		
	X-Sectional Area [m2]	2.207		
	Section Height [m]	5.486		

FIGURE 7.49 UniSim/Hysys calculated the diameter and height of the bubble cap tray column for the absorption process described in Example 7.8.

PRO/II SIMULATION

The selection of a suitable thermodynamic fluid package is essential to get the correct answer. NRTL was selected as the property estimation system using Henry's law constant, as shown in Figure 7.50. Figure 7.51 predicts the process flowsheet and the stream summary. It is observed that 9 bubble cap trays are required to achieve the desired separation, and the exit mole fraction of SO_2 in the exit stream is 323 ppm. The bubble cap diameter and the column's height are captured from the Output generated text report (Figure 7.52).

ASPEN PLUS SIMULATION

Using the converged AspenPlus file developed for Example 7.3, in this case, the packing is replaced by bubble cap trays. The stream summary result is shown in Figure 7.53. The tray tower diameter using the spread options is shown in Figure 7.54. Use the following information:

- Property method: NRTL
- Number of stages: 9 (bubble cap tray)

SUPERPRO DESIGNER

In SuperPro designer, there is no tray option. Changing the percent SO_2 removal (in Example 7.3) from 90% to 99% removal leads to an exit concentration of SO_2

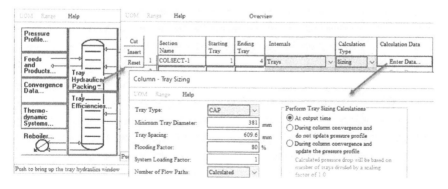

FIGURE 7.50 PRO/II selection of bubble cap trays for the case described in Example 7.8.

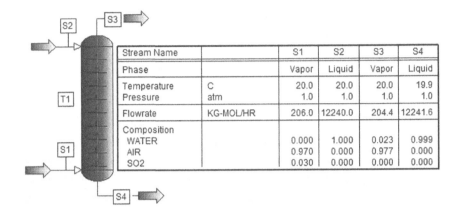

Stream Name		S1	S2	S3	S4
Phase		Vapor	Liquid	Vapor	Liquid
Temperature	C	20.0	20.0	20.0	19.9
Pressure	atm	1.0	1.0	1.0	1.0
Flowrate	KG-MOL/HR	206.0	12240.0	204.4	12241.6
Composition					
WATER		0.000	1.000	0.023	0.999
AIR		0.970	0.000	0.977	0.000
SO2		0.030	0.000	0.000	0.000

FIGURE 7.51 PRO/II generated the process flow diagram and stream summary for the absorption system described in Example 7.8.

```
TRAY SELECTION FOR TRAY RATING

  BUBBLE CAP DIAMETER    101.6 MM
  BUBBLE CAP SPACING      25.4 MM


                  DESIGN                      NUMBER    -----
       SECTION     TRAY    DIAMETER   NP    OF VALVES   SIDE
                  NUMBER      MM             OR CAPS     MM
  ------------    ------   --------   --    ---------   ------
     COLSECT-1       1       1829      1        64      543.582
```

FIGURE 7.52 PRO/II calculated bubble cap diameter and spacing for the tray tower absorption case presented in Example 7.8.

	Units	S1	S2	S3	S4
+ Mole Flows	kmol/hr	12240	206	188.198	12257.8
− Mole Fractions					
SO2		0	0.03	4.88778e-06	0.000504094
WATER		1	0	0.023084	0.998193
AIR		0	0.97	0.976911	0.00130264

FIGURE 7.53 Aspen Plus generated process flows diagram and streams summary of tray towers simulated by the case presented in Example 7.8.

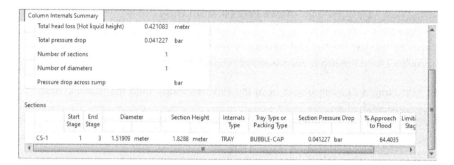

FIGURE 7.54 Tray column diameters generated by Aspen Plus for the case described in Example 7.8.

as 310 ppm, which is within the acceptable range. Figure 7.55 shows the process flowsheet and stream summary.

The calculated NTUs are nine close to those previously found using Hysys, PRO/II, and Aspen Plus (Figure 7.56). Right click on the column and select equipment Data.

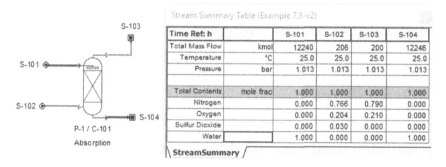

FIGURE 7.55 PFD and steam summary depicted with SuperPro designer for the absorber described in Example 7.8.

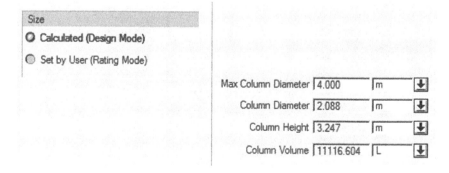

FIGURE 7.56 Column diameter and height predicted with SuperPro software for the case described in Example 7.8.

AVEVA PROCESS SIMULATION

Select the DefFluid from Fluid's menu and choose the NRTL fluid package. Fully specify the gas stream and freshwater feed stream (W, T, P, and composition). Connect feed streams to the column inlet. Edit the configuration page:

1. Drag a Column object from the Process library onto the Canvas, next to feed.
2. Rename it to Example 7.8 by overwriting its default name directly on the Canvas.
3. Connect feed streams (S1 and S2).
4. Expand the column icon to an appropriate size by acting on its corner blocks.
5. Set the number of stages to 3.
6. Set the two feeds' location stages; 1 for liquid solvent (top) and 3 for gas feed stream (bottom).
7. Setup Condenser and Reboiler to "None" for Absorbers.
8. Set Setup to Solve.
9. Set Contact between 0 to 1; here, it is 0.6 (otherwise, vapor and liquid will bypass each other).
10. The column is squared and solved. The result predicts the successful removal of SO_2 from the air with a 9.0 trays column.

CONCLUSION

The tray tower diameters obtained with four software packages were around 2 m above the minimum required diameter estimated with hand calculation (1.2 m), which is acceptable.

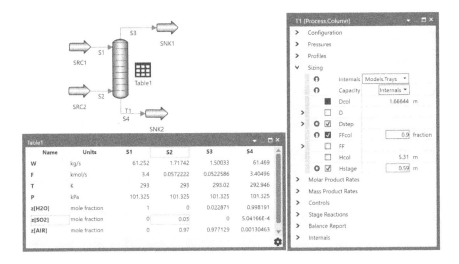

FIGURE 7.57 Aveva Process Simulation software generated the process flow diagram and estimated column diameter and height predicted for the case described in Example 7.8.

SUMMARY OF DESIGN PROCEDURE

The following steps are typical in the design of gas absorbers:

1. Calculate Henry's law constant from the slope of the equilibrium line y–x diagram.
2. Estimate the minimum liquid-to-gas ratio from the operating line equation.
3. Determine the actual liquid flow rate by multiplying the ratio with a predefined factor.
4. Determine the mass flow rate of the gas per unit cross-sectional area of the tower using the generalized flooding correlation.
5. Determine the required cross-sectional area, and the absorption tower's diameter is at the gas operating point velocity (50–75% of the flooding velocity).
6. The tower height is calculated from the NTU times the HTUs.
7. The tower height strongly depends on the pollutant gas concentration in the inlet gas and removal efficiency.
8. The tower height also depends on the size of the packing material.

PROBLEMS

7.1 ABSORPTION OF AMMONIA IN A PACKED TOWER

Water absorbs ammonia (NH_3) in a packed column. The inlet polluted air stream is 20,000 ppm NH_3. The air stream flows at a 7.0 ft³/min rate, and the column operates at 70°F and 1 atm. The inlet water is pure flowing at a rate of 500 mL/min. The concentration of ammonia in the exit air should not exceed 250 ppm. The packing consists of ceramic Raschig rings, length 3/8 inch, width 3/8 in., wall thickness 1/16 inch, weight 15 lbs/cubic foot, equivalent spherical diameter is 0.35 inch, 0.68 void fraction. The packed column consists of 4.0 inches ID × a 36 inch-long section of borosilicate pipe. Determine the NTUs, column diameter, and column height, and then verify the answer with simulation results obtained from Hysys, PRO/II, Aspen, and SuperPro designer software packages. The suitable fluid package is NRTL.

7.2 ABSORPTION OF ACETONE FROM AIR USING WATER

Acetone is being absorbed from the air by pure water in a packed column designed for 80% of flooding velocity at a temperature of 293 K and 1 atm pressure. The inlet air to the absorber contains 2.6 mol% acetone and outlet 0.5 mol% acetone. The total gas inlet flow rate is 14.0 kmol/h. The pure water inlet flow is 45.36 kmol/h. The column is packed randomly with ceramic Raschig rings, 1.5 inches nominal diameter. Calculate the packed-column height and verify the answer using Hysys/UniSim, PRO/II, Aspen, and SuperPro designer. The suitable fluid package for such a system is NRTL.

7.3 STRIPPING OF ETHANE FROM A HYDROCARBON MIXTURE

A measure of 100 kgmol/h of feed gas at 17 atm and 100°C, containing 3% ethane, 20% propane, 37% n-butane, 35% n-pentane, 5% n-hexane, is to be separated such that 100% ethane, 95% propane, and 1.35% n-butane of the feed stream are to be recovered in the overhead stream. Use stripper to find the molar flow rates and compositions of the bottom stream. Use UniSim to simulate the absorption column (Use Peng–Robinson as the suitable fluid package).

7.4 ABSORPTION OF CO$_2$ FROM GAS STREAM IN A FERMENTATION PROCESS

Ethanol is absorbed from a gas stream in a fermentation process. The gas stream contains 2.0 mol% ethanol and the remaining CO$_2$. All streams enter at 30°C, and the process is isobaric at 1 atm. The entering gas flow rate is 1,000 kgmol/h, and the water flow rate is 2,000 kgmol/h (no ethanol). Use 60% Murphree Tray efficiencies. Determine the number of stages required to absorb 95% of the ethanol from the air stream using water as the absorption media. Use UniSim of any other software package to simulate the absorption column. The appropriate fluid package is NRTL.

7.5 ABSORPTION OF CO$_2$ FROM A GAS STREAM USING METHANOL

A gas stream at 100°C and 6,000 kPa pressure enters a gas absorber. The primary objective of the CO$_2$ absorber is to absorb CO$_2$ contained in the feed stream by contacting countercurrently with methanol solvent in an absorber. The gas stream (0.35 CO, 0.002 H$_2$O, 0.274 CO$_2$, 0.37 H$_2$, 0.002 CH$_4$, and 0.002 N$_2$) is flowing at a rate of 100 kmol/h. Methanol at 30°C and 6,000 kPa is used as an absorbent solvent. The molar flow rate of methanol liquid is 330 kmol/h. Determine the number of theoretical trays required to achieve a fraction of 0.06 CO$_2$ in the exit stream, column diameter, and height. Use UniSim or any other software package to simulate the absorption column. The suitable fluid package is PRSV.

7.6 ABSORPTION OF SO$_2$ FROM AIR USING PURE

Polluted air is fed to the bottom of an absorption column at a rate of 100 kmol/h (0.03 SO$_2$, 0.97 air) at 30°C and 1 atm. Pure water at a flow rate of 10,000 kgmol/h and 30°C and 1 atm is used as absorbent. Use UniSim/Hysys (fluid package Peng–Robinson) to estimate the composition of SO$_2$ in the clean air.

7.7 ABSORPTION OF H$_2$S FROM NATURAL GAS USING PURE WATER

Polluted air is fed to the bottom of an absorption column at a rate of 100 kmol/h (0.05 H$_2$S, 0.95 CH$_4$) at 30°C and 1 atm. Pure water at a flow rate of 10,000 kgmol/h and 30°C and 1 atm is used as absorbent. Use UniSim/Hysys (fluid package Sour PR) to estimate the composition of H$_2$S in the clean air.

7.8 ABSORPTION OF H$_2$S FROM NATURAL GAS USING PROPYLENE CARBONATE

Sour natual gas is fed to the bottom of an absorption column (10 stages) at a rate of 100 kmol/h (0.05 H$_2$S, 0.95 CH4) at 30°C and 1 atm. The solvent composed of 80% propylene carbonate and 20% water at a flow rate of 10,000 kgmol/h and 30°C and

1 atm is used as absorbent. Use UniSim/Hysys (fluid package Sour PR) to estimate the composition of H_2S in the treated natural gas.

7.9 ABSORPTION OF CO_2 FROM NATURAL GAS USING DIETHANOLAMINE (DEA)

Sour natural gas is fed to the bottom of an absorption column (10 stages) at a rate of 100 kmol/h (0.15 CO_2, 0.85 CH_4) at 30°C and 10 atm. The solvent is composed of 10 wt% DEA in water at a flow rate of 7,000 kgmol/h and 30°C and 10 atm (absorbent). Use UniSim/Hysys (fluid package Amine Pkg) to estimate the composition of CO_2 in the treated natural gas. Compare the exit treated methane with the one treated in 7.8 using the different absorbents.

7.10 ABSORPTION OF ACETIC ACID

A feed stream (100 kg/h) enters the bottom of an absorption column (5 stages) at 40°C and 10 atm. The absorber uses pure water (150 kg/h) at 25°C and 10 atm to absorbed acetic acid from a gas mixture, a mass fraction of 98% CO_2, 1.4% N_2, and 0.6% acetic acid. Use UniSim/Hysys (NRTL) to simulate the absorption process and find the treated stream's composition.

REFERENCES

1. Geankopolis, C. J, 1993. Transport Processes and Unit Operations, 3rd edn, Prentice-Hall, New Jersey, NJ.
2. Calvert, S., J. Goldschmid, D. Leith and D. Mehta, 1972. Wet Scrubber System Study, Scrubber Handbook, EPA-R2-72-118a, Vol. 1, U.S. Environmental Protection Agency, Washington, DC.
3. Sherwood, K. T. and R. L. Pigford, 1952. Absorption and Extraction, McGraw-Hill, New York, NY.
4. Peytavy, J. L., M. H. Huor, R. Bugarel and A. Laurent, 1990. Interfacial area and gas-side mass transfer coefficient of a gas-liquid absorption column: Pilot-scale comparison of various tray types, Chemical Engineering and Processing, 27(3), 155–163.
5. Manyele, S. V, 2008. Toxic acid gas absorber design considerations for air pollution control in process industries, Educational Research and Review, 3(4), 137–147.
6. Cheremisinoff, P. N. and R. A. Young, 1977. Air Pollution Control and Design Handbook, Marcel Dekker, New York, NY.
7. McCabe, W. L. and C. J. Smith, 1967. Unit Operations of Chemical Engineering, McGraw-Hill, New York, NY.
8. Perry, J. H, 1973. Chemical Engineers' Handbook, 5th edn, McGraw-Hill, New York, NY.
9. Sinnott, R. K, 1999. Coulson & Richardson's Chemical Engineering, Vol. 6, 3rd edn, Butterworth Heinemann, Oxford.
10. Treybal, R. E, 1968. Mass Transfer Operations, 2nd edn, McGraw-Hill, New York, NY.
11. Onda, K., H. Takeushi and Y. Okumoto, 1968. Mass transfer coefficients between gas and liquid in packed columns, Journal of Chemical Engineering of Japan, 1(1), 56–62.
12. Diab, Y. S. and R. N. Maddox, 1982. Absorption, Chemical Engineering, 89, 38–56.
13. Zenz, F. A. 1972. Designing gas absorption towers, Chemical Engineering, 79, 120–138.

8 Liquid–Liquid Extraction

At the end of this chapter, students should be able to:

1. Explain the principles of liquid–liquid extraction (LLE).
2. Estimate the equilibrium number of stages using a ternary equilibrium diagram.
3. Compute the flow rate and compositions of Extract and Raffinate streams.
4. Justify the manual calculations using five software packages; UniSim/Hysys, PRO/II, Aspen Plus, SuperPro Designer, and Aveva Process Simulation.

8.1 INTRODUCTION

Liquid–liquid extraction (LLE) is an essential unit operation that allows one to separate fluids based on solubility differences of solutes in various solvents. In liquid extraction, separation of liquid solution occurred as a result of contact with another insoluble liquid. If the original solution's components are distributed differently between the two fluids, separation will result (Figure 8.1). Chemical differences drive extraction, and it can be used in situations when distillation is impractical, such as the separation of compounds with similar boiling points in which distillation is not viable or mixtures containing temperature-sensitive components. The solution to be extracted is called the feed, and the liquid used in the extraction is called the solvent. The enriched solvent product is the Extract, and the depleted feed is called the Raffinate [1]. In the design of a LLE column, there are two preliminary calculations: the number of stages and the amount of solvent needed to make a separation.

Since liquid–liquid equilibrium is seldom available in algebraic form, the calculations tend to be iterative or graphical. Use a modified McCabe–Thiele approach if y data versus x data are available. The coordinates for the diagram are the mass fraction of solute in the Extract phase and the Raffinate mass fraction for the other. When one has a convenient equilateral triangle diagram (Figure 8.2), construction can be done directly on the triangle. Some authors refer to this as the Hunter–Ash method [2]. The figure shows the loci of the M1 mixture of components A, B, and C. Try to find the composition of M2.

Figure 8.2 shows the ternary equilibrium phase diagram (A/B/C). Each apex of the triangle represents the pure component. The mass fraction of the species at point M1 is 0.33 A, 0.33 B, and 0.34 C. The mass fraction of compounds at point M2 is 0.6 C, 0.2 A, and 0.2 B. The mass fraction at point M3 is 0.3 B and 0.7 A.

Design of extraction column; the general graphical approach:

1. Determine operating lines using the ternary diagrams.
2. Estimate the number of theoretical stages or plates.
3. Determine the actual equipment size by converting the theoretical stages using the following relation.

DOI: 10.1201/9781003167365-8

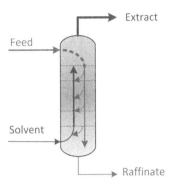

FIGURE 8.1 The schematic diagram represents a multistage countercurrent liquid–liquid extraction column.

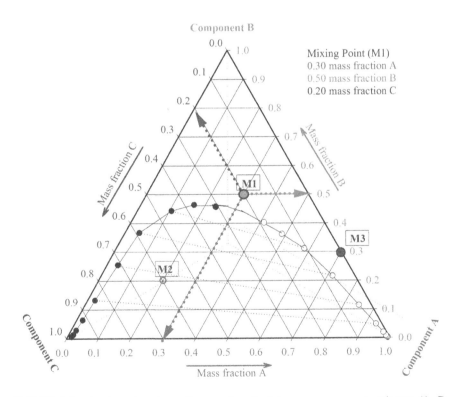

FIGURE 8.2 A triangular phase diagram described ternary components mixture (A, B, and C).

The unit size $\big(\text{or height}\big)$ = number of theoretical stages × HETS/stage efficiency.

HETS is the Height Equivalent to a Theoretical Stage provided by vendors. Suppose we calculated seven theoretical stages ($N = 7$). It means we need a column with the equivalent operation of seven countercurrent mixer-settlers that reach equilibrium entirely.

8.2 MATERIAL BALANCE

The feed stream containing the solute A extracted at one end of the process, and the solvent stream enters at the other end. The Extract and Raffinate streams flow counter currently from stage to stage, and the final products are then extracted stream V_1 leaving stage 1 and the Raffinate stream L_N leaving stage N. The overall material balance on the column [3, 4]:

$$L_0 + V_{N+1} = L_N + V_1 = M \tag{8.1}$$

M represents the total mass flow rate and is constant. L_o represents the feed mass flow rate, V_{N+1} the solvent mass flow rate, V_1 is the exit extract stream, and the L_N is the Raffinate exit stream. Mass flow rates in kg/h or lb/h. The overall component balance on component C:

$$L_0 x_{C0} + V_{N+1} y_{CN+1} = L_N x_{CN} + V_1 y_{C1} = M x_{CM} \tag{8.2}$$

Dividing Equation 8.2 by Equation 8.1 and rearranging:

$$\frac{L_0 x_{C0} + V_{N+1} y_{CN+1}}{L_0 + V_{N+1}} = \frac{L_N X_{CN} + V_{1y C1}}{L_N + V_1} = \frac{M\ X_{CM}}{M} = x_{CM} \tag{8.3}$$

The component balance on component A:

$$L_o x_{A0} + V_{N+1} y_{AN+1} = L_N x_{AN} + V_1 y_{A1} = M x_{AM} \tag{8.4}$$

Dividing Equation 8.4 by Equation 8.1:

$$\frac{L_o X_{A0} + V_{N+1} y_{AN+1}}{L_o + V_{N+1}} = \frac{L_N X_{AN} + V_1 y_{A1}}{L_N + V_1} = \frac{M\ X_{AM}}{M} = x_{AM} \tag{8.5}$$

$$x_{CM} = \frac{L_o X_{C0} + V_{N+1} y_{CN+1}}{L_o + V_{N+1}} \tag{8.6}$$

$$x_{AM} = \frac{L_o X_{A0} + V_{N+1} y_{AN+1}}{L_o + V_{N+1}} \tag{8.7}$$

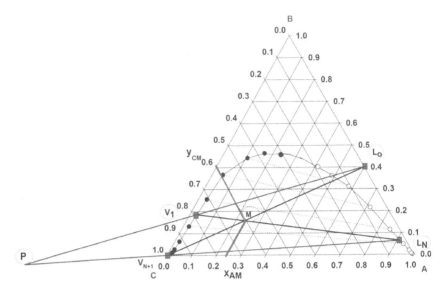

FIGURE 8.3 Equilateral triangle ternary equilibrium diagrams described that the operating lines' intersection is the operating point (p).

Use Equations 8.6 and 8.7 to calculate point M's coordinates on the phase diagram that ties together the two entering streams L_o (0.6 B, 0.4 A) and V_{N+1}(pure C), and the two exit streams V_1 and L_N (0.9 A). Figure 8.3 shows the operating lines. The intersection of the operational lines is the operating point (P).

The following steps allow estimating the number of stages utilizing the countercurrent graphical solution:

1. Commonly, the feed (F) and solvent (S) composition is known; hence, locate F and S on the ternary equilibrium diagram and connect with a straight line.
2. Locate mixture (M) using Equations 8.6 and 8.7.
3. Either specify Extract (E1) or Raffinate (RN); one of them is always known.
4. Connect the one specified with a straight line through M.
5. Solve for unspecified one via tie lines.
6. Connect S through RN and extrapolate, and E1 through F and extrapolate; the lines cross at P.
7. Locate operating point (P) by the intersection of the two operating lines (Figure 8.3).
8. Connect En and Rnvia equilibrium tie lines.

Example 8.1: Extraction of Acetone from Water by Methyl Isobutyl Ketone

Use a countercurrent LLE to remove acetone from a feed stream which contains 43% acetone (A), 50 wt% water (B), and 7 wt% methyl isobutyl ketone (MIBK) (C). Use pure solvent (MIBK) at a flow rate of 820 kg/h in this separation extraction. A feed flow rate of 1,000 kg/h to be treated. The Raffinate contains 4.0 wt% acetone. The operation takes place at 25°C and 1 bar. Perform the following:

1. Determine the number of stages necessary for the separation as specified and the final Extract's composition manually.
2. Determine the mass flow rate of Extract and Raffinate mass flow rate using the following five software packages: UniSim/Hysys, PRO/II, Aspen Plus, SuperPro, and Aveva Process Simulation.

SOLUTION

MANUAL CALCULATION

Plot the ternary equilateral triangle diagram using the equilibrium data in Table 8.1. Using overall material and the component balance, calculate the mixing point compositions:

$$M = L_N + V_1 = V_{N+1} + L_o$$

Component balance for water,

$$Mx_c = L_N x_{CN} + V_1 y_{C1} = V_{N+1} y_{CN+1} + L_o x_{CO}$$

TABLE 8.1
Equilibrium Data for System Acetone–MIBK–Water

Raffinate Layer, Mass Fraction			Extract Layer, Mass Fraction		
Water	Acetone	MIBK	Water	Acetone	MIBK
0.02	0.02	0.96	0.97	0.01	0.02
0.025	0.06	0.915	0.95	0.03	0.02
0.03	0.10	0.87	0.91	0.06	0.03
0.035	0.16	0.805	0.88	0.09	0.03
0.04	0.20	0.76	0.83	0.13	0.04
0.045	0.25	0.705	0.79	0.17	0.04
0.05	0.30	0.65	0.745	0.20	0.055
0.07	0.36	0.57	0.68	0.26	0.06
0.09	0.40	0.51	0.62	0.30	0.08
0.14	0.48	0.38	0.49	0.40	0.11
0.33	0.49	0.18	0.33	0.49	0.18

Source: Data from Christie John Geankoplis, Daniel H. Lepek, Allen Hersel, (2018). Transport Processes and Separation Process Principles, 5th edn, Prentice-Hall, Boston, MA.

Rearranging the above equations,

$$x_{CM} = \frac{y_{CN+1}V_{N+1} + x_{co}L_o}{V_{N+1} + L_o}$$

From the overall material balance and acetone component balance, the mass fraction of acetone at the mixing point is

$$x_{AM} = \frac{y_{AN+1}V_{N+1} + x_{A0}L_o}{V_{N+1} + L_o} = \frac{0 \times 820 + 0.43 \times 1000}{820 + 1000} = 0.24$$

The mass fraction of the MIBK (C) component at the mixing point:

$$x_{CM} = \frac{y_{CN+1}V_{N+1} + x_{co}L_o}{V_{N+1} + L_o} = \frac{1 \times 820 + 0.07 \times 1000}{820 + 1000} = 0.49$$

To calculate the Raffinate and Extract streams mass flow rate,

$$M = 1820 = L_N + V_1$$
$$M = 1820 - V_1 = L_N$$

Determine MIBK compositions at Extract and Raffinate from Figure 8.4.

$$Mx_{CM} = (1820 - V_1)x_{NC} + V_1 y_{1C}$$

Substitute known values

$$1820 \times 0.49 = (1820 - V_1) \times 0.03 + V_1 \times 0.67$$

Solving for V_1 (Extract), $V_1 = 1308$ kg/h
  Solving for L_N (Raffinate), $L_N = 512$ kg/h
  Locate the mixing point on the ternary diagram from calculated compositions. Draw a line starting from L_N through the mixing point (M) to the exit equilibrium concentrations of the Extract (V_1) (Figure 8.4). There are two operating lines. The first is drawn by connecting the L_o and V_1 points and extending the line to the right. The second operating line connects L_N and V_{N+1}, then broadens the line until both lines cross. The operating point (P) is the operational line joint, as shown in Figure 8.5. From V_1, draw the equilibrium tie line that ends at the Raffinate equilibrium line. From the end of the equilibrium, the Raffinate line draws a line to the operating point (P). From the point that crosses the equilibrium Extract curve, draw a tie line. Repeat until lines cross the exit Raffinate

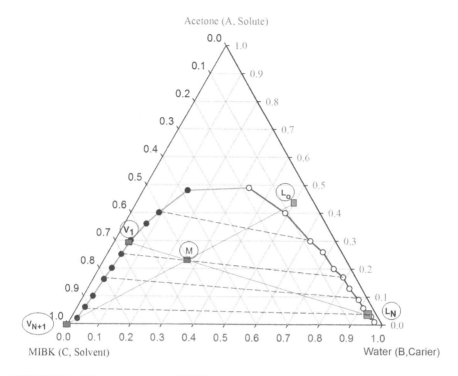

FIGURE 8.4   The acetone–water–MIBK ternary equilibrium diagram (25°C and 1 bar).

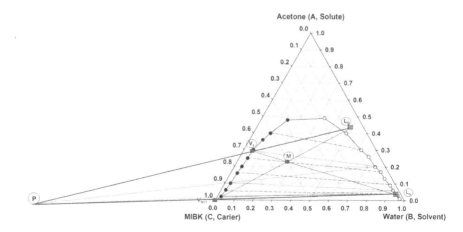

FIGURE 8.5   The manually calculated number of equilibrium stages for the liquid–liquid extraction case is described in Example 8.1 (2.5 stages).

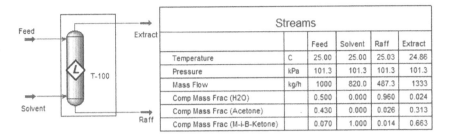

FIGURE 8.6 UniSim/Hysys process flow diagram and streams summary generated for the case presented in Example 8.1.

concentration (L_N). Figure 8.6 shows that seven equilibrium stages are necessary to reach a Raffinate concentration of 4% acetone. The Extract composition is the composition at V_N.

$$y_c = 0.67,\ y_A = 0.29,\ \text{and}\ y_B = 0.04$$

UniSim/Hysys Calculation

In UniSim/Hysys, to simulate an extraction column, one has to specify the number of equilibrium stages, and feed streams. UniSim/Hysys will calculate exit flow rates and components mass fractions, following the steps:

1. Start a new case in UniSim; select the involved components (Acetone, Water, and MIBK). Use the non-random two liquids (NRTL) as the suitable fluid package. For LLE cases, check the binary coefficients of the compounds, if some of the coefficients are not specified, a fatal error will occur, and the process cannot converge. To resolve the problem, simply click on the Binary Coeffs tab. If some of the coefficients are not specified, select UNIFAC LLE, and click on Unknowns only. Close and return to the simulation environment.
2. From the object palette, select the LLE icon (Liquid-liquid extraction column). Double click on the column and in the connection page; name the streams as Feed, Solvent, Extract, and Raffinate, then click on Next to continue.
3. Specify a constant pressure of 1 atm by entering it in the pressure columns.
4. Click on done to complete this section.
5. Specify the number of trays as three trays, the same as those obtained by hand calculation.
6. Double click on the Feed stream. Make an entry of the desired feed properties; the temperature is 25°C, and the pressure is 1.0 atm, a flow rate of 1,000 kg/h, and a feed mass fraction of 0.43 acetone, 0.5 water, 0.07 MIBK.
7. The solvent stream is pure MIBK, at a flow rate of 820 kg/h, at 25°C, 1.0 atm.
8. Double click on the column and click on Run.

9. Close the window and see that the Extract and Raffinate streams have turned from light to dark blue, signifying that UniSim/Hysys has successfully solved those streams of unknown properties.

10. After this step, click on the Reset option, if any of the LLE column specifications are to be changed, and then click the Run button to generate the new results (Figure 8.6).

PRO/II CALCULATIONS

Start PRO/II, and create a new simulation. Add a new unit from the object palette. The Column/Distillation unit operation simulates any distillation or LLE process. A column must contain at least one equilibrium stage or theoretical tray. For purposes of this discussion, the term "trays" denotes "equilibrium stages." The condenser, when present, is always numbered tray number one, and the reboiler, when existing, is assigned the highest tray number in the model.

1. Create a New Simulation in PRO/II (click New), and select the column from the PFD palette. Double click on Units/Streams for input. LLE does not require a condenser or reboiler. Identify the number of stages to 3.0 to verify the manual calculation results. The number of theoretical stages is close to the value obtained using manual calculations (2.5).

2. Click on Streams, and then add two feed streams and two product streams.

3. Click on the Component Selection icon (the benzene ring) in the toolbar, select Acetone, Water, and MIBK, and then click on OK to close the window.

4. Click on Thermodynamic Data, under Category select liquid activity, and then NRTL under Primary Method. Click on Add, and in the pop-up window, select Enable Two Liquid-Phase Calculations. Click on OK to close the current window and then on OK to close the following window.

5. Double click on the Feed stream and specify the total flow rate and compositions. Click on Flow rate and compositions, then select Total Flow Rate. Specify the total flow rate as 1,000 kg/h and specify the compositions as 0.43 acetone, 0.5 for water, and 0.07 mass fractions MIBK. Click OK to close the window. Under Thermal specification, select Temperature and Pressure, specify the temperature as 25°C and pressure as 1 atm. Use UOM at the left top corner to change the units when needed.

6. Double click on solvent feed stream and specify total flow rate as 820 kg/h; for compositions, set mass fraction of MIBK as 1 (pure solvent).

7. Double click on the Column icon, under Algorithm and Calculated Phases, select Liquid-Liquid from the pull-down menu.

8. Click on Pressure Profile. Set the pressure to 1 atm; the pressure drop per tray is 0.0, and then click on OK. The red button turns black.

9. Double click on Feeds and Products, and specify the feed and product tray and an initial estimate of the product stream. Click on OK, and then once again, click on OK.

10. Click on Run for the column to be converted, and the color of the PFD will change to light blue, as shown in Figure 8.7.

11. Click Output in the toolbar, and then click on Generate report to generate the results shown in Figure 8.7.

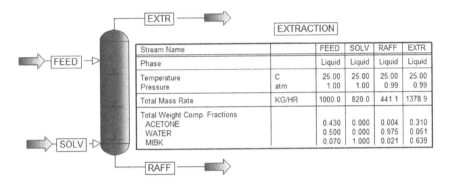

FIGURE 8.7   PRO/II generated the process flow diagram and streams summary for the extraction case described in Example 8.1.

ASPEN PLUS CALCULATIONS

The following steps help in simulating the LLE process using Aspen Plus:

1. Start by opening up a new Aspen file, creating and adding water acetone and MIBK components. A bit of a challenge finding MIBK; use the find. For the method, choose the NRTL-RK, and this setup is ready to go. Once accepting the binary interaction parameters, one thing that is going to be important in this particular case is that we want to see if, with these three components, we will be able to get separation, so the ternary diagram is handy for this (click on Ternary Diagram, then click Continue to Aspen Plus Ternary diagram, and Run Analysis button). Aspen could create those triangle diagrams (Figure 8.8). The two-phase region that we would have expected exists. So it seems like this is going to be the right choice.

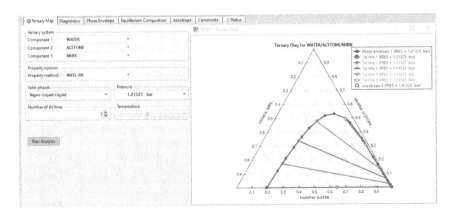

FIGURE 8.8   Aspen Plus generated the ternary diagram for water/acetone/MIBK at 1 atm and 25°C.

2. We are going to only focus on the Column/Extract; add this column extractor. Water will be the bottom product there, and the rich solvent goes on to the top of the column. The feed stream 1.0 is at 25°C and 1 bar and flows at 1,000 kg/h; make sure to switch to a mass fraction (0.43 acetone, 0.5 water, and 0.07 MIBK). Stream 2.0 is a pure solvent (pure MIBK), 25°C, and 1 bar; the mass flow rate is 820 kg/h.

3. Click on the Block setup; the first thing to tell Aspen the number of stages (3.0 stages). Then, we are going to come to our key components, and in this case, we want to choose water as the key component for the first liquid phase and the MIBK as the other key component for the second liquid phase; the acetone is the one that will be traveling back and forth between stages. Suppose we come to our pressure tab. We are going to choose to keep this at a constant pressure of one.

4. Furthermore, now, we have such a red block, and we are going to choose a temperature estimate that will be approximately 25°C, matching our string temperature.

5. This next step is something we have not done before, and this is to go to the convergence tab, and convergence will allow us to override this LLE. The calculations are challenging for the computer. So we need to increase the maximum number of iterations to 200 and error tolerance, 1×10^{-7}. So at this point, we are ready to run this. Look at the summary.

Summary

1. To begin, open the Aspen Plus, and click New. After prompting with the pop-up menu, click on Create on the bottom right corner to create a blank simulation; start in the Properties tab of the Aspen plus program.

2. Under the Components ID, input the components used in the process (acetone, water, and MIBK).

3. The property tab's input is probably the most critical input required to run a successful simulation. This key input is the Base Method found under the Specifications option. The Base Method is the thermodynamic basis for all simulation calculations, select NRTL-RK.

4. Select Extraction (Extract) under the Column tab from the Model Palette and then transfer it to the flowsheet window (click in the Main Flowsheet area).

5. Users can add streams by clicking on the process flowsheet where the stream begins by clicking one more time on the stream to end (first select the Material button).

6. Under the Streams tab, enter all of the specifications for each of the feed streams one at a time.

7. Click on Next, for Specs: for the Number of stages type 3, click on Next, and for the 1st liquid phase, select water as the key component, and for the second liquid phase, select MIBK. Click on Next and for pressure type 1 for stage 3, click on Next, and for stage 3.0, the temperature is 25.

8. To run the simulation, the user could select Next in the toolbar, which tells that all the required inputs are complete, and run the simulation. The user can also choose the toolbar's Run button (this is the button with

FIGURE 8.9 Aspen Plus generated a process flow diagram and stream summary for the case described in Example 8.1.

  a black arrow pointing to the right). Finally, the user can go running on the menu bar and select Run.

9. The Results Summary tab on the Data Browser window has a blue checkmark. Clicking on that tab will open up the Run Status. If the simulation has been converged, the run status should state, "Calculations were completed normally."

10. Adding stream tables to the process flowsheet is a simple process; the two options for varying the stream table: Display and Format on the current screen. Under the Display drop-down menu, there are two options, All Streams or Streams. The Streams option allows the user to choose which stream they would like to present, one by one. Under the Format drop-down menu, there are several types of stream tables. Each of the options offers the data in a slightly different fashion, depending on the intended application. Use the CHEM_E option. To add a stream table, click on the Stream Table button, and a stream table is added to the process flowsheet (Figure 8.9).

11. There is one other location where the user can modify the appearance and content of stream tables. In the Data Browser window, under the Setup tab, there is an option titled Report Options.

12. If one prefers some changes, first reinitialize the simulation to delete the current results and click Run/Reinitialize in the menu bar.

SUPERPRO DESIGNER SIMULATION

SuperPro Designer is different from the previous three commercial software packages, Hysys, PRO/II, and Aspen Plus, so that the Partition coefficient and solubility of the solvent are required in this program and the user needs to provide the value. These values are either found from searching the literature or done experimentally in SuperPro. The following steps help in constructing the process flow diagram in SuperPro Designer:

1. Under Unit Procedure, go to extraction, then Liquid Extraction, select Differential Extractor.

2. Click on Connection Mode; add two inlet streams (heavy liquid from the top, usually the feed stream, and light liquid in the bottom, for example, solvent. Now there are two product streams: Extract (top) and Raffinate (bottom). Figure 8.10 shows the SuperPro predicted results.

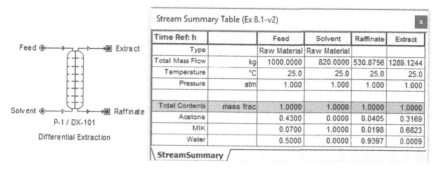

Feed ●——————→▣ Extract

Solvent ●——————→▣ Raffinate
P-1 / DX-101

Differential Extraction

Stream Summary Table (Ex 8.1-v2)						
Time Ref: h			Feed	Solvent	Raffinate	Extract
Type			Raw Material	Raw Material		
Total Mass Flow		kg	1000.0000	820.0000	530.8756	1289.1244
Temperature		°C	25.0	25.0	25.0	25.0
Pressure		atm	1.000	1.000	1.000	1.000
Total Contents	mass frac		1.0000	1.0000	1.0000	1.0000
Acetone			0.4300	0.0000	0.0405	0.3169
MIK			0.0700	1.0000	0.0198	0.6823
Water			0.5000	0.0000	0.9397	0.0009
StreamSummary						

FIGURE 8.10   SuperPro generates the process flow diagram and streams summary table for the case described in Example 8.1.

3. Partition coefficients of solute should be obtained experimentally or found from the literature. Accurate values of partition coefficients are necessary to get the number of transfer units (NTUs) in this example, and while in the Material Balance menu, set the value of Ki for acetone to 3.0 and zero for the rest of the components.
4. Solubility of the light phase (water) and heavy phase (MIBK) and should be specified (1 g/L, 10 g/L).

The distribution coefficient (K) is the ratio of mass fraction of solute in the Extract phase to that in the Raffinate phase. For example, for a feed stream consists of acetone (solute)/water/chloroform (solvent), the distribution coefficient (K) is:

$$K = \frac{\text{Mass fraction acetone in chloroform phase}}{\text{Mass fraction acetone in the water phase}}$$

$$K = \frac{y}{x} = 1.72$$

This means acetone is preferentially soluble in the chloroform phase.

AVEVA PROCESS SIMULATION

Follow the following steps:

1. Create a new simulation and rename it (e.g., Example 8.1).
2. Copy DefFluid from the Process library to the Example model library.
3. Edit the DefFluid by right clicking on the DefFluid icon and selecting Edit. Click on the method and choose NRTL fluid package.
4. Click on the Select Component and add the components involved in the extraction system: acetone, water, and MIBK.
5. From the Process library, drag and drop the Source icon twice for Feed and Solvent streams, and then drag and drop the Extractor icon into the Canvas (the simulation area).

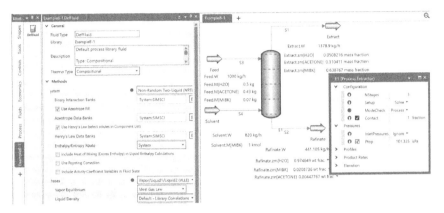

FIGURE 8.11 Aveva Process Simulation produces the process flow diagram and stream conditions for the case described in Example 8.1.

6. Click on the feed and solvent icons and fully specify the two streams by adding values of temperature, pressure, flow rate, and composition, T, P, F, and M, respectively.

7. Connect the two streams to the Extractor column and enter the required values.

8. While the column is still in Configure mode and contact is zero, the system should be squared and solved even this is not a realistic solution.

9. Connect the product stream (Extract and Raffinate), drag and drop the sink icons twice and rename both Extract and Raffinate.

10. Change the Configure to solve and contact 0 to 1 (if not converged, increase it gradually, 0, 0.1, 0.2 up to 1).

11. The process should be squared and solved (Figure 8.11).

12. To plot the ternary diagram, select the triangle from the Tools library and drag it to Canvas. Double click and enter the three components; acetone, water, and MIBK (Figure 8.12).

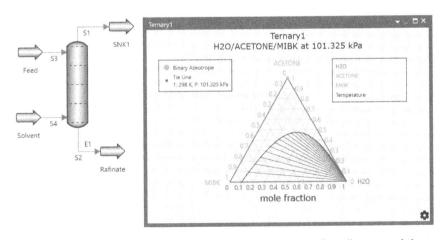

FIGURE 8.12 Aveva Process Simulation generates the process flow diagram and the ternary diagram of the component involved in Example 8.1.

TABLE 8.2
Results of Example 8.1

	Hand Calculations	Hysys	PRO/II	Aspen Plus	SuperPro Designer	Aveva Process Simulation
Feed (kg/h)	1,000	1,000	1,000	1,000	1,000	1,000
Solvent (kg/h)	820	820	820	820	820	820
Extract (kg/h)	1,308	1,333	1,379	1,327	1,289	1,379
Raffinate (kg/h)	512	487	441	429	531	441

CONCLUSIONS

Table 8.2 shows the results of hand calculations and the five simulation software packages. Results reveal the hand calculation and those obtained from the software packages within the acceptable range.

Example 8.2: Extraction of Acetone Using Pure Trichloroethane

In a continuous countercurrent extraction column, 100 kg/h of 40 wt% acetone, 60 wt% water solution is to be reduced to 10 wt% acetone by a LLE process using pure 1,1,2 trichloroethane (TCE) at 25°C and 1 atm; (see Table 8.3). Perform the following:

1. Manually determine the minimum solvent rate and at a solvent rate equal to 1.8 times the minimum solvent/feed rate.
2. Find the required number of mixer settlers.
3. Compare manually calculated values with the software packages: UniSim/ Hysys, PRO/II, Aspen Plus, SuperPro, and Aveva Process Simulation.

TABLE 8.3
Equilibrium Data for the Ternary Liquid Mixture, Acetone– Water–TCE Required for Example 8.2

Water Phase (Mass Fraction)			Water Phase (Mass Fraction)		
Water	Acetone	$C_2H_3Cl_3$	Water	Acetone	$C_2H_3Cl_3$
0.822	0.170	0.007	0.011	0.251	0.738
0.721	0.269	0.010	0.023	0.385	0.592
0.680	0.309	0.012	0.031	0.430	0.539
0.627	0.357	0.016	0.043	0.482	0.475
0.570	0.409	0.021	0.061	0.540	0.400
0.502	0.461	0.038	0.089	0.574	0.337
0.417	0.518	0.065	0.134	0.603	0.263

Source: Data from McCabe, W. L., J. C. Smith and P. Harriott, 1993. Unit Operations of Chemical Engineering, 5th edn, McGraw-Hill, Boston, MA.

SOLUTION

MANUAL CALCULATIONS

Plot the ternary equilibrium diagram and locate inlet and outlet stream compositions as shown in Figure 8.12. To calculate the minimum amount of solvent, one needs to connect the feed stream location to the equilibrium point on the Extract line employing the equilibrium tie line. The line crosses the Extract equilibrium line at E_{min}. From Figure 8.13, the mass fraction of water at minimum solvent feed rate, $X_{B@Mmin}$.

$$X_{B,min} = \frac{x_B F + x_B S_{min}}{F + S_{min}} = \frac{0.6 \times 100 \text{ kg/h} + 0}{100 + S_{min}} = 0.48$$

The minimum solvent flow rate

$$S_{min} = \frac{(60 - 48)}{0.48} = 25 \text{ kg/h}$$

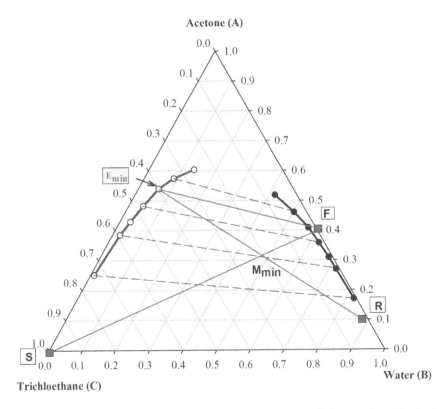

FIGURE 8.13 The location of feed and product streams at a minimum solvent feed rate for the ternary liquid mixture (trichloroethane, water, and acetone). Example 8.2 presents the case.

The actual inlet solvent is

$$S = 1.8 \times S_{min} = 1.8 \times 25 \text{ kg/h} = 45 \text{ kg/h}$$

To locate the mixing point on the ternary equilibrium diagram, the water (B) mass fraction at the mixing point,

$$x_{B,M} = \frac{x_B F + x_B S}{F + S} = \frac{0.6 \times 100 \text{ kg/h} + 0}{100 + 45} = 0.41$$

The mass fraction of acetone (A) at the mixing point,

$$x_{A,M} = \frac{x_A F + x_A S}{F + S} = \frac{0.4 \times 100 \text{ kg/h} + 0}{100 + 45} = \frac{40}{145} = 0.276$$

To calculate product streams flow rate and compositions (Extract and Raffinate streams), locate the calculated mixing point compositions point (M_{min}) on the ternary diagram as shown in Figure 8.14. Connect E_{min} with M_{min}, and then extend the line to cross the Raffinate equilibrium line to point R.

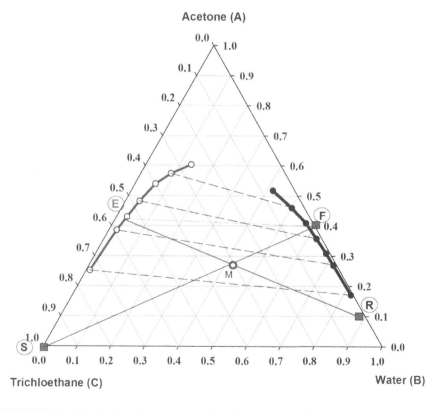

FIGURE 8.14 Feed and product streams at the solvent rate are 1.8 minimum solvent rate for the case described in Example 8.2.

The overall material balance:

$$R + E = F + S = M = 145$$

Acetone component balance:

$$0.1\,R + 0.42\,E = 0.276\,M = 0.276(145)$$

Substitute R = 145 − E:

$$0.1(145 - E) + 0.42\,E = 0.276\,M = 0.276(145)$$

Rearranging:

$$14.5 - 0.1\,E + 0.42\,E = 0.276(145) = 39.15$$

Simplifying and rearranging,

$$0.32\,E = 25.5 \rightarrow E = 77.75\ kg/h,\ R = 145 - E\ R = 67.25\ kg/h$$

From Figure 8.15, the equilibrium number of stages is around 3.0.

UNISIM/HYSYS SIMULATION

1. Select the components (water, acetone, and trichloroethane) involved in the extraction process.
2. Use UNIQUAQ for the fluid package (note that NRTL did not work for this example).
3. Make sure that the binary coefficients exist; otherwise, use UNIQUAQ LLE to calculate unknown values. Figure 8.16 predicted the required values.

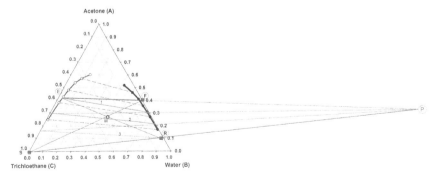

FIGURE 8.15 The manual calculated the total number of equilibrium stages for the extraction case described in Example 8.2.

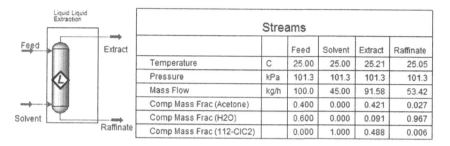

Streams					
		Feed	Solvent	Extract	Raffinate
Temperature	C	25.00	25.00	25.21	25.05
Pressure	kPa	101.3	101.3	101.3	101.3
Mass Flow	kg/h	100.0	45.00	91.58	53.42
Comp Mass Frac (Acetone)		0.400	0.000	0.421	0.027
Comp Mass Frac (H2O)		0.600	0.000	0.091	0.967
Comp Mass Frac (112-ClC2)		0.000	1.000	0.488	0.006

FIGURE 8.16 UniSim/Hysys generates the extraction process flow diagram and streams summary for the case described in Example 8.2.

PRO/II CALCULATIONS

1. Start PRO/II and create a new simulation. Add the component involved in the extraction process (acetone, water, and trichloroethane) and the thermodynamic package (NRTL).
2. From the PFD Palette, click on the Column and select the Distillation column.
3. Uncheck the condenser and reboiler since both are not needed.
4. Set the number of equilibrium stages to three stages.
5. Note: to display streams in weight fraction, click on Output in the toolbar, then Report format, Stream properties: stream component flow rate and composition report.
6. Figure 8.17 shows the PRO/II predicted results.

ASPEN PLUS RESULTS

Using Aspen Plus, Figure 8.18 shows that the process flowsheet and stream table compositions. The fluid base method is NRTL. The extraction column consists of three equilibrium stages.

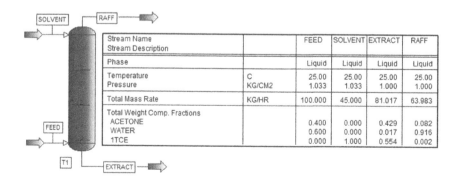

Stream Name Stream Description		FEED	SOLVENT	EXTRACT	RAFF
Phase		Liquid	Liquid	Liquid	Liquid
Temperature	C	25.00	25.00	25.00	25.00
Pressure	KG/CM2	1.033	1.033	1.000	1.000
Total Mass Rate	KG/HR	100.000	45.000	81.017	63.983
Total Weight Comp. Fractions ACETONE		0.400	0.000	0.429	0.082
WATER		0.600	0.000	0.017	0.916
1TCE		0.000	1.000	0.554	0.002

FIGURE 8.17 PRO/II generates the extraction process flow diagram and streams summary for the case described in Example 8.2.

FIGURE 8.18 Aspen Plus generates the extraction process flow diagram and streams summary for the case described in Example 8.2.

SuperPro Designer Simulation

In SuperPro Designer, specify the solvent's partition factor along with light- and heavy-phase solvents' solubility, as previously shown in Example 8.1. Figure 8.19 depicts the process flowsheet and components mass flow rates.

Aveva Process Simulation

Following the same procedure of Example 8.1, Figure 8.20 shows Aveva Process Simulation software's predictions using UNIQUAC for the fluid package and LLE. Figure 8.21 predicts the generated ternary diagram for the system water/acetone/trichloroethane.

Conclusion

Table 8.4 shows the results of hand calculations and predictions of the five software. Results reveal that hand calculations and those obtained from the software packages were in good agreement. The product obtained by PRO/II is close to hand calculations and Aveva Process Simulation; in contrast, the Hysys results were far from hand calculations.

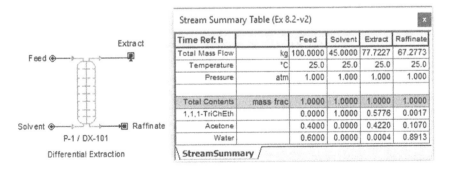

FIGURE 8.19 SuperPro generates the extraction process flow diagram with stream table for the case described in Example 8.2.

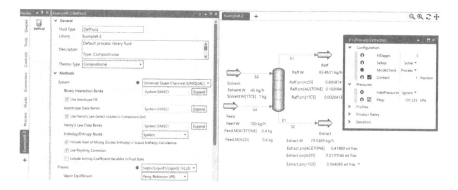

FIGURE 8.20 Aveva Process Simulation generates the extraction process flow diagram and stream conditions for the case described in Example 8.2.

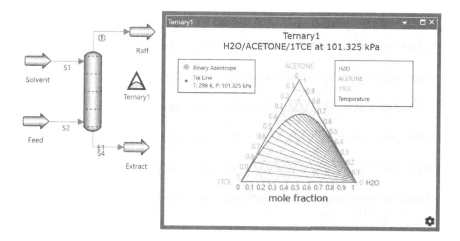

FIGURE 8.21 Aveva Process Simulation generates the process flow diagram and the ternary diagram of the component involved in Example 8.2.

TABLE 8.4

Results of Example 8.2 Performed Manually and Five Software Packages

	Manual Calculations	UniSim/ Hysys	PROII	Aspen ONE	SuperPro Designer	Aveva Process Simulation
Feed (kg/h)	100.0	100	100	100	100	100
Solvent (kg/h)	45.00	45	45	45	45.0	45.0
Extract (kg/h)	77.75	91.58	83.66	76	77.7	79.5
Raffinate (kg/h)	67.25	53.42	61.34	69	67.3	65.5

Example 8.3: Extraction of Acetone from Water Using MIBK

Figure 8.22 shows a countercurrent extraction column engaged in extracting acetone from 100 kg/h of feed mixture. The feed consists of 40 wt% acetone (A) and 60 wt% water (w) using MIBK to Extract acetone at a temperature of 25°C and 1 atm. The extracting liquid is pure solvent (MIBK) fed at 100 kg/h [3]. Perform the following calculations:

1. Manually determine how many ideal stages are required to Extract 99% of the acetone fed.
2. Compare the manually calculated Extract, Raffinate mass flow rates, and the compositions with the five software packages: UniSim, Aspen, PRO/II, SuperPro, and Aveva Process Simulation.

SOLUTION

MANUAL CALCULATION

The composition and flow rate of inlet streams (Feed and Solvent):

Feed: 100 kg/h, 0.4 mass fraction acetone, and 0.6 mass fraction water.
Solvent: 100 kg/h of pure MIBK.

First, determine stream compositions:

Assume that the mass flow rate of water in the Extract stream is w. The mass flow rate of MIBK in the Raffinate stream is m. Since 99% of the acetone is recovered in the Extract.

$$0.99\,(0.4 \times 100) = 39.6 \ kg/h$$

Mass flow rate of acetone in the Raffinate = $40 - 39.6 = 0.4$ kg/h. Stream's flow rates:

$$L_0 = 100 \ kg/h, V_{N+1} = 100 \ kg/h$$
$$V_N = 39.6 + (100 - m) + w$$

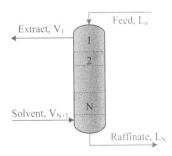

FIGURE 8.22 Process flowsheet of an extraction system in Example 8.3.

$$L_N = 0.4 + (60 - w) + m = 60.4 + m - w$$

As the first assumption neglect the amount of w and m,

$$V_N = 140 \text{ kg/h}$$

The mass fraction of acetone in the exit Extract stream

$$y_{A,N} = \frac{39.6}{140} = 0.283$$

The corresponding mass fraction of water from the equilibrium diagram is, $y_{W,N} = 0.049$

$$w = \frac{0.049}{1 - 0.049}(39.6 + 100 - m)$$

The amount of m is minimal and neglected as in the first trial: $w = 7.2$ kg/h
The mass fraction of acetone in the exit Raffinate stream, $x_{A,N}$

$$x_{A,N} = \frac{0.4}{60} = 0.007$$

The corresponding mass fraction of MIBK from the equilibrium diagram is

$$x_{M,N} = 0.02$$

The amount of MIBK:

$$m = \frac{0.02}{1 - 0.02}(0.4 + 60 - w)$$

Simplifying further,

$$m = \frac{0.02}{1 - 0.02}(0.4 + 60 - 7.2) = 1.1 \text{ kg/h}$$

The revised value of water,

$$w = \frac{0.049}{1 - 0.049}(39.6 + 100 - 1.1) = 7.1 \text{ kg/h}$$

SUMMARY

$L_o = 100$ kg/h and the compositions are $x_{A,0} = 0.4$, $x_{w,o} = 0.6$

$V_{N+1} = 100$ kg/h and the compositions are $y_{A,N+1} = 0$, $y_{M,N+1} = 1.0$

$L_N = 0.4 + (60 - w) + m = 60.4 + m - w = 60.4 + 1.1 - 7.1 = 54.4$ kg/h

$$x_{A,N} = \frac{0.4}{54.4} = 0.0074$$

$V_N = 39.6 + (100 - m) + w = 139.6 - 1.1 + 7.1 = 145.6$ kg/h

$$y_{A,N} = \frac{39.6}{145.6} = 0.272$$

RESULTS

Figure 8.23 locates the inlet and exit stream compositions in the ternary diagram. Performing stage-by-stage calculations reveals that 4.0 equilibrium stages are required.

$L_o = 100$ kg/h, $V_{N+1} = 100$ kg/h, $L_N = 54.4$ kg/h, $V_{N+1} = 145.6$ kg/h

UNISIM CALCULATIONS

Start a new case in Hysys. Select acetone, water, and MIBK for the components. Use the UNIQUAQ Fluid Package. Check the binary coefficients; simply click the Binary Coeffs tab. To specify the undefined values, select UNIFAC LLE, and click on Unknowns Only.

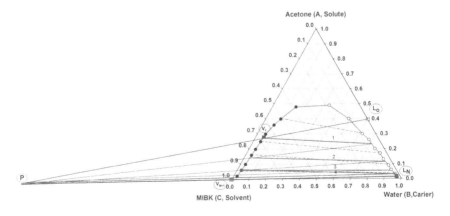

FIGURE 8.23 Manually calculates the equilibrium stages of the case described in Example 8.3.

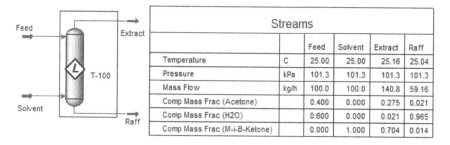

FIGURE 8.24 UniSim/Hysys generates the extraction process flow diagram and streams summary for the case described in Example 8.3.

1. Close and return to the simulation environment. Name the streams Feed, Solvent, Extract, and Raffinate, and then click on Next. Specify a constant pressure of 1 atm by typing it into the pressure columns. Click on Done button to complete the Liquid-Liquid Extractor Input screen.

2. Specify the number of stages as 4.0 stages to compare with hand calculation.

3. Double click on the Feed stream. Enter the desired feed properties; a temperature of 25 °C and the input pressure is 1.0 atm, the feed flow rate is 100 kg/h, and a feed composition: 40 wt% acetone and 60 wt% water.

4. The solvent stream contains pure MIBK entering at 25°C and 1.0 atm, and flowing at 100 kg/h.

5. Double click on the column and click on Run. Close the window and see that the Extracted Product and Raffinate streams have turned from light blue to dark blue, signifying that UniSim/Hysys has successfully solved those streams. After this stage, if any of the specifications on the LLE column are changed, press the Reset button, and then the click the Run button again to rerun the distillation column to the solved mode with new specifications (Figure 8.24).

PRO/II CALCULATIONS

Start PRO/II and create a new simulation, and from the components menu, select Acetone, Water, and MIBK. Use the UNIQUAQ Fluid Package. From the PFD Palette, choose Column and then Distillation column. Uncheck the reboiler and condenser (i.e., not needed) and specify the stages as four equilibrium stages. Repeat the PRO/II procedure used in Example 8.1. Figure 8.25 shows the process flowsheet and the generated Output.

ASPEN PLUS SIMULATION

Use NRTL-RK for the fluid package to measure liquid property; the first liquid-phase key component is water, and the second liquid-phase key component is MIBK, acetone is moving back and forth between the two phases. Using 4.0 stages and the Extractor is operating at 25°C and 1 atm, the results should look like that shown in Figure 8.26.

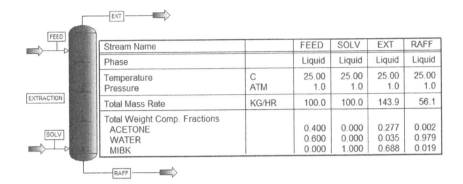

FIGURE 8.25 PRO/II generates the extraction process flow diagram and streams summary for the case described in Example 8.3.

SuperPro Designer Simulation

Using SuperPro Designer, as previously mentioned, the partition coefficient of solute should be specified either by trial and error to achieve specific separation or by using experimental values. Figure 8.27 shows the predicted results.

Aveva Process Simulation

Following the same procedure of Example 8.1, Figure 8.28 shows the predictions of Aveva Process Simulation software. Use NRTL for the fluid package. Figure 8.29 shows the process flow diagram or the results of acetone/water/MIBK. Figure 8.28 shows the Aveva Process Simulation generates the process flow diagram and the ternary diagram of the component involved in Example 8.3.

Conclusions

Table 8.5 shows the results of hand calculations and the five simulation software packages. Results reveal that hand calculations and those obtained from the software packages were in good agreement. The PRO/II generated output is close to hand calculations; in contrast, the results obtained using SuperPro Designer were

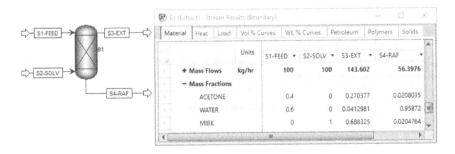

FIGURE 8.26 Aspen Plus generates the process flowsheet with the stream table for the extraction case described in Example 8.3.

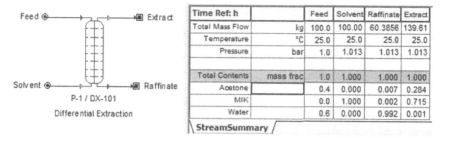

Time Ref: h			Feed	Solvent	Raffinate	Extract
Total Mass Flow		kg	100.0	100.00	60.3856	139.61
Temperature		°C	25.0	25.0	25.0	25.0
Pressure		bar	1.0	1.013	1.013	1.013
Total Contents		mass frac	1.0	1.000	1.000	1.000
Acetone			0.4	0.000	0.007	0.284
MIK			0.0	1.000	0.002	0.715
Water			0.6	0.000	0.992	0.001

FIGURE 8.27 SuperPro Designer generates the process flowsheet and stream table for the extraction case described in Example 8.3.

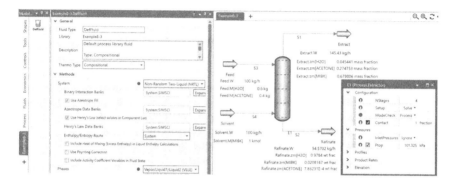

FIGURE 8.28 Aveva Process Simulation software predicts the outlet stream conditions for the extraction case described in Example 8.3.

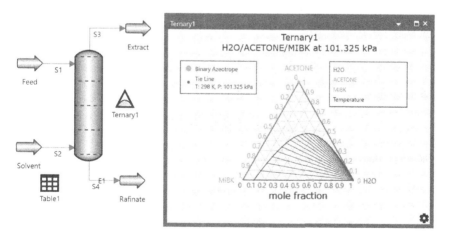

FIGURE 8.29 Aveva Process Simulation generates the process flow diagram and the ternary diagram of the component involved in Example 8.3.

TABLE 8.5
Comparison of the Manual Calculation and the Software Predicted Results of the Case in Example 8.3

	Manual Calculations	UniSim	PRO/II	Aspen Plus	SuperPro Designer	Aveva Process Simulation
Feed (kg/h)	100.0	100.0	100.0	100.0	100.0	100.0
Solvent (kg/h)	100.0	100.0	100.0	100.0	100.0	100.0
Extract (kg/h)	54.4	58.2	56.1	58.4	60.4	54.6
Raffinate (kg/h)	145.6	141.8	143.9	141.6	139.6	145.4

far from manual estimates and other software packages, accurate values of partition coefficients may give better results.

Example 8.4: Extraction of Acetone from Water using Two Feed Streams

A countercurrent extraction column was employed to Extract acetone from water in a two feed stream consuming 5,000 kg/h of the solvent trichloroethane to give a Raffinate containing 10 wt% acetone. The feed stream F_1 is at a rate of 7,500 kg/h (50% acetone and 50 wt% water). The second feed stream, F_2, is 7,500 kg/h (25 wt% acetone and 75 wt% water). The column is operating at 25°C and 1 atm [3]. Manually calculate the required equilibrium number of stages and the stages of the location of the feed streams. Use equilibrium data of Example 8.2. Compare the manually calculated values with the predictions of the available software packages.

SOLUTION

MANUAL CALCULATIONS

Draw the ternary diagram and locate feed streams (F_1, F_2, and S). Find the mixing point compositions (Figure 8.30). Mixing point compositions,

$$F_1 + F_2 + V_{N+1} = L_N + V_1 = M$$

Substitution

$$7500 + 7500 + 5000 = L_N = V_1 = M$$

Rearranging,

$$M = 20,000 \ kg/h$$

Component balance of acetone (A),

$$0.5 \times 7500 + 0.25 \times 7500 + 0 \times 5000 = L_N = V_1 = x_{AM} \times 20,000, \ x_{AM} = 0.28$$

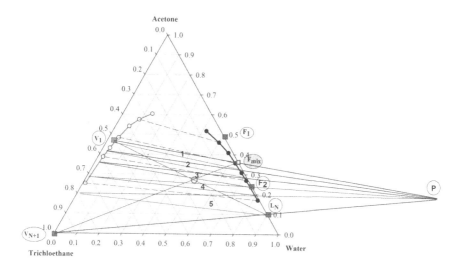

FIGURE 8.30 The manual calculation of the extraction process generates five stages for the case described in Example 8.4.

Component balance of 1,1,2 trichloroethane (C),

$$0.0 \times 7500 + 0.0 \times 7500 + 1 \times 5000 = x_{CM} \times 20{,}000$$

Simplifying, $x_{CM} = 0.25$

The Raffinate (L_N) and Extract (V_1) mass flow rate: $L_N + V_1 = M = 20{,}000$
Component balance (acetone): $0.1 \times L_N + 0.47 \times V_1 = 0.28 \times 20{,}000$
Substituting total balance: $0.1 \times (20{,}000 - V_1) + 0.47 \times V_1 = 0.28 \times 20{,}000$

$$V_1 = 9{,}729 \text{ kg/h, and } L_N = 10{,}271 \text{ kg/h}$$

UniSim Simulation

To verify the hand calculations using Hysys, open a new case in Hysys, select the components (water, acetone, and TCE). Select UNIQUAQ as the fluid package. Ensure that the binary coefficients exist, otherwise estimate unknown values, and set the number of stages to 5. Figure 8.31 depicts the Hysys process flow-sheet and the result of the exit stream.

PRO/II Solution

Select distillation column from object palette; no condenser or reboiler is required, so uncheck reboiler and condenser radio button. Set equilibrium number of stages to five; enter first feed at stage 1 and the second feed at stage 3. Add components (water, acetone, and trichloroethane); select NRTL for a fluid package under the Liquid Activity menu. Figure 8.32 shows the converged process flow diagram and the generated report. The report shows the flow rate and composition of inlet and exit streams.

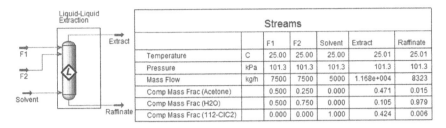

Streams						
		F1	F2	Solvent	Extract	Raffinate
Temperature	C	25.00	25.00	25.00	25.01	25.01
Pressure	kPa	101.3	101.3	101.3	101.3	101.3
Mass Flow	kg/h	7500	7500	5000	1.168e+004	8323
Comp Mass Frac (Acetone)		0.500	0.250	0.000	0.471	0.015
Comp Mass Frac (H2O)		0.500	0.750	0.000	0.105	0.979
Comp Mass Frac (112-ClC2)		0.000	0.000	1.000	0.424	0.006

FIGURE 8.31   UniSim/Hysys generates the process flow diagram and streams summary for the case described in Example 8.4.

ASPEN PLUS CALCULATIONS

Start a new case in Aspen Plus; add the component involved in the extraction process (acetone, water, 1,1,2 trichloroethane). The NRTL-RK is a suitable fluid property measurement package. There are five stages, and the feed stream enters stage 1, the F2 stream enters stage 2, and the S2-SOLV stream to stage 5. Figure 8.33 shows the Aspen Plus generated process flowsheet and stream table compositions.

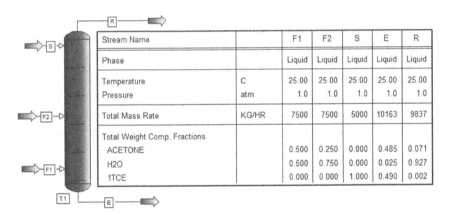

Stream Name		F1	F2	S	E	R
Phase		Liquid	Liquid	Liquid	Liquid	Liquid
Temperature	C	25.00	25.00	25.00	25.00	25.00
Pressure	atm	1.0	1.0	1.0	1.0	1.0
Total Mass Rate	KG/HR	7500	7500	5000	10163	9837
Total Weight Comp. Fractions						
ACETONE		0.500	0.250	0.000	0.485	0.071
H2O		0.500	0.750	0.000	0.025	0.927
1TCE		0.000	0.000	1.000	0.490	0.002

FIGURE 8.32   PRO/II generates the process flow diagram and streams summary for the extraction case described in Example 8.4.

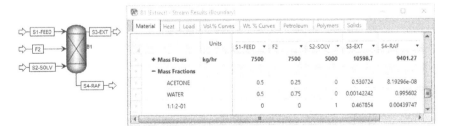

	Units	S1-FEED	F2	S2-SOLV	S3-EXT	S4-RAF
+ Mass Flows	kg/hr	7500	7500	5000	10598.7	9401.27
− Mass Fractions						
ACETONE		0.5	0.25	0	0.530724	8.19296e-08
WATER		0.5	0.75	0	0.00142242	0.995602
1:1:2-01		0	0	1	0.467854	0.00439747

FIGURE 8.33   Aspen Plus generates the process flow diagram and streams summary for the extraction case described in Example 8.4.

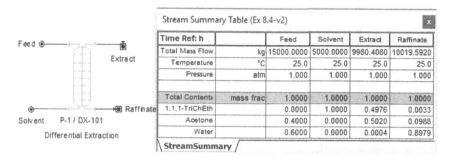

Stream Summary Table (Ex 8.4-v2)						
Time Ref: h			Feed	Solvent	Extract	Raffinate
Total Mass Flow	kg	15000.0000	5000.0000	9980.4080	10019.5920
Temperature	°C	25.0	25.0	25.0	25.0
Pressure	atm	1.000	1.000	1.000	1.000
Total Contents	mass frac	1.0000	1.0000	1.0000	1.0000
1,1,1-TriChEth		0.0000	1.0000	0.4976	0.0033
Acetone		0.4000	0.0000	0.5020	0.0988
Water		0.6000	0.0000	0.0004	0.8979

StreamSummary

FIGURE 8.34 SuperPro Designer generates the process flow diagram and streams summary for the extraction case described in Example 8.4.

SUPERPRO DESIGNER SIMULATION

The SuperPro generated process flowsheet and stream table for the extraction case in Example 8.4 are shown in Figure 8.34.

AVEVA PROCESS SIMULATION

Design an Extractor column for Example 8.4 with one feed stream and one solvent stream. Figure 8.35 shows how this would solve and deliver identical results of PRO/II. Use the NRTL method with the vapor and liquid-liquid equilibrium (VLLE) option and the default interaction parameters. These results show the extraction principles. Figure 8.36 shows the Aveva Process Simulation generated ternary diagram of the extraction process described in Example 8.4.

CONCLUSIONS

Table 8.6 shows the results of hand calculations and the predictions of the five simulation software packages. Results reveal that hand calculations and those

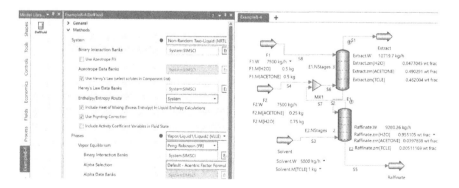

FIGURE 8.35 Aveva Process Simulation generates the process flow diagram and streams summary for the extraction case described in Example 8.4.

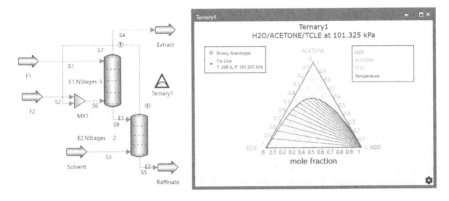

FIGURE 8.36 Aveva Process Simulation generated the ternary equilibrium diagram of the extraction process described in Example 8.4.

obtained from the software packages were in good agreement. The product obtained by PROII is closer to hand calculations than others; in contrast, the results obtained using Hysys were far from hand calculations and other software packages; trying other fluid packages from Hysys will give better results.

Example 8.5: Extraction of Acetic Acid Using Isopropyl Ether [5]

Pure isopropyl ether at 20°C and 1 atm and a mass flow rate of 45 kg/h Extract acetic acid from a mixture flowing at 15 kg/h contain 30 wt% acetic acids in water. Calculate the number of equilibrium stages required to achieve a concentration of 2 wt% acetic acids in the product Raffinate stream (Table 8.7)

TABLE 8.6
Comparison of Manual Calculation with Software Predicted Results of Example 8.4

		Manual Calculations	Hysys	PROII	Aspen Plus	SuperPro Designer	Aveva Process Simulation
Feed (kg/h)	F_1	7,500	7,500	7,500	7,500	7,500	7,500
	F_2	7,500	7,500	7,500	7,500	7,500	7,500
Solvent (kg/h)		5,000	5,000	5,000	5,000	5,000	5,000
Extract (kg/h)		9,729	11,679	10,500	11,752	10,020	10,720
Raffinate (kg/h)		10,271	8,321	9,500	8,248	9,980	9,280

TABLE 8.7

Example 8.5 Required the Equilibrium Data at 20°C for the Acetic Acid–Water–Isopropyl Ether

Water Layer (Mass Fraction)			Isopropyl Ether Layer (Mass Fraction)		
Water	Acetic Acid	Isopropyl Ether	Water	Acetic Acid	Isopropyl Ether
0.981	0.007	0.012	0.005	0.002	0.993
0.971	0.014	0.015	0.007	0.004	0.989
0.955	0.029	0.016	0.008	0.008	0.984
0.917	0.064	0.019	0.010	0.019	0.971
0.844	0.133	0.023	0.019	0.048	0.933
0.711	0.255	0.034	0.039	0.114	0.847
0.589	0.367	0.044	0.069	0.216	0.715
0.451	0.443	0.106	0.108	0.311	0.581
0.371	0.464	0.165	0.151	0.362	0.487

Source: Data from Christie John Geankoplis, Daniel H. Lepek, Allen Hersel, 2018. Transport Processes and Separation Process Principles, 5th edn, Prentice-Hall, Boston, MA.

SOLUTION

MANUAL CALCULATIONS

Figure 8.37 shows the schematic diagram of a multistage extraction process flow diagram.
The mixing point of the second stage is determined as follows:

$$S_1 + L_o = V_1 + L_N = M_1 = 60 \ kg/h$$

The mass fraction of isopropyl ether (C) at the mixing point

$$x_{CM} = \frac{y_{c1}S_1 + x_{c0}L_o}{S_1 + L_o} = \frac{1 \times 45 + 0 \times 15}{45 + 15} = 0.75$$

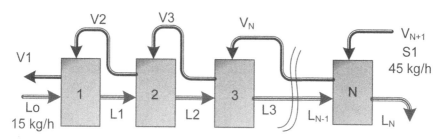

FIGURE 8.37 The multi-stages countercurrent extraction process is employed to extract acidic acid from acidic acid/water mixture using isopropyl ether in Example 8.5.

The mass fraction of acetic acid at the mixing point

$$x_{AM} = \frac{y_{A1}S_1 + x_{A0}L_o}{S_1 + L_o} = \frac{0 \times 45 + 0.3 \times 15}{45 + 15} = 0.075$$

Locate the mixing point on the ternary diagram; the mass fraction of the Extract stream leaving stage 1, V_1, is

$$x_{A1} = 0.1, \ x_{B1} = 0.02, \ x_{C1} = 0.88$$

Calculate the amount of L_N: $L_N + V_1 = 60 \rightarrow V_1 = 60 - L_N$
   Component balance (isopropyl ether, C): $0.02 \times L_N + 0.88(60 - L_N) = 0.75 \times 60$
   The mass flow rate of streams leaving unit: $L_N = 9.1 \ kg/h$, $V_1 = 50.9 \ kg/h$
   Figure 8.38 shows the ternary equilibrium diagram, and the number of stages required for the extraction process is four stages.

UNISIM/HYSYS SIMULATIONS

The fluid package used in the simulation is NRTL, and the binary coefficients were calculated using UNIFAC LLE. Built-in default values of binary coefficients with the NRTL method did not give good results (Figure 8.39).

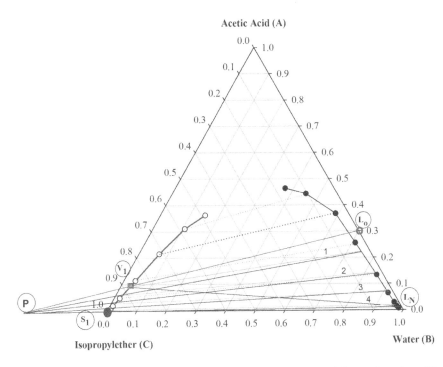

FIGURE 8.38 Manually calculated ternary equilibrium diagram and the number of equilibrium stages for water–acetic acid–isopropyl ether (4 stages) for the extraction case described Example 8.5.

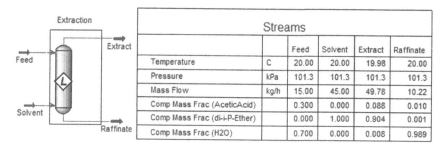

FIGURE 8.39   UniSim/Hysys generates the process flow diagram and streams summary for the extraction case described in Example 8.5.

PRO/II SIMULATIONS

Figure 8.40 shows the PRO/II generated process diagram and streams summary for the case of extraction described in Example 8.5. The fluid package used to calculate the physical properties is the NRTL. The number of trays is 4, the top tray pressure of the column is at 0.968 atm. The selected algorithm and calculated phases are the liquid–liquid.

ASPEN PLUS CALCULATIONS

Create a blank case in the Aspen Plus, add the component involved in the extraction process (Acetic acid, water, and diisopropyl ether). The suitable fluid property measurement package is the NRTL-RK. The number of stages is 4. Figure 8.41 shows the Aspen Plus generated process flow diagram and stream table with component mass fractions.

SUPERPRO DESIGNER SIMULATION

Figure 8.42 depicts the process flow diagram and table stream summary with total contents and mass of each component generated with SuperPro Designer for the case described in Example 8.5.

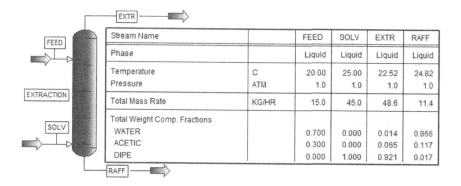

FIGURE 8.40   PRO/II generates the process flow diagram and streams summary for the extraction case described in Example 8.5.

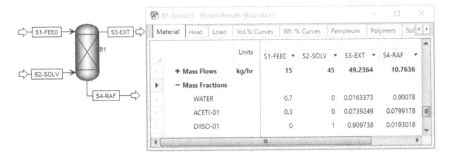

FIGURE 8.41   Aspen Plus generates the process flow diagram and streams summary for the extraction case described in Example 8.5.

AVEVA PROCESS SIMULATION

Following the same procedure of Example 8.1, Figure 8.43 shows the predictions of Aveva Process Simulation software. Use the NRTL method with the VLLE option and the default interaction parameters. The Aveva Extractor column in this version is with one feed stream and one solvent stream. Figure 8.44 shows the Aveva Process Simulation generated the process flow diagram and the ternary diagram of the component involved in Example 8.5.

Stream Summary Table (Ex. 8.5-v2)					
Time Ref: h			Feed	Solvent	Extract	Raffinate
Type			Raw Material	Raw Material		
Total Mass Flow	kg	15.0000	45.0000	49.4439	10.5561
Temperature	°C	20.0	20.0	20.0	20.0
Pressure	atm	1.000	1.000	1.000	1.000
Total Contents	**mass frac**	1.0000	1.0000	1.0000	1.0000
Acetic-Acid		0.3000	0.0000	0.0901	0.0043
Diethyl Ether		0.0000	1.0000	0.9086	0.0070
Water		0.7000	0.0000	0.0013	0.9887

StreamSummary

FIGURE 8.42   SuperPro Designer generates the process flow diagram and streams summary for the extraction case described in Example 8.5.

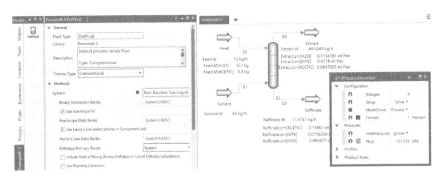

FIGURE 8.43   Aveva Process Simulation generates the process flow diagram and streams summary for the extraction case described in Example 8.5.

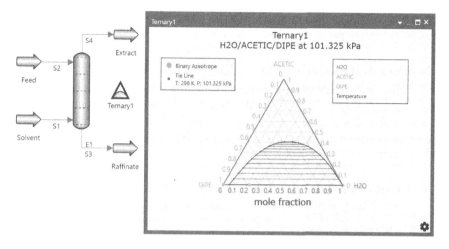

FIGURE 8.44 Aveva Process Simulation generates the process flow diagram and the ternary diagram of the component involved in Example 8.5.

TABLE 8.8
Results of Example 8.5

	Hand Calculations	Hysys	PROII	Aspen Plus	SuperPro Designer	Aveva Process Simulation
Feed (kg/h)	15	15	15	15	15	15
Solvent (kg/h)	45	45	45	45	45	45
Extract (kg/h)	50.9	49.8	48.6	49.2	49.4	48.6
Raffinate (kg/h)	9.1	10.2	11.4	10.8	10.6	11.4

CONCLUSIONS

Table 8.8 shows the results of hand calculations and the four simulation software. The table reveals that the results obtained from the four software packages were very close to each other, while there is also a discrepancy in that obtained by hand calculations due to the accuracy in drawing of the equilibrium tie lines in the equilibrium ternary diagram.

PROBLEMS

8.1 EXTRACTION OF ACETONE FROM WATER BY MIBK

Utilize a countercurrent liquid-liquid extraction column to extract acetone from a feed that contains 50% acetone (A), 50 wt% water (B). Pure MIBK, at a flow rate of 80 kg/h, is used as the solvent in this separation. The Raffinate composition should be around 4.0 wt% acetone at a feed flow rate of 100 kg/h. The operation is taking place at 25°C and 1 atm. Determine the number of stages necessary for the separation. Simulate the extraction process using UniSim/Hysys (NRTL) or any available software package.

8.2 EXTRACTION OF ACETONE USING PURE TCE

In a continuous countercurrent extraction column, 100 kg/h of 30 wt% acetone and 70 wt% water solution is to be reduced to 10 wt% acetone by extraction with 100 kg/h of pure TCE at 25°C and 1 atm. Find the number of mixer settlers required. Verify the manual calculations using UniSim/Hysys (UNIQUAC) predictions or any available software package.

8.3 EXTRACTION OF ACETONE FROM WATER USING MIBK

A countercurrent extraction plant was employed to extract acetone from 100 kg/h of feed mixture. The feed consists of 20 wt% acetone (A) and 80 wt% water (w) using MIBK at a temperature of 25°C and 1 atm. 100 kg/h of pure solvent MIBK is the extracting liquid. How many ideal stages are required to extract 90% of the acetone fed? What are the Extract and Raffinate mass flow rates, and what are the compositions? Compare the manual calculations with UniSim/Hysys (NRTL-ideal) predictions or any available software package (e.g., PRO/II, Aspen Plus, SuperPro, and Aveva Process Simulation).

8.4 EXTRACTION OF ACETONE FROM WATER USING 1,1,2 TRICHLOROETHANE

Practice the countercurrent extraction column to extract acetone from water in a two-feed stream using 50 kg/h of TCE to give a Raffinate containing 10.0 wt% acetone. The feed stream F_1 is 75 kg/h, including 50% acetone and 50 wt% water. The second feed, F_2, is 75 kg/h (25 wt% acetone and 75 wt% water). The column is operating at 25°C and 1 atm. Calculate the required equilibrium number of stages and the feed stage. Validate manual results with the predictions of UniSim/Hysys (UNIQUAC-Ideal) or any available software packages.

8.5 EXTRACTION OF ACETIC ACID FROM WATER USING ISOPROPYL ETHER

A mixture of 100 kg/h of acetic acid and water containing 30 wt% acids is to be extracted in a three-stage extractor with isopropyl ether at 20°C using 40 kg/h of the pure solvent. A fresh solvent at 40 kg/h enters the second stage. Determine the composition and quantities of the Raffinate and Extract streams. Perform the extraction process by using UniSim/Hysys (NRTL) or any available software package.

8.6 EXTRACTION OF ACETONE IN TWO STAGES

The liquid mixture with a flow rate of 100 kg/h comprises 0.24 mass fraction acetone, and 0.78 water is extracted through 50 kg/h of MIBK in two stages using a continuous countercurrent extractor. Determine the amount and composition of the Extract and Raffinate phases. Validate manual results with the predictions of UniSim/Hysys (NRTL) or any available software packages.

8.7 THREE-STAGE EXTRACTOR CONTINUOUS EXTRACTOR

Use a three-stage countercurrent extractor to extract acetic acid from water by isopropyl ether. A solution of 40 kg/h containing 35 wt% acetic acids in water comes into direct contact with 40 kg/h of pure isopropyl ether. Calculate the amount and composition of the Extract and Raffinate layers manually by using UniSim/Hysys (NRTL) or any other available software packages.

8.8 EXTRACTION OF ACETIC ACID WITH PURE ISOPROPYL ETHER

An aqueous feed solution of acetic acid contains 30 wt% acetic acids and 70 wt% water is to be extracted in a continuous counter-current extractor with pure isopropyl ether to reduce the acetic acid concentration to 5 wt% in the final Raffinate. If 2,500 kg/h of pure isopropyl ether is used [1], determine the number of theoretical stages required. Use UniSim/Hysys (NRTL) to validate the manually calculated results.

8.9 EXTRACTION OF ACETIC ACID WITH PURE ISOPROPYL ETHER

An aqueous feed solution of acetic acid contains 30 wt% acetic acids and 70 wt% water is to be extracted in a continuous countercurrent extractor with pure isopropyl ether to reduce the acetic acid concentration to 5 wt% in the final Raffinate using 1,000 kg/h of pure isopropyl ether. Determine the number of theoretical stages required. Compare the number of theoretical stages with Example 8.3 and the effect of the amount of solvent on the necessary theoretical stages. Compare the manually calculated results with software prediction using UniSim/Hysys (NRTL) or any other available software packages.

8.10 EXTRACTION OF ACETIC ACID WITH NONE PURE ISOPROPYL ETHER

An aqueous feed solution of acetic acid contains 30 wt% acetic acids and 70 wt% water to be extracted in a continuous countercurrent extractor with pure isopropyl ether to reduce the acetic acid concentration to 5 wt% in the final Raffinate, using 1,000 kg/h of 90 wt% isopropyl ether and 10.0 wt% acetic acids. Determine the number of theoretical stages required. Compare the number of theoretical stages with Problem 8.3 and the effect of the amount of solvent on the necessary theoretical stages. Simulate the extraction process using UniSim/Hysys (NRTL) software or any other available software packages (e.g., PRO/II, Aspen Plus, SuperPro, and Aveva Process Simulation).

REFERENCES

1. Geankoplis, C. J., D. H. Lepek and A. Hersel, 2018. Transport Processes and Separation Process Principles, 5th edn, Prentice-Hall, Boston, MA.
2. McCabe, W. L., J. C. Smith and P. Harriott, 2005. Unit Operations of Chemical Engineering, 7th edn, McGraw-Hill, Boston, MA.
3. Seader, J. D. and E. J. Henley, 2016. Separation Process Principles with Applications using Simulators, 4th edn, John Wiley & Sons, New York, NY.
4. Seider, W. D., J. D. Seader, D. R. Lewin and S. Widagdo, 2016. Product and Process Design Principles: Synthesis, Analysis and Design, 4th edn, Wiley, New York, NY.
5. Treybal, R. E., 1987. Mass-transfer Operations, 3rd edn, McGraw-Hill, Boston, MA.

9 Simulation of Entire Processes

At the end of this chapter, students should be able to:

1. Apply the knowledge gained from previous chapters together.
2. Select appropriate process units.
3. Build the process flow diagram.
4. Simulate an entire process using UniSim/Hysys, PRO/II, Aspen Plus, SuperPro Designer, and Aveva Process Simulation.

9.1 INTRODUCTION

Industrial processes rarely involve one process unit. Keeping track of material flow for all individual units' overall operations and material flow requires simulating the overall plant [1]. A chemical plant is an industrial process plant that manufactures chemicals, usually on a large scale. The general objective of a chemical plant is to create new material [2, 3]. The simulation software packages are the key to know and control the full-scale industrial plant used in the chemical, oil, gas, and electrical power industries [4]. Simulation software simulates equipment so that the final product will be as close to design specifications as possible without further expenses in process modification. Simulation software gives helpful training experience without the panic of a catastrophic outcome. Process simulation is used to design, develop, analyze, and optimize technical processes and applied to chemical plants and chemical processes. Process simulation is a model-based representation of chemical, physical, biological, and other specialized processes and unit operations in software. The necessary fundamentals are a thorough knowledge of the chemical and physical properties of pure components and mixtures, reactions, and mathematical models that allow the calculation of a computer process. Process simulation software describes methods in flow diagrams where unit operations are positioned and connected by-product or inlet streams. The software has to solve the mass and energy balance to find a stable operating point [5]. In this chapter, the previous chapters' knowledge gained help in simulating an entire process flow diagram.

Example 9.1

The gas-phase reaction of HCl with ethylene over a copper chloride produces ethyl chloride with the catalyst supported on silica. The feed stream comprises 50% HCl, 48 mol% C_2H_4, and 2.0 mol% N_2 at 100 kmol/h, 25°C, and 1 atm. Since the reaction achieves only 90 mol% conversions, separate the ethyl chloride product from the unreacted reagents, and the latter is recycled. A distillation column

DOI: 10.1201/9781003167365-9

reaches the required separation. A purge stream withdrew a portion of the distillate. Design a process for this purpose using Hysys, PRO/II, Aspen Plus, SuperPro software packages, and Aveva Process Simulation to prevent the accumulation of inert in the system.

METHODOLOGY

The reaction of ethylene (C_2H_4) and hydrogen chloride (HCl) over a copper chloride catalyst supported on silica to produce ethylene chloride (C_2H_5Cl) is highly exothermic. In this example, the reaction operates in an isothermal conversion reactor (90% conversion). The heat evolved from the chemical reaction is removed from the reactor to keep the chemical reaction at a constant temperature. The reactor effluent stream is compressed, cooled, and then separated in a flash unit followed by a distillation column. The flash and distillation top products are collected and then recycled to the reactor after a portion of the stream is purged to avoid the accumulation of an inert component (N_2). The recycled stream is depressurized and heated to the fresh feed stream conditions. The liquid from the bottom of the flash enters a distillation column to separate ethyl chloride from the unreacted HCl and ethylene. Simulate the entire process using Hysys/UniSim, PRO/II, Aspen, and SuperPro Designer software packages.

Hysys/UniSim Simulation

Aspen Hysys is a process simulator used mainly in the oil and gas industry. Hysys is a highly interactive process flow diagram for building and navigating through extensive simulations. In this section, UniSim simulated the entire process. Use the shortcut method for the separation of ethylene chloride from unreacted gases. Start a new case in Hysys and add all components involved in the process (ethylene, hydrogen chloride, ethyl chloride, and nitrogen). Select the Peng–Robinson equation of state as the base property package. While in the simulation environment, select mixer from the object palette and place it on the process flow diagram (PFD) environment. Use the same procedure for a conversion reactor, compressor, cooler, flash, distillation column, mixer, tee, recycle, throttling valve, and heater, as shown in Figure 9.1.

The second step is to utilize the knowledge gained from previous chapters and define each unit operation. Describing reaction, from Flowsheet/Reaction Package,

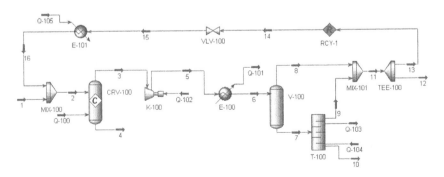

FIGURE 9.1 UniSim generated the process flow diagram using the UniSim shortcut distillation column for the case described in Example 9.1.

then click Add Rxn, and choose conversion. Next, add the three components involved in the chemical reaction and set the stoichiometric coefficient accordingly. Click Basis page, and type 90 for Co as percent conversion, where ethylene is the basis component. Close the reaction page and create a new set, suppose named conversion set and then send the conversion set to available reaction sets. Double click the conversion reactor, click the Reaction tab and then choose conversion set from the pull-down menu. Double click on the compressor, put exit pressure to 20 atm, and then cool the stream to 20°C to liberate unreacted components nitrogen, hydrogen chloride, and ethylene. The flash unit's bottom is separated in a distillation column where almost pure ethylene chloride is collected in the bottom, and unreacted compounds with negligible ethylene chloride are recycled. The shortcut method is employed for this purpose. The Light Key in the Bottom is HCl, its mole fraction is 0.001, and the Heavy Key in Distillate is C_2H_5Cl with mole fraction 0.001. The External Reflux Ratio is 2. The top product from the flash unit and distillate from the separation column was collected using a mixer. The exit of the mixer is recycled after a portion is purged (10%). The recycled streams pressured dropped through the throttling valve to 1 atm and then heated to 25°C (the new feed conditions). Figure 9.2 shows the process flowsheet and stream summary table. UniSim generated the stream summary table using the UniSim Workbook from the toolbar.

Figure 9.3 shows the UniSim predictions of the second case, wherein in this case, the rigorous distillation column replaced the shortcut column. Figure 9.4 shows the stream summary table of the simulation predictions. The distillation column required providing the condenser and reboiler pressure as 20 atm, the reflux ratio is 2, and the initial estimate of distillate molar flow rate is 10.0 kmol/h, results from the UniSim Shortcut column can provide the required data.

PRO/II SIMULATION

PRO/II has the power and flexibility to simulate a wide range of processes at a steady-state; it provides robust and accurate results based on industry-standard

Streams									
		1	2	3	4	5	6	7	8
Temperature	C	25.00	24.99	25.00	25.00	255.4	20.00	20.00	20.00
Pressure	kPa	101.3	101.3	101.3	101.3	2026	2026	2026	2026
Molar Flow	kgmole/h	100.0	144.5	97.09	0.0000	97.09	97.09	61.42	35.68
Comp Mole Frac (Ethylene)		0.4800	0.3647	0.0543	0.0029	0.0543	0.0543	0.0314	0.0937
Comp Mole Frac (HCl)		0.5000	0.4905	0.2416	0.0214	0.2416	0.2416	0.1881	0.3337
Comp Mole Frac (ClC2)		0.0000	0.0162	0.5128	0.9750	0.5128	0.5128	0.7681	0.0734
Comp Mole Frac (Nitrogen)		0.0200	0.1285	0.1913	0.0007	0.1913	0.1913	0.0124	0.4993
		9	10	11	12	13	14	15	16
Temperature	C	-178.5	126.5	-35.79	-35.79	-35.79	-35.81	-92.27	25.00
Pressure	kPa	2026	2026	2026	2026	2026	2026	101.3	101.3
Molar Flow	kgmole/h	14.21	47.20	49.89	4.989	44.90	44.53	44.53	44.53
Comp Mole Frac (Ethylene)		0.1356	0.0000	0.1057	0.1057	0.1057	0.1058	0.1058	0.1058
Comp Mole Frac (HCl)		0.8096	0.0010	0.4692	0.4692	0.4692	0.4693	0.4693	0.4693
Comp Mole Frac (ClC2)		0.0010	0.9990	0.0528	0.0528	0.0528	0.0527	0.0527	0.0527
Comp Mole Frac (Nitrogen)		0.0538	0.0000	0.3724	0.3724	0.3724	0.3722	0.3722	0.3722

FIGURE 9.2 UniSim generated stream summaries, conditions, and compositions using the shortcut methods for the case described in Example 9.1.

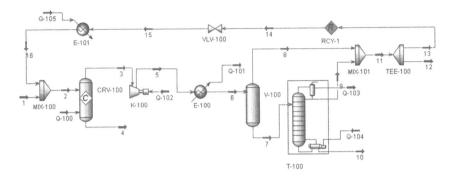

FIGURE 9.3 UniSim generated the process flow diagram using the rigorous distillation column for the case described in Example 9.1.

thermodynamic methods and physical property data. PRO/II is a valuable tool allowing engineers and management to enhance their plant's bottom line. Example 9.1 is prepared in PRO/II by building the process flowsheet shown in Figure 9.5 using the PFD palette. Connect each process component with the following streams. The streams connect the process components by first clicking on the Streams icon on the PFD palette, then dragging the other end of the stream to the next process component.

The following components are selected: a Mixer, conversion reactor, compressor, cooler, shortcut column, Mixer, splitter, pressure release valve, and simple heater; all are chosen from the PFD palette and placed on the screen. Click on the first Mixer and set the pressure drop to zero. Click the Conversion Reactor and then select the conversion in the Reaction Set Name. Click on the Extent of Reaction and Pressure and then input the contents for each component. Under Thermal Specification, select fixed temperature and set the temperature as 25°C.

Streams									
		1	2	3	4	5	6	7	8
Temperature	C	25.00	24.99	25.00	25.00	248.2	20.00	20.00	20.00
Pressure	kPa	101.3	101.3	101.3	101.3	2026	2026	2026	2026
Molar Flow	kgmole/h	100.0	138.2	90.77	0.0000	90.77	90.77	58.22	32.55
Comp Mole Frac (Ethylene)		0.4800	0.3815	0.0581	0.0029	0.0581	0.0581	0.0339	0.1013
Comp Mole Frac (HCl)		0.5000	0.4618	0.1804	0.0151	0.1804	0.1804	0.1413	0.2503
Comp Mole Frac (ClC2)		0.0000	0.0161	0.5473	0.9813	0.5473	0.5473	0.8107	0.0763
Comp Mole Frac (Nitrogen)		0.0200	0.1406	0.2142	0.0007	0.2142	0.2142	0.0141	0.5721
		9	10	11	12	13	14	15	16
Temperature	C	-180.5	120.4	-35.67	-35.67	-35.67	-35.69	-87.62	25.00
Pressure	kPa	2026	2026	2026	2026	2026	2026	101.3	101.3
Molar Flow	kgmole/h	9.997	48.23	42.55	4.255	38.29	38.23	38.23	38.23
Comp Mole Frac (Ethylene)		0.1965	0.0002	0.1237	0.1237	0.1237	0.1237	0.1237	0.1237
Comp Mole Frac (HCl)		0.7215	0.0210	0.3610	0.3610	0.3610	0.3618	0.3618	0.3618
Comp Mole Frac (ClC2)		0.0000	0.9787	0.0583	0.0583	0.0583	0.0583	0.0583	0.0583
Comp Mole Frac (Nitrogen)		0.0820	0.0000	0.4569	0.4569	0.4569	0.4562	0.4562	0.4562

FIGURE 9.4 UniSim generated stream summaries, conditions, and compositions using the rigorous distillation column for the case described in Example 9.1.

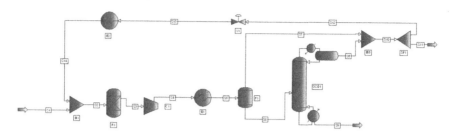

FIGURE 9.5 PRO/II generated process flowsheet using the shortcut distillation method for ethyl chloride production, as described in Example 9.1.

The pressure drop is zero; the extent of the reaction for the ethylene component is 0.9. Click on the compressor and enter the outlet pressure 20 bar. Click on the Simple HX (heat exchanger) and set the pressure drop to zero. Click on the Specification tab and select the exit temperature and type in value 20°C. Click on the flash column and classify the pressure drop and heat duty as zero. Click on the Shortcut distillation and then click on the specification button and set the mole fraction of HCl in the distillate to 0.7 and the mole fraction of ethyl chloride to 0.99 at the bottom. Click on Products, and the initial estimate sets the molar flow rate of the top stream to 10 kmol/h. After the splitter, connect the recycled stream to a pressure release valve followed by heat to bring the recycle stream to feed stream conditions. Construct a table to insert on the process screen, click on stream properties in the PFD palette, and place a table on the process screen. Double click on the table. A window will open. Select Stream Summary and click on Add All, then click on OK (Figure 9.6).

In the second method, replace the shortcut column with the rigorous distillation column (Figure 9.7). Click on the distillation column; the trays are 10, and the feed tray is the fifth tray. Click on Feed and Products and set the initial estimate of the bottom stream flow rate as 10 kmol/h. In the performance specifications, set the reflux ratio to 2, and the molar fraction of ethyl chloride in the bottom stream is 0.99. Figure 9.8 shows the stream summary.

ASPEN PLUS SIMULATION

Aspen Plus includes the world's largest database of pure component and phase-equilibrium data for conventional chemicals, electrolytes, solids, and polymers. In this example, connect the Mixer, conversion reactor, compressor, cooler, flash,

Stream Name		S1	S2	S3	S4	S5	S6	S7	S8	S9	S10	S11	S12	S13	S14
Stream Description															
Phase		Vapor	Vapor	Vapor	Vapor	Mixed	Liquid	Vapor	Liquid	Liquid	Mixed	Mixed	Mixed	Mixed	Vapor
Temperature	C	25.000	24.990	25.000	370.233	20.000	20.000	20.000	-113.013	124.716	8.895	8.895	8.895	-40.975	25.000
Pressure	KG/CM2	1.033	1.033	1.033	20.665	20.665	20.665	20.665	20.665	20.665	20.665	20.665	1.033	1.033
Flowrate	KG-MOL/HR	100.000	525.216	477.222	477.222	477.222	82.991	394.231	34.782	46.208	429.013	0.100	428.913	428.913	428.913
Composition															
ETHYLENE		0.480	0.102	0.011	0.011	0.011	0.004	0.013	0.010	0.000	0.012	0.012	0.012	0.012	0.012
HCL		0.500	0.413	0.354	0.354	0.354	0.216	0.383	0.501	0.010	0.392	0.392	0.392	0.392	0.392
CLE		0.000	0.083	0.192	0.192	0.192	0.766	0.072	0.458	0.990	0.103	0.103	0.103	0.103	0.103
N2		0.020	0.402	0.443	0.443	0.443	0.014	0.533	0.032	0.000	0.492	0.492	0.492	0.492	0.492

FIGURE 9.6 PRO/II generated the stream summary using shortcut columns for separation, the case described in Example 9.1.

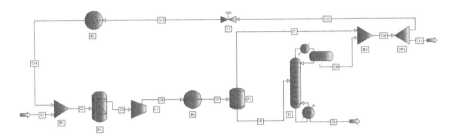

FIGURE 9.7 PRO/II generated process flowsheet using the rigorous distillation method for ethyl chloride production, as described in Example 9.1.

Stream Name		S1	S2	S3	S4	S5	S6	S7	S8	S9	S10	S11	S12	S13	S14
Stream Description															
Phase		Vapor	Vapor	Vapor	Vapor	Mixed	Liquid	Vapor	Liquid	Liquid	Mixed	Mixed	Mixed	Vapor	Vapor
Temperature	C	25.000	24.986	25.000	388.702	20.000	20.000	20.000	-203.079	124.742	11.730	11.730	11.730	-22.050	25.000
Pressure	KG/CM2	1.033	1.033	1.033	20.665	20.665	20.665	20.665	20.665	20.665	20.665	20.665	20.665	1.033	1.033
Flowrate	KG-MOL/HR	100.000	531.625	483.627	483.627	483.627	62.550	421.077	14.339	48.211	435.416	0.100	435.316	435.316	435.316
Composition															
ETHYLENE		0.480	0.100	0.011	0.011	0.011	0.004	0.012	0.018	0.000	0.012	0.012	0.012	0.012	0.012
HCL		0.500	0.424	0.367	0.367	0.367	0.219	0.389	0.924	0.010	0.407	0.407	0.407	0.407	0.407
CLE		0.000	0.056	0.161	0.161	0.161	0.763	0.071	0.000	0.990	0.069	0.069	0.069	0.069	0.069
N2		0.020	0.419	0.461	0.461	0.461	0.013	0.528	0.058	0.000	0.512	0.512	0.512	0.512	0.512

FIGURE 9.8 PRO/II generated the stream summary using the distillation column for separation, the case described in Example 9.1.

shortcut column mixer, splitter, throttling valve, and heater (Figure 9.9). The Mixer requires no information. Double click on the reactor and specify the reaction stoichiometry; the fractional conversion of ethylene is 0.9. The reactor product is compressed to 20 bar and then cooled to 20°C before flashed.

The flash is operating adiabatically at 20 bar. The bottom of the flash is separated in a shortcut column. In the shortcut column, the condenser and reboiler pressure is 20 bar. The light key component in the distillate is HCl, and its mole fraction at the bottom is 0.001. The heavy key is ethyl chloride, and its composition in the distillate is minimal; in this example, it is 0.001. The reflux ratio is 2. The bottom product is mostly pure ethylene chloride. The purge (stream 11) split fraction in the distillate is set at 0.1. The pressure of the recycle stream is reduced to 1 atm using a throttling valve. The exit stream from the valve is heated in a heater to match the feed stream temperature (recycle). The recycle stream is

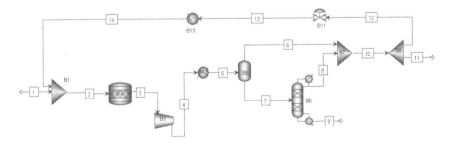

FIGURE 9.9 Aspen Plus generated a process flowsheet using the shortcut distillation (DSTWU) method for ethyl chloride production, as described in Example 9.1.

Stream ID		1	2	3	4	5	6	7	8	9	10	11	12	13	14
								Example 9.1							
Temperature	K	298.0	298.0	827.8	1603.3	293.1	293.2	293.2	262.4	400.8	286.2	286.2	286.2	264.0	298.1
Pressure	atm	1.00	1.00	0.07	20.00	20.00	20.00	20.00	20.00	20.00	20.00	20.00	20.00	1.00	1.00
Vapor Frac		1.000	1.000	1.000	1.000	0.385	1.000	0.800	1.000	0.000	0.996	0.996	0.996	1.000	1.000
Mole Flow	kmol/hr	100.000	147.943	100.470	100.470	100.470	38.670	61.800	14.600	47.201	53.270	5.337	47.943	47.943	47.943
Mass Flow	kg/hr	3225.639	4853.384	4853.384	4853.384	4853.384	1297.154	3556.231	311.437	3044.773	1808.605	180.861	1627.745	1627.745	1627.745
Volume Flow	l/min	40507.814	59981.223	1.67163E+6	11073.269	787.139	721.047	66.097	214.345	71.838	945.574	94.563	851.066	17214.737	19472.429
Enthalpy	MMkcal/hr	-1.990	-4.000	-4.000	2.643	-8.553	-1.312	-7.241	-0.906	-5.553	-2.297	-0.230	-2.068	-2.068	-2.018
Mole Flow	kmol/hr														
ETHYL-01		48.000	52.747	5.275	5.275	5.275	3.416	1.859	1.858	<0.001	5.274	0.527	4.747	4.747	4.747
HYDRO-01		50.000	72.641	25.169	25.169	25.169	13.239	11.929	11.917	0.012	25.157	2.516	22.641	22.641	22.641
ETHYL-02			2.555	50.027	50.027	50.027	2.792	47.236	0.047	47.188	2.839	0.284	2.555	2.555	2.555
NITRO-01		2.000	20.000	20.000	20.000	20.000	19.223	0.777	0.777	trace	20.000	2.000	18.000	18.000	18.000

FIGURE 9.10 Stream summary table using shortcut distillation methods.

tied with the fresh feed stream. Figure 9.10 shows the stream conditions and compositions.

In the second case, replace the shortcut column by the distillation column (Figure 9.11). In the distillation column, the number of trays is 10.0 stages, the distillate rate is 10 kmol/h, the reflux ratio is 2, and the column pressure is 20 atm. Figure 9.12 shows the generated table of the stream conditions and compositions.

SuperPro Designer Software

SuperPro Designer facilitates modeling, evaluation, and optimization of integrated processes in a wide range of industries. The combination of manufacturing and environmental operation models in the same package enables the user to concurrently design and evaluate manufacturing and end-of-pipe treatment processes and practice waste minimization via pollution prevention and pollution control.

In a new case in SuperPro Designer, select all active components. The user adds ethyl chloride since this component does not exist in the SuperPro library. Insert the process units in the process flowsheet diagram; connect the streams as shown in Figure 9.13. In the plug flow (PFR) conversion reactor, the chemical reaction is taking place in the vapor phase. Set the Exit Temperature to 25°C and extent to 90% based on reference component ethylene. Pressure the reactor's exit to 20 bar in a compressor (compressor's efficiency is 70%). The pressurized gases are cooled to 20°C and fed into a shortcut column. The relative volatilities and percent in the distillate of each component are to be provided by the user. The Light key is HCl, and the Heavy key is ethyl chloride. In the splitter, set 90% to the top stream.

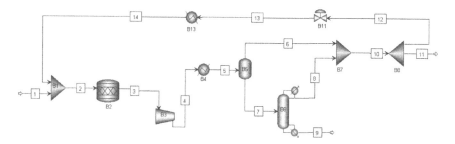

FIGURE 9.11 Aspen Plus generated the process flow diagram using the rigorous distillation (RedFrac) method for ethyl chloride production, as described in Example 9.1.

		1	2	3	4	5	6	7	8	9	10	11	12	13	14
Stream ID		1	2	3	4	5	6	7	8	9	10	11	12	13	14
Temperature	K	298.0	298.0	877.2	1591.5	253.2	253.2	253.2	103.2	389.4	198.1	198.1	198.1	159.0	299.1
Pressure	atm	1.00	1.00	0.87	20.00	20.00	20.00	20.00	20.00	20.00	20.00	20.00	20.00	1.00	1.00
Vapor Frac		1.000	1.000	1.000	1.000	0.283	1.000	0.000	0.000	0.000	0.713	0.713	0.713	0.803	1.000
Mole Flow	kmol/hr	100.000	130.052	82.629	82.629	82.629	23.391	59.238	10.000	49.238	33.391	3.339	30.052	30.052	30.052
Mass Flow	kg/hr	3225.630	4138.947	4138.947	4138.947	4138.947	687.125	3451.822	327.665	3124.157	1014.787	101.479	913.308	913.308	913.308
Volume Flow	l/min	40507.814	52737.631	1.45668E+6	9040.878	447.766	386.883	60.879	5.419	71.215	299.319	29.932	269.387	3194.777	12236.647
Enthalpy	MMBtu/hr	-1.990	-2.407	-2.407	3.207	-7.123	-0.171	-6.932	-0.535	-5.811	-0.706	-0.071	-0.636	-0.636	-0.417
Mole Flow	kmol/hr														
ETHYL-01		48.000	52.691	5.269	5.269	5.269	1.903	3.207	3.230	0.056	5.213	0.521	4.691	4.691	4.691
HYDRO-01		50.000	57.037	9.614	9.614	9.614	2.207	7.407	3.611	1.796	7.819	0.782	7.037	7.037	7.037
ETHYL-02			0.324	47.746	47.746	47.746	0.360	47.366	< 0.001	47.366	0.360	0.036	0.324	0.324	0.324
NITRO-01		2.000	19.999	19.999	19.999	19.999	18.841	1.158	1.158	trace	19.999	2.000	17.999	17.999	17.999

FIGURE 9.12 Aspen Plus generated the stream summary tables using a rigorous distillation method, the case described in Example 9.1.

AVEVA PROCESS SIMULATION

Aveva Process Simulation is the first commercially available platform to take advantage of developing web-based and cloud technologies to deliver an enjoyable user experience so that engineers will be more productive, collaborative, creative, and inspired.

Example 9.1 is prepared in Aveva Process Simulation by building the process flowsheet shown in Figure 9.14. Connect each process component with the following streams. The streams connect the process components by first clicking on the Streams icon on the process library, then dragging the other end of the stream to the next process component. The following members are selected: a mixer, conversion reactor, compressor, cooler, distillation column, Mixer, splitter, pressure release valve, and simple heater; all are selected from the Process model library and placed on the screen. Click on the first Mixer and set the pressure drop to zero. Click the Conversion Reactor and then select the Conversion reactor (CNVR). Set the fractional conversion to 0.9. Click on the compressor and enter the outlet pressure 20 bar. Click on the Simple HX (heat exchanger) and set the pressure drop to zero. Click on the Specification tab and select the exit temperature and type in value 20°C. Click on the flash column, classify the pressure drop, and heat duty as zero (adiabatic operation).

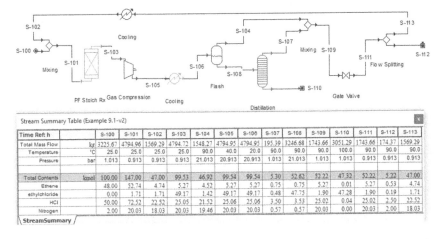

Stream Summary Table (Example 9.1-v2)															
Time Ref: h		S-100	S-101	S-102	S-103	S-104	S-105	S-106	S-107	S-108	S-109	S-110	S-111	S-112	S-113
Total Mass Flow	kg	3225.67	4794.96	1569.29	4794.72	1548.27	4794.95	4794.95	195.39	3246.68	1743.66	3051.29	1743.66	174.37	1569.29
Temperature	°C	25.0	25.0	25.0	25.0	90.0	40.0	20.0	90.0	90.0	90.0	100.0	90.0	90.0	90.0
Pressure	bar	1.013	0.913	0.913	0.913	21.013	20.913	20.913	1.013	21.013	1.013	1.013	0.913	0.913	0.913
Total Contents	kmol	100.00	147.00	47.00	99.53	46.92	99.54	99.54	5.30	52.62	52.22	47.32	52.22	5.22	47.00
Ethene		48.00	52.74	4.74	5.27	4.52	5.27	5.27	0.75	0.75	0.75	0.01	5.27	0.53	4.74
ethylchloride		0.00	1.71	1.71	49.17	1.42	49.17	49.17	0.48	47.75	1.90	47.28	1.90	0.19	1.71
HCl		50.00	72.52	22.52	25.05	21.52	25.06	25.06	3.50	3.53	25.02	0.04	25.02	2.50	22.52
Nitrogen		2.00	20.03	18.03	20.03	19.46	20.03	20.03	0.57	0.57	20.03	0.00	20.03	2.00	18.03
\StreamSummary/															

FIGURE 9.13 SuperPro Designer generated process flowsheet and stream summary using the distillation method for ethyl chloride production, as described in Example 9.1.

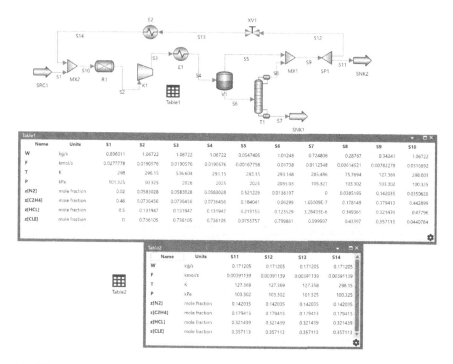

Name	Units	S1	S2	S3	S4	S5	S6	S7	S8	S9	S10
W	kg/s	0.896011	1.06722	1.06722	1.06722	0.0547405	1.01248	0.724806	0.28767	0.34241	1.06722
F	kmol/s	0.0277778	0.0190576	0.0190576	0.0190576	0.00167758	0.01738	0.0112348	0.00614521	0.00782278	0.0316892
T	K	298	298.15	536.604	293.15	293.15	293.148	285.486	75.7694	127.369	298.003
P	kPa	101.325	50.325	2026	2025	2025	2033.03	105.321	103.302	103.302	100.325
z[N2]	mole fraction	0.02	0.0583028	0.0583028	0.0583028	0.521229	0.0136197	0	0.0385195	0.142035	0.0350628
z[C2H4]	mole fraction	0.48	0.0736456	0.0736456	0.0736456	0.184041	0.06299	1.65009E-7	0.178149	0.179413	0.442899
z[HCL]	mole fraction	0.5	0.131947	0.131947	0.131947	0.219155	0.123529	3.28433E-6	0.349361	0.321439	0.47796
z[CLE]	mole fraction	0	0.736105	0.736105	0.736105	0.0755757	0.799861	0.999997	0.43397	0.357113	0.0440784

Name	Units	S11	S12	S13	S14
W	kg/s	0.171205	0.171205	0.171205	0.171205
F	kmol/s	0.00391139	0.00391139	0.00391139	0.00391139
T	K	127.369	127.369	127.358	298.15
P	kPa	103.302	103.302	101.325	100.325
z[N2]	mole fraction	0.142035	0.142035	0.142035	0.142035
z[C2H4]	mole fraction	0.179413	0.179413	0.179413	0.179413
z[HCL]	mole fraction	0.321439	0.321439	0.321439	0.321439
z[CLE]	mole fraction	0.357113	0.357113	0.357113	0.357113

FIGURE 9.14 Aveva Process Simulation generated the process diagram and stream summary using the distillation method for ethyl chloride production, as described in Example 9.1.

Click on the distillation column, set the condenser and reboilers to internal, click on the specification button, and select the mole fraction of HCl in the distillate to 0.7 and the mole fraction of ethyl chloride to 0.99 at the bottom.

Click on Products, and the initial estimate sets the molar flow rate of the top stream to 10 kmol/h. After the splitter, the recycled stream is connected to a pressure release valve followed by heat to bring the recycle stream to feed stream conditions. To construct a table including the process streams summary, drag the Table module to the canvas, then right click on any stream, add to table 1 (streams 1 to 10), and add the streams from 11 to 14 to table 2. Double click on the tables to display the results (Figure 9.14).

PROBLEMS

9.1 ETHYL CHLORIDE PRODUCTION IN AN ADIABATIC REACTOR

Ethyl chloride is produced by the gas-phase reaction of HCl with ethylene over a copper chloride catalyst supported on silica. The feed stream is composed of 50% HCl, 48 mol% C_2H_4, and 2 mol% N_2 at 100 kmol/h, 25°C, and 1 atm. Since the reaction achieves only 80 mol% conversions, the ethyl chloride product is separated from the unreacted reagents, and the latter is recycled. The chemical reaction takes place in an adiabatic reactor. The separation is achieved using a distillation column. A portion of the distillate is withdrawn in a purge stream to prevent the accumulation of inert in the system. Suggest a process flowsheet for this purpose and simulate the process using the accessible software package. Compare results with those obtained from Example 9.1. For hydrocarbon gases, the recommended fluid package is Peng–Robinson.

9.2 ETHYLENE PRODUCTION IN AN ISOTHERMAL REACTOR

Ethane enters a furnace and decomposed at 800°C to produce ethylene. Assume that the reaction takes place in an isothermal conversion reactor where ethane single-pass conversion is 65%. Develop a process flowsheet for the production of ethylene from pure ethane. Use the existing software package in your university to perform the material and energy balance of the entire process. The recommended fluid package is Peng–Robinson.

9.3 AMMONIA SYNTHESIS PROCESS IN AN ADIABATIC REACTOR

Ammonia is synthesized through the reaction of nitrogen and hydrogen in an adiabatic conversion reactor. The feed stream to the reactor is at 400°C. Nitrogen and hydrogen are fed in stoichiometric proportions. The single-pass fractional conversion is 0.15. The product from the converter is condensed and produced as ammonia liquid. The unreacted gases are recycled. The reactor effluent gas is used to heat the recycled gas from the separator in a combined reactor effluent/recycle heat exchanger. Construct a process flow diagram and use one of the available software packages to simulate the entire process. The suggested fluid package is Peng–Robinson.

9.4 AMMONIA SYNTHESIS PROCESS IN AN ISOTHERMAL REACTOR

Repeat Problem 9.3; in this case, the reaction occurs at 400°C in an isothermal reactor. Compare the results with Problem 9.3. The suggested fluid package is Peng–Robinson.

9.5 METHANOL DEHYDROGENATION

Methanol at 675°C and 1.0 bar at a rate of 100 kmol/h enter an adiabatic reactor where 25% of it is dehydrogenated to formaldehyde (HCHO). Calculate the temperature of the gases leaving the reactor and separate the component where unreacted methanol is recycled and almost pure formaldehyde is produced. The advised fluid package is PRSV.

REFERENCES

1. Ghasem, N. M. and R. Henda, 2015. Principles of Chemical Engineering Processes, 2nd edn, CRC Press, New York, NY.
2. Ghasem, N. M., 2018. Modeling and Simulation of Chemical Process Systems, CRC Press, New York, NY.
3. Douglas, J. M., 1988. Conceptual Design of Chemical Processes, McGraw-Hill, New York, NY.
4. Thomas, P. J., 1999. Simulation of Industrial Processes for Control Engineers, Butterworth Heinmann, Oxford.
5. Rhodes, C. L., 1996. The process simulation revolution: Thermophysical property needs and concerns, Journal of Chemical Engineering Data, 41, 947–950.

Appendix A: Introduction to UniSim/Hysys

UniSim is a commercial simulation software package. Simulation is the imitation of the operation of a process or system in the real world over time. This means that we can create a complete plant, like a real plant, in our software, and we can see how the plant is working before we turn it on in the current plant. The program allows checking the parameters that were overdue. What purity appears as a product, what elements we need here before entering the actual factory, that is, the primary uses of the UniSim software package.

Example A.1: Methane Steam Reforming

Consider the steam reforming reaction where 100 kmol/h of methane fed to a conversion reactor with 200 kmol/h of steam. The chemical reaction took place at 900°C and 25 bars. The single-pass conversion of methane is 98%. Simulate the process with UniSim and get the product molar flow rate.

SOLUTION

Start UniSim software will launch the blank menu shown in Figure A.1.

1. To create a new one, go to File, click New case. The first thing that appears is the simulation basis manager; within this basis manager, we do the first three steps of the simulation workflow to set the chemical species, the thermodynamic description, and chemical reactions (Figure A.2).
2. Within the simulation basis manager, the components tab is the first tab to open up, and in the box or left-hand side, you have what is termed the master component list where all the chemical species of the simulation are held.
3. Next is to populate this master component list with all the chemical species within our simulation. If we look at the UniSim database components, we will see that methane is the first component. There are three ways to specify names within UniSim's component database; you can look at what you might term the common name or the Simulation name (first list), searching by Full name/synonym (second list), or search by an empirical formula (third list). Single left-click the formula radio button and add the rest of the components (CO, hydrogen, and water) as shown in Figure A.3.
4. So the next tab to components is Fluid Packages. If we single, left click on that, we will see that the fluid package is the UniSim or Hysys phrase for the thermodynamic model. Again, we will see that by default, none is selected, so a single left click on Add and see what we have available.

FIGURE A.1   UniSim launched graphical user interface.

When this new window opens, we will see many thermodynamic models implemented within UniSim, making it a handy tool for process simulation and physical property calculation, subject to a suitably validated model. The thermodynamic model should be validated with the available experimental data before being used; otherwise, it leads to wrong results. For this tutorial, we are going to use the non-random two liquid or NRTL activity coefficient model. We chose that because it is good at describing mixtures with polar molecules, of which, of course, water is one. So I'm going to single left click on the activity models selection box to narrow the number of packages I have, and then single left click on NRTL. Before we do that, watch this indicator down here that is currently red. Single left click and NRTL, it turns green. This is an excellent introduction to the UniSim traffic light color-coding system. Close that box.

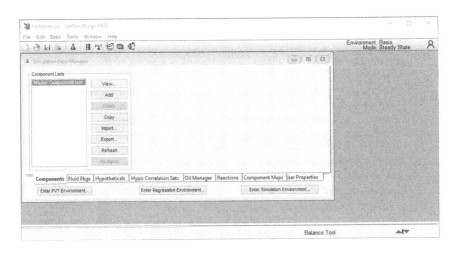

FIGURE A.2   UniSim lunched simulation manager window.

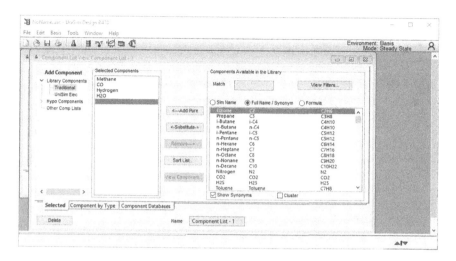

FIGURE A.3 UniSim completed the component selection.

5. When you choose to make, they will typically be a status bar, like that red or green bar. If it's red, it means you need to do something. If it is amber, it means you've started to do something, but you haven't completed it, as in the property package, or it isn't correctly and wholly set up. If it's green, it means that the property package or model is fully classified, but because it's fully specified, it doesn't necessarily mean that it is correct for your purposes, so always make sure that all the information that is pertinent to your simulation has been entered, we are going to tune in on random to liquid model any further. We're going to assume that the database coefficients, which will be here on the binary coefficients, will

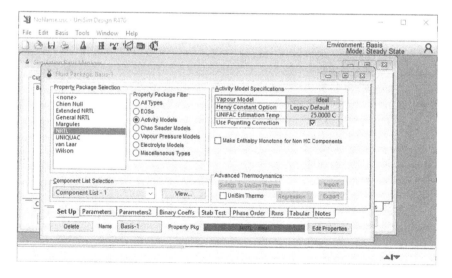

FIGURE A.4 UniSim selected fluid package from the property package selection interface.

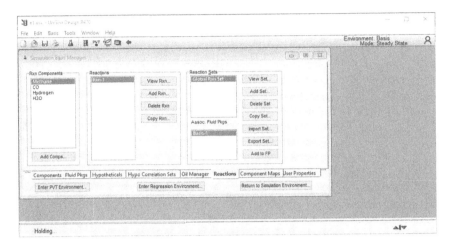

FIGURE A.5   UniSim load up simulation basis manager, reactions interface menu.

be part of your validation process if you're setting this up independently, but we've already assumed that the standard NRTL model is fit for purpose. So having chosen that, we're going to close that box.

6. The final thing we're going to do is to set the chemical reactions. So if we skip over the following three tabs to reactions and single left click that, we will see that we have a list of components on the left-hand side, a list of chemical reactions in the middle, and then on the right-hand side and global reactions sets and associated fluid packages so associated fluid package is the thermodynamic model that underpins the reaction.

7. Let's go through this in sequence; we need to define a reaction so that with a single left click on Add reaction, we will see a selection box appear with the five essential reactions.

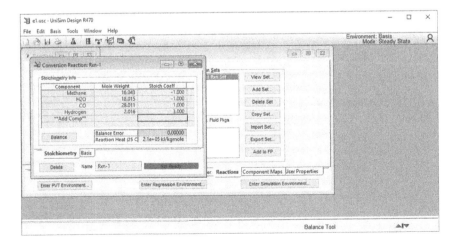

FIGURE A.6   UniSim inputted reaction stoichiometry completed.

8. Single left-click conversion for this tutorial and then Add reaction. The next screen that appears allows you to add a component to your chemical reaction, adding methane as a reaction raw material, water as a new reactant material. Add the two products, carbon monoxide, and hydrogen.

9. What UniSim requires from us are the stoichiometric coefficients to describe their place within the reaction. Stoichiometric coefficients are negative for species consumed by reaction and positive for species produced (consumption is with negative sign and production is positive). The stoichiometric coefficient of methane is minus one (reactant) the stoichiometric coefficient for water is again minus one because it is consumed. The stoichiometric coefficient for carbon monoxide is one because it is produced with the same stoichiometry, as consumption of both water, methane, and it's positive because the reaction produces it. The stoichiometric coefficient for hydrogen is three because the chemical reaction produces it.

10. Note that our traffic light system still says, not ready. But the first thing that we check, rather than the traffic light system in the reaction, is the balance error. The balance error is zero, which means my stoichiometry is correct. Note that the reaction heat at 25°C also has now calculation status. Again this is an excellent example of how UniSim can help figure various data pieces that you can then use, for example, in hand calculation.

11. The UniSim imitated traffic light says not ready because we haven't looked at this second tab yet, so that Single left click basis. And we will see the various parameters here that he used to calculate the conversion reaction. Now, always read the definition on the box conversion; here is a percentage conversion, and it is not fractional conversion; please don't make that mistake. If you need to enter, the temperature dependence is measured in Kelvin, not Celsius, not Fahrenheit. So, we choose how we define the extent of reaction concerning the component, so we're going to say that our extent of the chemical

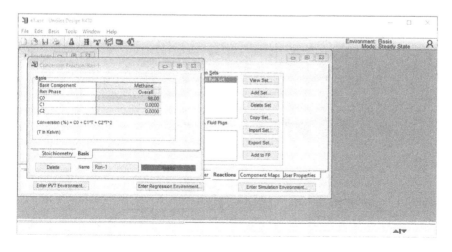

FIGURE A.7   UniSim inputs the 98% conversion on the basis page.

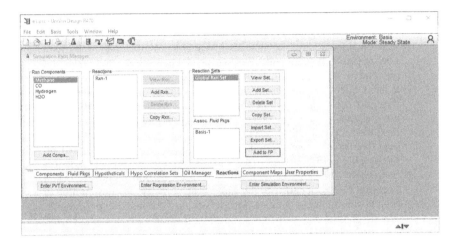

FIGURE A.8   Enter the simulation environment, start to draw up the process flowsheet.

reaction is against methane. If we look at the reaction phase, we can say, well, the reaction extent pertains to either the vapor phase, the liquid phase, or some combined phase, or just everything that we've got reacting. The methane steam reforming reaction is a gas phase. We could set the vapor phase option here, but there will be no other phase other than vapor when we define the simulation, so I'm just going to keep it Overall defined. Remember that we said we want a 98% conversion, and we didn't specify temperature dependence, put 98%, and see the traffic light now gone green.

12. Associate the reaction with thermodynamic description. With the global reaction set selected, single left click on it and then single left click on Add fluid package at a thermodynamic description. When we do that,

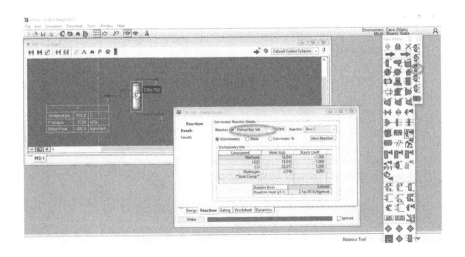

FIGURE A.9   UniSim Simulation basis environment.

the global reaction set box opened, and we going to the single left click and set it to fluid package.

13. The first three steps of the simulation workflow are complete. We have set components, a thermodynamic description, and the chemical reaction.

LOGICAL OPERATORS IN UNISIM/HYSYS

Four logical unit operations are used primarily in steady-state mode. These are ADJUST, SET, RECYCLE, and SPREADSHEET.

A.1 THE SET OPERATION

The SET operator is used to set the value of a specific process variable (PV) in relation to another PV. The linear relationship of the Y and X, form $Y = mX + b$ and the process variables must be of the same type.

Example A.2: Use of Set Logical Operator

Based on the following reaction

$$CH_4 + 2H_2O \rightarrow CO_2 + 2H_2$$

Set the molar flow rate of steam twice that of methane. The flow rate of methane is 100 kgmol/h at 2 atm, 25°C.

SOLUTION

The user opens a new case in UniSim, should first select the components (CH_4 and H_2O), and then define and select a Fluid Package (Peng–Robinson). Once the user has specified the system's components and required fluid package, hence completed all necessary input to begin the simulation. Click on the Enter Simulation Environment button and then select two material streams, rename the first to Natural gas, and the second to H_2O. Fully specify the Natural gas stream (pure CH_4) (*Four* variables needed for input stream are composition, flow rate, and two from temperature, pressure, or vapor/phase fraction). Undefine the flow rate for the H_2O stream (assume pure water). The user selects "Set" from the object pallet. Because we are using Natural gas stream information to set the properties of the stream of H_2O vapor, natural gas is the "Source" stream, and steam is the "Target" stream. Figure A.10 shows the Connections and Parameters pages.

THE ADJUST OPERATION

The ADJUST operator is used to adjust one variable until a target variable reaches a user-specified value.

Example A.3: Adjust Logical Operator

The feed stream contains a mixture of acetone, and water needs to manipulate the feed temperature until the bottom stream is pure liquid. The feed stream contains

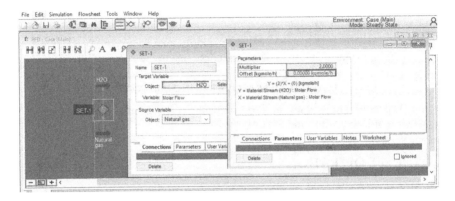

FIGURE A.10   UniSim set operator process flow diagram.

50 mol% acetone in water. Stream temperature, pressure, and flow rate are 70 C, 1 bar, and 100 kmol/h.

SOLUTION

Start new case, select Components (acetone, water), define and select a Fluid Package (NRTL). Once you have specified the components and fluid package, you have now completed all necessary input to begin your simulation. Click on the Enter Simulation Environment button. Add the Separator, connect feed, top, and bottom streams. Fully specify the feed stream (temperature, pressure, flow rate, and composition). Then add the Adjust operator. Click the Select Variable button and choose stream 1 as the Object adjusted variable and temperature as the variable. Set the Target Variable, press the Select Var. Button. Choose stream 2 as an Object and the component mole fraction of H_2O as the variable. Enter 0.95 as the Specified Target value.

RECYCLE OPERATION

The Recycle block operation is used to connect the two streams. Once the Recycle is attached and running, UniSim compares the two values, adjusts

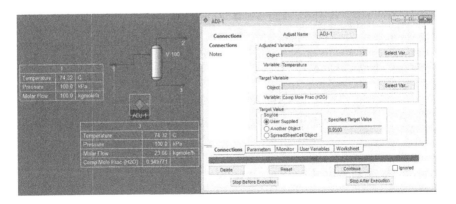

FIGURE A.11   UniSim completed adjust operator diagram.

the assumed stream, and reruns the flowsheet. A Recycle operation has an inlet (calculated) stream and an outlet thought stream. Hysys/UniSim uses the supposed stream's conditions (outlet) and solves the flowsheet up to the computed inlet. Hysys/UniSim then compares the values of the calculated stream to those in the assumed stream. Based on the difference between the values, Hysys/UniSim modifies the calculated stream values and passes the changed values to the supposed stream.

Example A.4: Recycle Operation

This section aims to understand recycling by using Recycle operator in the Hysys/UniSim object palette. Build the following flowchart using the feed stream data in Table A.1. For Separator, assign the pressure drop to 0 atm, Expander feed pressure 2,000 kPa, second separator pressure drop to 0, tee flow ratio 0.5, pump outlet pressure 4,000 kPa.

SOLUTION PROCEDURE

1. Start a new case, add the components, select a Fluid Package (PR), and click on Enter Simulation Environment.
2. Add feed stream and specify all conditions and composition in Table A.1.
3. Add separator turbine and the rest of the units shown in Figure A.3.
4. The user needs to create a new material stream, rename it as Recycle, double click on the Recycle stream from the stream property view, click on Define from other Stream buttons, and then select stream 9 from available steams like box on the Source Stream.
5. Attach this stream as an inlet to Separator V-100.
6. Add the RECYCLE icon from the object palette and attach 9 as feed and Recycle stream as output. The PFD will look like Figure A.3. The flowsheet will now solve, and the RECYCLE should quickly converge.
7. Double click on the RECYCLE icon and click the composition. The assumed and calculated recycles will be almost identical once a solution is reached.

TABLE A.1
Feed Stream Conditions

Material Stream

Conditions	Temperature	16°C
	Pressure	4,000 kPa
	Molar flow	2,000 kmol/h
Composition	Nitrogen	0.01
	CO_2	0.015
	Methane	0.89
	Ethane	0.06
	Propane	0.025

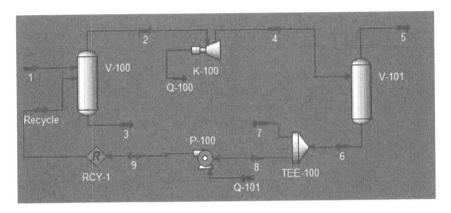

FIGURE A.12 Process flow diagram with recycling.

THE SPREADSHEET

The Spreadsheet is doing the same function as Excel.

Example A.5: Spreadsheet

In a process where there are two streams (Natural gas and steam), the flow rate of steam is equal to three times the molar flow rate of natural gas (methane plus two times the ethane flow plus three times the propane flow plus four times i-butane and n-butane). The natural gas contains 0.5 mole fraction methane, 0.3 ethane, 0.05 propane, 0.025 i-butane, 0.025 n-butane, and 0.1 nitrogen. Set Operation with a multiplier equal to three times the carbon atom. The natural gas stream flow rate is 1 kgmol/h, pressure is 2 bar, and temperature is 200°C.

The required steam molar flow rate = $3 \times (C1 + 2 \times C2 + 3 \times C3 + 4 \times i\text{-}C4 + 4 \times n\text{-}C4)$

The required steam molar flow rate = $3 \times (0.5 + 2 \times 0.3 + 3 \times 0.0.05 + 4 \times 0.025 + 4 \times 0.025) = 4.35$

SOLUTION

1. Create a material stream, and rename it Natural Gas. Input the temperature, pressure, molar flow rate, and mole fractions.
2. Create a stream called steam and define it exactly the way you did with Natural gas stream, though this time leave the Molar Flow fields empty, as shown in Figure A.4.
3. From the Object Palette, select the Spreadsheet. Open the Property view.
4. Press Add Import and make the selections to all hydrocarbon components (Figure A.5).
5. In cells C2 through C6 enter the following cells respectively: =B2, =2 × B3, = 3 × B4, = 4 × B5, = 4 × B6. Then, in cell B8, write "Total Carbon Flow" and in B9 "Required Steam Flow." In C8, enter = C2 + C3 + C4 + C5 + C6. In C9, enter = 3 × C8 (required steam molar flow rate).

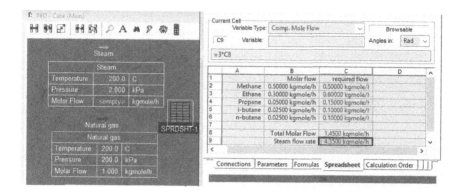

FIGURE A.13 Spreadsheet of the calculated variable.

6. Return to the Connections Page, and press Add Export and select steam and then Molar Flow. Change the cell to C9 to match the one to be exported.

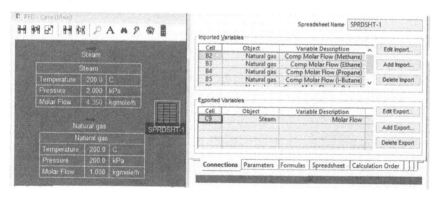

FIGURE A.14 Spreadsheet exported variables.

Appendix B: Introduction to PRO/II Simulation

PRO/II is a steady-state process simulator intended for process design and operational study, primarily for process engineers in the chemical and natural gas industries.

To start PRO/II, either click on the icon on the desktop or open it from the start menu. Once launched, the graphical user interface should look like in Figure B.1. To build a simulation, the first thing to do is to click on the New button. There are various ways a user can access the multiple functionalities of PRO/II Tools. One of them is using the ribbon. The other one would be to go to the specific tab and then look for the particular feature, but the primary tab a user usually will be working within PRO/II is the "Express" tab.

Click on New and create a new simulation. Three windows appear on the message window on the left, the flowsheeting area in the middle, and the process flow diagram (PFD) Pallete right (Figure B.2). This PFD Pallete consists of all the unit operations in PRO/II, divided into various tabs. For example, any unit operation across which there is a pressure variation located under the pressure change tab; a compressor, an expander, a pump, and a valve, all these unit operations across which there is a pressure change. The heat exchanger represents all types of heat exchanges. The left side is the Messages window. When the simulation is running, it shows the number of iterations taken and if there are any errors or warnings, these types of messages can be seen during the simulation too. The flow sheeting area in the middle is where the user can build the flowsheet. PRO/II simulation is a steady-state simulation, which means that the controllers, the transmitters, or any instrumentation are irrelevant for the software because it doesn't do any level controls.

There are several steps to build a simulation in PRO/II: unit of measurement, component selection, thermodynamics, flowsheet construction, stream specifications, specifying unit operations, running the system, and presenting results. They are described in more detail as follows:

1. Unit of measure: The first step was selecting the correct units of measure. I have English units of measure, metrics, and SI. Let us use SI, if the user does not want to work with Kelvin, you would like to work with Celsius, and then you can specify only that particular parameter (Figure B.3).
2. Component selection: The second step is to select the components. Add whatever raw material, intermediate products, finished products, or any species involved in the process by clicking on the Component selection box (the benzene ring). When you click on that, it opens up the components selection box to select the component's name and add them. So for our demonstration. If the compound is longer than eight characters, the user has to choose Select from the list (Figure B.4).

FIGURE B.1   PRO/II launched graphical user interface and the Express tab.

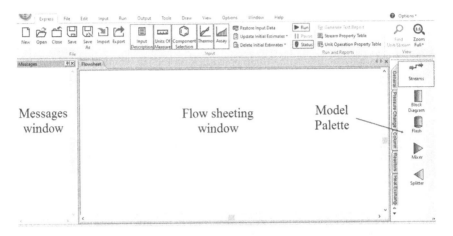

FIGURE B.2   PRO/II three main windows.

3. Thermodynamic: Since we are working with light hydrocarbons, Peng–Robinson is a suitable choice, good enough for light hydrocarbons. Click on the Thermodynamic button, and in the popup thermodynamic Data menu, select the appropriate property method (Figure B.5).
4. The fourth step is to build a flowsheet; it is not advisable to create the entire process flowsheet at one goal. The best practice would be to make sections of that simulated plant, or the process in steps, converge, and then move on further. The following simple example explains the fourth step (Figure B.6).

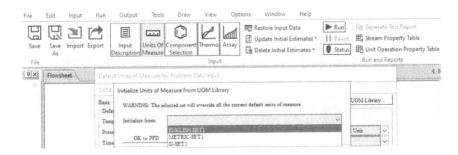

FIGURE B.3   Unit measurement system and the default unit of measure for problem data input.

FIGURE B.4 Component selection window.

Example B.1: Building a Flowsheet

Natural gas stream flows at 100 kgmol/h, contains 0.4 methane, 0.3 ethane, 0.2 propane, and 0.1 butane (in mole fraction) at 5.0 bars and 25°C flashed into an adiabatic flash drum with zero pressure drop. The vapor from the top stream fed to a compressor (75% efficiency), enables doubling the inlet pressure. The bottom liquid stream enters a pump where the pressure is rising twice the inlet pressure.

SOLUTION

Start by choosing a separator, which is a simple flash drum. Then we have a compressor and a pump. Connect the units with streams. First, streaming would require the feed stream on the feed stream and connect with the separator. As soon as one clicks on the stream, all the unit operations' borders have a red color. From all over this red border is we are going to connect to this green stream.

1. Streams information data: PRO/II simulation does not require providing any information for the internal streams or streams connected to operating units. Users do not have to give any information for the product stream, calculated once the simulation run. So the only thing required

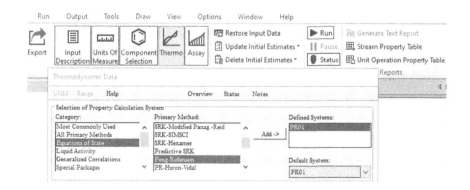

FIGURE B.5 Thermodynamic data and selection of the property calculation system.

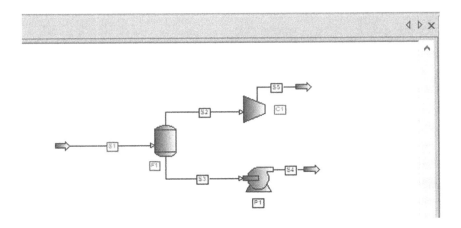

FIGURE B.6   Building the required flowsheet.

to provide is the stream information for the feed. Double click on the feed stream and provide the needed information (total molar flow rate, pressure, temperature, and composition). It is easy to navigate because wherever there is a red border, tap it to give the required data, so you do not miss a thing. For example, even if you look at the blue sheet, you know which are the streams or unit operations that need to provide the required information. The feed stream is 100 kgmol/h, temperature is 25°C, pressure is 5 bar, and composition in mole fractions is 0.4 methane, 0.2 ethane, 0.2 propane, and 0.2 butane (Figure B.7).

2. Unit operation data: click the flash drum and set the pressure drop to zero and the duty to zero (Adiabatic flash drum). By default, the pump takes a pressure rise of zero, which means that whatever pressure comes in for the feed stream is getting delivered, so the pump is not doing any work. The information required for the flash unit is that there is no pressure drop, the duty is also zero (Figure B.8). Double click on the compressor, and set a pressure ratio of 2, and the efficiency at excellent efficiency is 75%. These are the six steps done and nothing left on the flowsheet to specify.

3. Run the simulation and view the results. If the color of the unit process turns blue in color, it indicates that the problem with the simulation has converged successfully. If it turns red in color, it is an indication that it

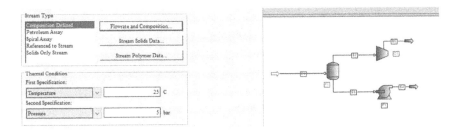

FIGURE B.7   Provide the feed stream information.

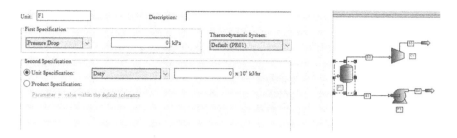

FIGURE B.8   Specification of the unit operations.

has failed to solve. Sometimes the unit appears green in color for some time, such as the pump. The feed rate to the pump is zero, which is just an indication that it is still calculating a particular unit, and sometimes even after a similar converge remains in yellow, a sign that there are some warnings to have a look at it. That's all for in terms of running the simulator (Figure B.9).

DISPLAY OF RESULTS

There are various ways of looking at results; looking at your results takes your cursor onto the screen and throws a tooltip. This particular small window that comes up with a tooltip tells you the temperature, pressure, and flow rate. If you take it to a unit operation, it shows the parameters associated with that particular platform. If you go to the compressor, it shows the shaft work and actual head.

The other way of looking at your results would be to build something called a stream property table. Click on the Output tab and then click on the Stream property table. And then we can come and place it somewhere on the flowsheet, and you have a box attached to the cursor; when you double click on that box, it shows the different streams you have available. So we can say we select all this. The result gives me the material balance for this process, and when you click OK, you know that stream property table or products appear (Figure B.10).

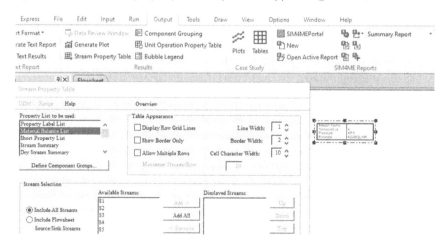

FIGURE B.9   Adding stream table summary below the process flowsheet.

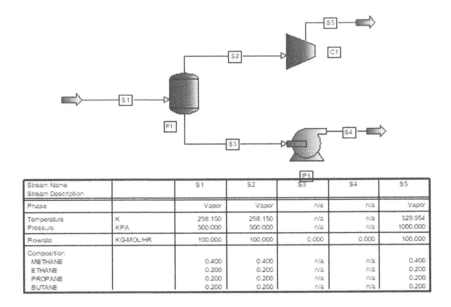

Stream Name Stream Description		S1	S2	S3	S4	S5
Phase		Vapor	Vapor	n/a	n/a	Vapor
Temperature	K	298.150	298.150	n/a	n/a	329.964
Pressure	KPA	500.000	500.000	n/a	n/a	1000.000
Flowrate	KG-MOL/HR	100.000	100.000	0.000	0.000	100.000
Composition						
METHANE		0.400	0.400	n/a	n/a	0.400
ETHANE		0.200	0.200	n/a	n/a	0.200
PROPANE		0.200	0.200	n/a	n/a	0.200
BUTANE		0.200	0.200	n/a	n/a	0.200

FIGURE B.10   PRO/II generated the material stream summary.

Generate the Excel-based results as follows (make sure to save the file before doing this step): Click on the Output tab, and while in the output windows, click on Generate Excel output. Save As the file to a specific directory. The summary report is under Detail report either these two types of reporting. The generation speed depends on how fast the computer is to start up Excel.

Appendix C: Introduction to Aspen Plus

Start Aspen Plus and create a new simulation. When running a new simulation, the default screen is set to the Properties tab and enters the system's relevant simulation information.

1. To begin, Click the components folder for this simulation, specify that the water, methane, ethane, and propane are the process chemicals, enter these individually into the component ID box, pressing Enter after each chemical (Figure C.1).
2. Aspen Plus does not recognize component ID that is longer than eight characters. You can search for it by clicking on the Find box, typing in part or all of the names, and pressing on Find Now. A list of relevant chemicals will appear below. Select the appropriate chemical and click Add selected components (Figure C.2).
3. After entering all process components, click on the Methods folder and select an appropriate property method (Figure C.3).
4. Next, click on the blue arrow at the top of the screen to be directed to a summary table of interaction effects. Click the blue arrow again to prompt a dialogue box asking if you want to Run the property analysis. Click OK, and Aspen will do so.
5. After Aspen generated the results, click the blue arrow at the screen's top to open the dialogue box again. This time, select to go to the simulation environment. At the bottom of the dialogue box and click OK (Figure C.4).
6. On the menu tree at the left of the screen, some folders have blue checkmarks on them, and some have red circles. Blue checkmarks indicate inputted sufficient information. By contrast, red circles indicate the lack of necessary process information. You can click on the arrow next to a folder to expand a menu tree. This screen allows you to enter more precisely the information needed. Alternatively, click the blue arrow at the top of the screen to be brought to the data input screens. Notice that the input screen's name on both the tab at the top of the input and the menu tree is the same.
7. While in the simulation environment, click on Exchangers in the Model Palette and select the shell and tube heat exchanger (HeatX). Click somewhere in the simulation area. Click on Material and connect cold inlet stream (S1), exit cold stream (S2), inlet hot stream (S3), and exit hot stream (S4) as shown in Figure C.5.
8. Click on the cold inlet stream, and suppose cooling water enters at 20°C at one bar with the mass flow rate of 500 kg/h. To enter the flow rate, change the total flow basis, and input the appropriate box's flow rate. In the

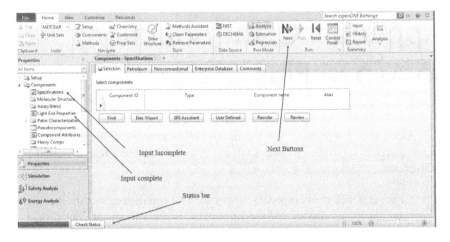

FIGURE C.1   Aspen Plus data browser and component selection windows.

composition box, change the left box to mass flow and specify all 500 kg/h water. Alternatively, you can select either mass or mole fraction (FRAC) and specify water as one. Click the blue next arrow to go to the hot inlet stream (Figure C.6).

9. Suppose the stream enters into 200°C, and one bar with the mass flow rate of 1,000 kg/h, in the composition box, change the left box to mass FRAC and specify the stream as 50 wt% methane 30 wt% ethane, and 20 wt% propane (Figure C.7).

10. Notice how the values are added in the bottom box to ensure no mistakes are made. Click the blue next arrow to go to the heat exchanger tab. Specify the type as design and under the exchanger specification tab, choose Hot stream outlet temperature from the drop-down menu. Enter 100°C in the value box. Pressing the blue next arrow will open a dialogue box asking if

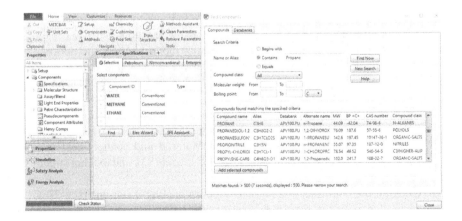

FIGURE C.2   Aspen Plus completed component selection and Find component window.

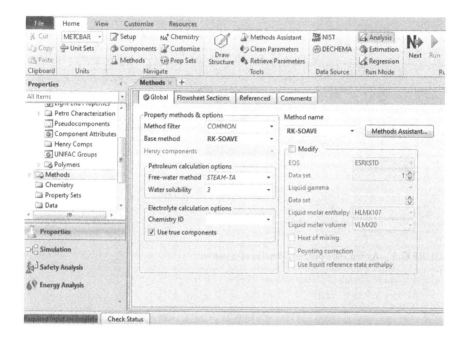

FIGURE C.3 Aspen Plus completed properties method and options.

you want to run the simulation. Click OK to have Aspen begin calculations, and the dialogue box will appear, indicating the simulation converges. Click Close to view the results (Figure C.8).

11. Running the Simulation: there are a few ways to run the simulation. Select either the Next button in the toolbar, which will tell that all of the required inputs are complete, and ask if you would like to run the simulation. The

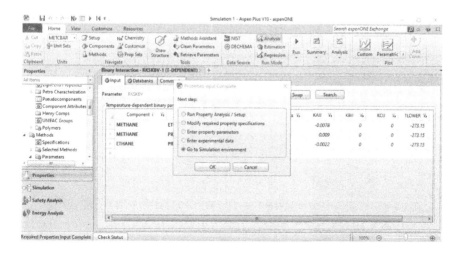

FIGURE C.4 Aspen Plus "Go to Simulation environment" dialog box.

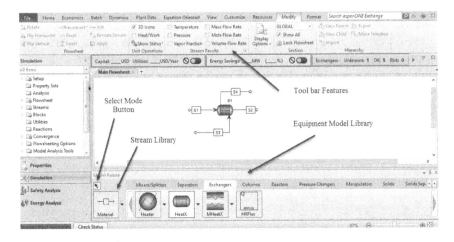

FIGURE C.5   Aspen Plus process flowsheet window.

FIGURE C.6   Aspen Plus completed cold feed stream input.

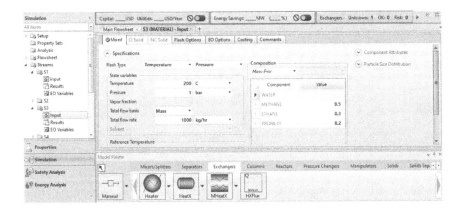

FIGURE C.7   Aspen Plus completed hot stream input.

FIGURE C.8 Aspen Plus required specifying the Model fidelity, specifications, and mode of calculation.

user can also simulate by selecting the toolbar's run button (this is the button with an arrow pointing to the right). The user can choose Run from the menu bar. After the simulation is converged, the Results Summary Tab on the Data Browser Window has a blue checkmark.

12. The results are displayed by clicking on the Stream Results under the block folder (Figure C.9).

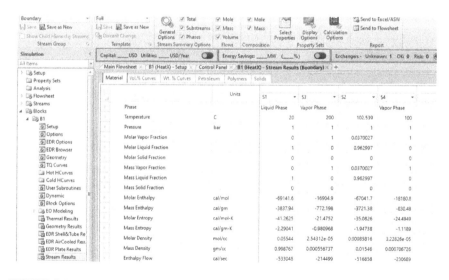

FIGURE C.9 Aspen Plus generated stream summary results.

Appendix D: Introduction to SuperPro Designer

This introduction focuses on the features and capabilities of SuperPro. SuperPro is used to evaluate and screen alternative technologies and to optimize processes under development as the user moves from development to manufacturing. They are used for ongoing process optimization and debottlenecking. In the language of SuperPro, the icons which look like equipment are under the Unit Procedures. The unit procedures are connected with streams that represent material flow operations. SuperPro enables users to model batch processes in detail. In continuous processes, each system includes just a single operation, and it behaves like a traditional unit operation. For every function within a unit procedure, SuperPro contains a mathematical model that performs material and energy balance calculations.

After selecting File≫New and before the simulation will open a new flowsheet, the user must define whether the process is batch or continuous (Figure D.1).

To register pure components, go to the menu bar and select tasks, compounds, register interview properties. Register your species, go to the menu bar, and select your components. Note that in every SuperPro file, nitrogen, oxygen, and water are existed by default. User can choose the source of the component database from either designer, user, or one of two other proprietary databases (Figure D.2).

Process Operating Mode

○ Batch
- Scheduling information is required.
- Process batch time is calculated.
- Stream flows are displayed on a per-batch basis.
- Inherently continuous processing steps can be included as unit operations in either continuous or semi-continuous mode.

○ Continuous
- Scheduling information is NOT required.
- Process batch time is NOT calculated.
- Stream flows are displayed on a per-hour basis.
- Inherently batch processing steps can be included ; user must specify process time and turnaround time for such steps.

FIGURE D.1   Specifying mode of operation.

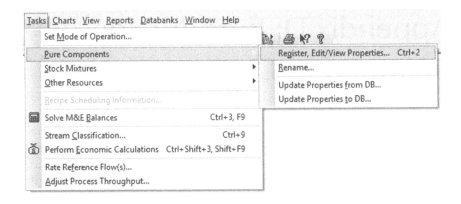

FIGURE D.2   Component selection menu.

Example D.1: Separation of a Binary Liquid Mixture using a Distillation Column

A mixture of 50 mol% ethyl alcohol and the balance is propanol fed to a distillation column at 100 kmol/h, 1 atm, and 25°C. The column has 20 trays, including a reboiler and condenser. The feed enters at tray 10, and the reflux ratio is 1.5. Simulate the process to achieve 93% of propanol in the bottom stream.

SOLUTION

Select a continuous distillation column. Unit Procedures≫Distillation≫Continuous (Rigors) (Figure D.3), and click in the work area, and it will appear

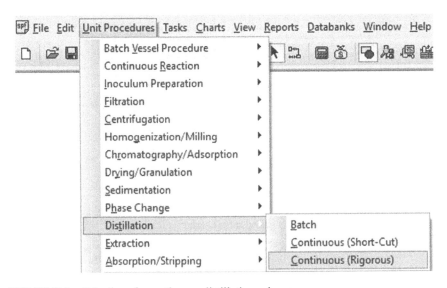

FIGURE D.3   Selection of a continuous distillation column.

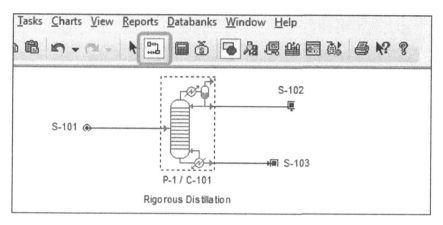

FIGURE D.4   Connecting of feed and product stream using the collect mode.

Add the feed stream, click on the Connects Mode next to the arrow point and click, and next click on the output, move to the right, and double-click to set the top product, repeat the same for the bottom product (Figure D.4).

Add the component: Tasks≫Pure component≫Register, Edit/View properties

Note that nitrogen, oxygen, and water are present by default; add ethyl alcohol and propanol and click OK to close the Pure Components Databank (Figure D.5).

To specify the feed stream, exit the connect mode (click on the arrow). Double click on the feed stream, put the feed details molar composition 50% ethyl alcohol and 50% propanol as per the question, the feed rate is 100 kgmol/h at 1 atm and 25°C (Figure D.6).

Right click on the distillation column, select Operation Data and input the number of trays and feed tray location (Figure D.7). Enter the number of trays (20), the feed tray (10), and reflux ration (2.0), as shown in Figure D.8.

To display results, click on the Toggle the Stream Summary Table (Figure D.9).

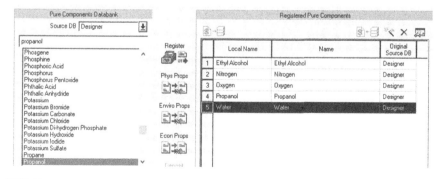

FIGURE D.5   Component selection menu.

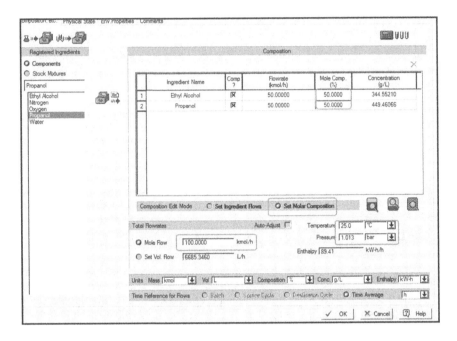

FIGURE D.6   Feed stream fully specified.

Right click on the empty docked window and select Edit Contents and the required streams shown in Figure D.10.

To delete undesired components, click on the Include/Exclude components and uncheck to excluded compounds (Figure D.11).

The final process flow diagram and stream summary (select float) is shown in Figure D.12.

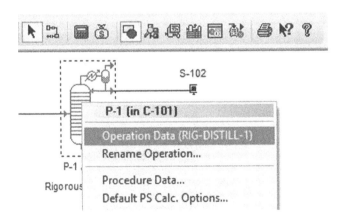

FIGURE D.7   Specifying distillation column operating data.

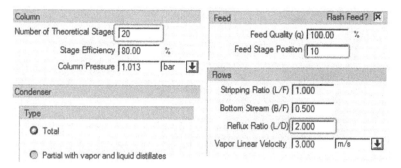

FIGURE D.8 Required distillation column specifications.

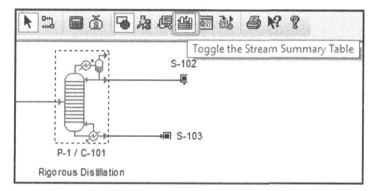

FIGURE D.9 Toggle the stream summary table to display the stream summary.

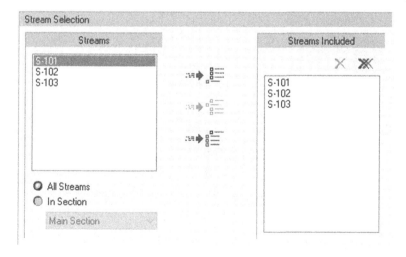

FIGURE D.10 Stream selection menu.

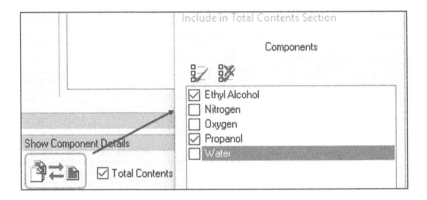

FIGURE D.11   Include the components to appear in the stream summary table.

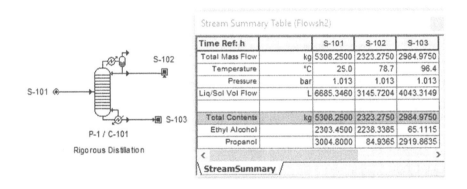

FIGURE D.12   SuperPro generated process flow diagram and stream summary.

Appendix E: Introduction to Aveva Process Simulation

The following brief introduction covers some best practices and handles mistakes, warnings, and errors. It tackles the basics of how to start building a new simulation in the Aveva Process Simulation platform.

1. To start a new simulation. Click on the New button in the simulation repository, or click on the plus sign above the canvas, the two options for choosing a template for a new simulation (Figure E.1).
2. This example will use the Process Model library, the most common library used in chemical and refining applications. Rename the simulation if desired.
3. To begin building a Process Flow Diagram (PFD), drag and drop models from the process model library to the canvas. When we start connecting the modules, make sure that we keep the three badges (Figure E.2) in the green solution status window before adding more models to the simulations, to make sure we have entered all the required data into the given simulation correctly.
4. Drag and drop the Source and valve from the model library to the canvas (Figure E.3). Fully specify the feed stream (W or F, T, P, and composition).
5. Suppose we run into a point or any of these badges are red; resolve those issues immediately before building the simulation. If you hover over the icons, you'll notice that information will pop up to help you better understand each icon's status for this red required input. It is telling us that there are models with data that need to be entered or confirmed. If we look at the PFD, we will also see the same badge in the icon's upper left for the models where this is applicable. That will also appear above the model icon for the spec. If you open the Model's Property Inspector, you will see the badges next to the variables affected. In each instance, we can hover over these badges for more information. To clear these input badges, we can change or confirm the value. Or hold the Shift key, and click on the icon next to the wonder.
6. We can also turn these badge applications off through the Edit, View tab, leave that just turned on while learning how to use Aveva Process Simulation for the first time.
7. Before adding any new models, we want to specify the existing models to the process conditions. Let's set the source flow rate to 100 kg/s, the source pressure to 200 kPa, and the source temperature to 300 K. We will accept the default pressure drop of 50 kPa. If you are unsure what a variable means,

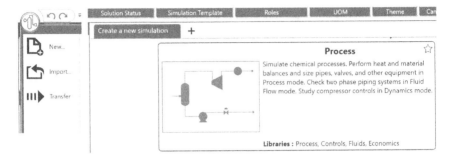

FIGURE E.1   Creating new simulation in Aveva Process Simulation software.

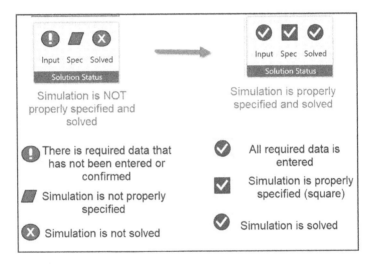

FIGURE E.2   Aveva Process Simulation solution status.

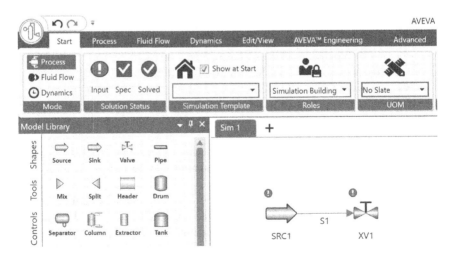

FIGURE E.3   Building process flowsheet in the canvas.

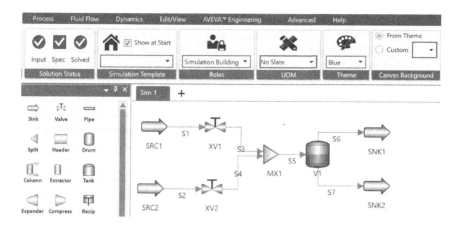

FIGURE E.4 Squared and solved process flow sheet.

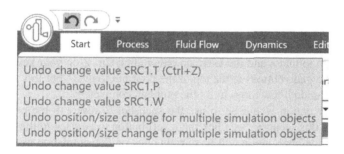

FIGURE E.5 Aveva Process Simulation Undo option.

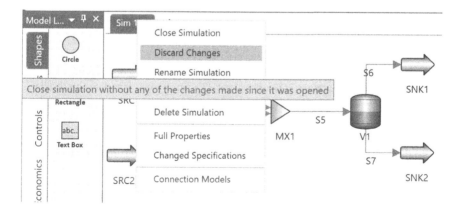

FIGURE E.6 Aveva Process Simulation discard changes option.

hover over the variable's name for a pop-up with a description. Notice that we did not have to enter thermodynamic information components, but the simulation is still solved. This is because all software sources start with a default fluid defined, with a specific set of components and composition information in thermodynamic method selection. You can change the values of the composition from the Source.

8. Add a mixer connected to the source stream. One can add another source going to this mixer either by dragging and dropping a new source and valve from the model library or selecting the existing Source and valve and doing a copy-paste.

9. Notice that the simulation becomes square and unsolved. Let's specify the source temperature. Now the simulation is back to being square installed, and we can continue building. Before we add more, let's set the vapor fraction value to 50%.

10. The simulation is continuously solving as we make. Since we have a correct solution, it is better to take a snapshot before making new changes and saving the current state's resolution. The snapshot allows returning to this solution in the future if we run into any conversion issues down the line or want to be able to return to this point to reference a result.

   Let's now see what happens when we do run into a problem. We can fix this mistake in several ways. The first would be to use the Undo feature program to undo, and the user can either click Ctrl Z on your keyboard or click the undo buttons at the top of the program to undo. For this issue, we have to undo it few times to fix the problem completely.

11. Another option would be to use the snapshot to revert to the prior solution. In this scenario, we click the revert button to go back to the snapshot selected; a final option will be to right click on the simulation name above the canvas, and we'll choose the discard changes option. This will completely close the simulation and undo any changes made to the simulation since it was the last chance. We are finished and ready to get our results from the simulation.

Example E.1: Separation of a Binary Liquid Mixture

A mixture of 50% benzene and 50% toluene is separated in a distillation column. The inlet stream flow rate is 100 kgmol/h at 25°C and 1 bar. Simulate the distillation column using Aveva Process Simulation.

SOLUTION

Use the Column from the Process library. Multistage fractionation of vapor and liquid mixtures. Single Model to simulate all possible Column types (i.e., distillation, absorber, and stripper). Right click and then select Help on the Column Model in the Model Library which opens extensive Help documentation. Starting Point (Initial State), using the default column set:

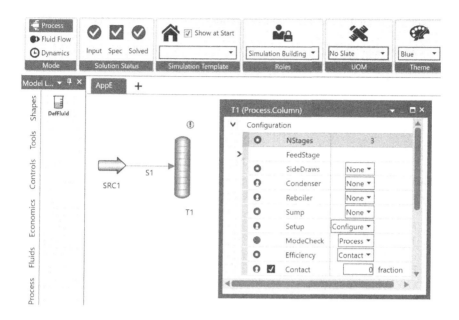

FIGURE E.7 Distillation column converged in the configuration mode (default setting).

- Defaults to 3 Stages
- Feed defaults to Stage 1
- Product streams not required
- Solved to a "no contact" solution
- No condenser
- No reboiler

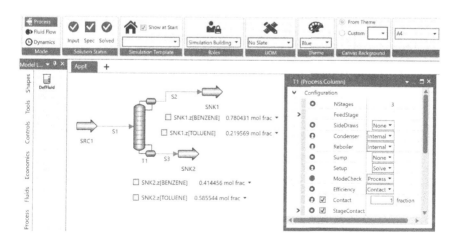

FIGURE E.8 Converged distillation column for benzene/toluene separation.

Achieving an Initial Rigorous Solution. Work Top-down from the Mini Inspector.

1. Set number of stages
2. Set feed tray location
3. Set Condenser type (commonly internal)
4. Set Reboiler (typically internal)
5. Set Setup to Solve
6. Set Contact to 1 (or vapor and liquid will bypass each other)

VARIABLE NOMENCLATURE

Naming conventions tend to follow standard engineering textbook names (hover over any variable name to get a full description):

P – Pressure
T – Temperature
L – Length
F – Mole Flow
Q – Volumetric Flow
W – Mass Flow
MW – Molecular Weight
L – Length
D – Diameter
VF – Vapor Fraction
z – Bulk Composition
y/x – Vapor/Liquid Composition

Greek letters get spelled out:

rho – Density
mu – Viscosity
eta – Efficiency
Duty – Heat transfer duty
Speed – Rotating speed
Power – Mechanical power
Level – Level

Index

Printed in the United States
by Baker & Taylor Publisher Services